Rockets and People

For sale by the Superintendent of Documents, U.S. Government Printing Office
Internet: bookstore.gpo.gov Phone: toll free (866) 512-1800; DC area (202) 512-1800
Fax: (202) 512-2250 Mail: Stop SSOP, Washington, DC 20402-0001

ISBN 0-16-073239-5

ISBN 0-16-073239-5

9 780160 732393 90000

Rockets and People

Volume I

Boris Chertok

The NASA History Series

National Aeronautics and Space Administration
NASA History Division
Office of External Relations
Washington, DC
January 2005
NASA SP-2005-4110

Library of Congress Cataloging-in-Publication Data

Chertok, B. E. (Boris Evseevich), 1912–
[Rakety i lyudi. English]
Rockets and people / by Boris E. Chertok ; [edited by] Asif A. Siddiqi.
p. cm. — (NASA History Series) (NASA SP-2005-4110)
Includes bibliographical references and index.
1. Chertok, B. E. (Boris Evseevich), 1912– 2. Astronautics—Soviet Union—Biography. 3. Aerospace engineers—Soviet Union—Biography. 4. Astronautics—Soviet Union—History.
I. Siddiqi, Asif A., 1966– II. Title. III. Series. IV. SP-2005-4110.
TL789.85.C48C4813 2004
629.1′092—dc22

2004020825

I dedicate this book
to the cherished memory
of my wife and friend,
Yekaterina Semyonova Golubkina.

Contents

Series Introduction

IN AN EXTRAORDINARY CENTURY, Academician Boris Yevseyevich Chertok lived an extraordinary life. He witnessed and participated in many important technological milestones of the twentieth century, and in these volumes, he recollects them with clarity, humanity, and humility. Chertok began his career as an electrician in 1930 at an aviation factory near Moscow. Thirty years later, he was one of the senior designers in charge of the Soviet Union's crowning achievement as a space power: the launch of Yuriy Gagarin, the world's first space voyager. Chertok's sixty-year-long career, punctuated by the extraordinary accomplishments of both Sputnik and Gagarin, and continuing to the many successes and failures of the Soviet space program, constitutes the core of his memoirs, *Rockets and People*. In these four volumes, Academician Chertok not only describes and remembers, but also elicits and extracts profound insights from an epic story about a society's quest to explore the cosmos.

Academician Chertok's memoirs, forged from experience in the Cold War, provide a compelling perspective into a past that is indispensable to understanding the present relationship between the American and Russian space programs. From the end of the World War II to the present day, the missile and space efforts of the United States and the Soviet Union (and now, Russia) have been inextricably linked. As such, although Chertok's work focuses exclusively on Soviet programs to explore space, it also prompts us to reconsider the entire history of spaceflight, both Russian and American.

Chertok's narrative underlines how, from the beginning of the Cold War, the rocketry projects of the two nations evolved in independent but parallel paths. Chertok's first-hand recollections of the extraordinary Soviet efforts to collect, catalog, and reproduce German rocket technology after World War II provide a parallel view to what historian John Gimbel has called the Western "exploitation and plunder" of German technology after the war.[1] Chertok describes how the Soviet design team under the famous Chief Designer Sergey Pavlovich Korolev

1. John Gimbel, *Science, Technology, and Reparations: Exploitation and Plunder in Postwar Germany* (Stanford: Stanford University Press, 1990).

quickly outgrew German missile technology. By the late 1950s, his team produced the majestic R-7, the world's first intercontinental ballistic missile. Using this rocket, the Soviet Union launched the first Sputnik satellite on 4 October 1957 from a launch site in remote central Asia.

The early Soviet accomplishments in space exploration, particularly the launch of Sputnik in 1957 and the remarkable flight of Yuriy Gagarin in 1961, were benchmarks of the Cold War. Spurred by the Soviet successes, the United States formed a governmental agency, the National Aeronautics and Space Administration (NASA), to conduct civilian space exploration. As a result of Gagarin's triumphant flight, in 1961, the Kennedy Administration charged NASA to achieve the goal of "landing a man on the Moon and returning him safely to the Earth before the end of the decade."[2] Such an achievement would demonstrate American supremacy in the arena of spaceflight at a time when both American and Soviet politicians believed that victory in space would be tantamount to preeminence on the global stage. The space programs of both countries grew in leaps and bounds in the 1960s, but the Americans crossed the finish line first when Apollo astronauts Neil A. Armstrong and Edwin E. "Buzz" Aldrin, Jr. disembarked on the Moon's surface in July 1969.

Shadowing Apollo's success was an absent question: What happened to the Soviets who had succeeded so brilliantly with Sputnik and Gagarin? Unknown to most, the Soviets tried and failed to reach the Moon in a secret program that came to naught. As a result of that disastrous failure, the Soviet Union pursued a gradual and consistent space station program in the 1970s and 1980s that eventually led to the Mir space station. The Americans developed a reusable space transportation system known as the Space Shuttle. Despite their seemingly separate paths, the space programs of the two powers remained dependent on each other for rationale and direction. When the Soviet Union disintegrated in 1991, cooperation replaced competition as the two countries embarked on a joint program to establish the first permanent human habitation in space through the International Space Station (ISS).

Academician Chertok's reminiscences are particularly important because he played key roles in almost every major milestone of the Soviet missile and space programs, from the beginning of World War II to the dissolution of the Soviet Union in 1991. During the war, he served on the team that developed the Soviet Union's first rocket-powered airplane, the BI. In the immediate aftermath of the war, Chertok, then in his early thirties, played a key role in studying and collecting captured German rocket technology. In the latter days of the Stalinist era, he worked to develop long-range missiles as deputy chief engineer of the main research institute, the NII-88 (pronounced "nee-88") near Moscow. In 1956,

---

2. U.S. Congress, *Senate Committee on Aeronautical and Space Sciences, Documents on International Aspects of the Exploration and Uses of Outer Space, 1954–1962, 88th Cong., 1st sess., S. Doc. 18* (Washington, DC: GPO, 1963), pp. 202-204.

Korolev's famous OKB-1 design bureau spun off from the institute and assumed a leading position in the emerging Soviet space program. As a deputy chief designer at OKB-1, Chertok continued with his contributions to the most important Soviet space projects of the day: Vostok, Voskhod, Soyuz, the world's first space station Salyut, the Energiya superbooster, and the Buran space shuttle.

Chertok's emergence from the secret world of the Soviet military-industrial complex, into his current status as the most recognized living legacy of the Soviet space program, coincided with the dismantling of the Soviet Union as a political entity. Throughout most of his career, Chertok's name remained a state secret. When he occasionally wrote for the public, he used the pseudonym "Boris Yevseyev."[3] Like others writing on the Soviet space program during the Cold War, Chertok was not allowed to reveal any institutional or technical details in his writings. What the state censors permitted for publication said little; one could read a book several hundred pages long comprised of nothing beyond tedious and long personal anecdotes between anonymous participants extolling the virtues of the Communist Party. The formerly immutable limits on free expression in the Soviet Union irrevocably expanded only after Mikhail Gorbachev's rise to power in 1985 and the introduction of *glasnost'* (openness).

Chertok's name first appeared in print in the newspaper *Izvestiya* in an article commemorating the thirtieth anniversary of the launch of Sputnik in 1987. In a wide-ranging interview on the creation of Sputnik, Chertok spoke with the utmost respect for his former boss, the late Korolev. He also eloquently balanced love for his country with criticisms of the widespread inertia and inefficiency that characterized late-period Soviet society.[4] His first written works in the *glasnost'* period, published in early 1988 in the Air Force journal *Aviatsiya i kosmonavtika* (Aviation and Cosmonautics), underlined Korolev's central role in the foundation and growth of the Soviet space program.[5] By this time, it was as if all the patched up straps that held together a stagnant empire were falling apart one by one; even as Russia was in the midst of one of its most historic transformations, the floodgates of free expression were transforming the country's own history. People like Chertok were now free to speak about their experiences with candor. Readers could now learn about episodes such as Korolev's brutal incarceration in the late 1930s, the dramatic story behind the fatal space mission of Soyuz-1 in 1967, and details of the failed and abandoned Moon project in the 1960s.[6] Chertok himself

---

3. See for example, his article "Chelovek or avtomat?" (Human or Automation?) in the book by M. Vasilyev, ed., *Shagi k zvezdam* (Footsteps to the Stars) (Moscow: Molodaya gvardiya, 1972), pp. 281–287.

4. B. Konovalov, "Ryvok k zvezdam" (Dash to the Stars), *Izvestiya*, October 1, 1987, p. 3.

5. B. Chertok, "Lider" (Leader), *Aviatsiya i kosmonavtika* no. 1 (1988): pp. 30–31 and no. 2 (1988): pp. 40–41.

6. For early references to Korolev's imprisonment, see Ye. Manucharova, "Kharakter glavnogo konstruktora" (The Character of the Chief Designer), *Izvestiya*, January 11, 1987, p. 3. For early revelations on Soyuz-1 and the Moon program, see L. N. Kamanin, "Zvezdy Komarova" (Komarov's Star), *Poisk* No. 5 (June 1989): pp. 4–5 and L. N. Kamanin, "S zemli na lunu i obratno" (From the Earth to the Moon and Back), *Poisk* no. 12 (July 1989): pp. 7–8.

shed light on a missing piece of history in a series of five articles published in *Izvestiya* in early 1992 on the German contribution to the foundation of the Soviet missile program after World War II.[7]

Using these works as a starting point, Academician Chertok began working on his memoirs. Originally, he had only intended to write about his experiences from the postwar years in one volume, maybe two. Readers responded so positively to the first volume, *Rakety i liudi* (Rockets and People) published in 1994, that Chertok continued to write, eventually producing four substantial volumes, published in 1996, 1997, and 1999, covering the entire history of the Soviet missile and space programs.[8]

My initial interest in the memoirs was purely historical: I was fascinated by the wealth of technical arcana in the books, specifically projects and concepts that had remained hidden throughout much of the Cold War. Those interested in dates, statistics, and the "nuts and bolts" of history will find much that is useful in these pages. As I continued to read, however, I became engrossed by the overall rhythm of Academician Chertok's narrative, which gave voice and humanity to a story ostensibly about mathematics and technology. In his writings, I found a richness that had been nearly absent in most of the disembodied, clinical, and often speculative writing by Westerners studying the Soviet space program. Because of Chertok's story-telling skills, his memoir is a much needed corrective to the outdated Western view of Soviet space achievements as a mishmash of propaganda, self-delusion, and Cold War rhetoric. In Chertok's story, we meet real people with real dreams who achieved extraordinary successes under very difficult conditions.

Chertok's reminiscences are remarkably sharp and descriptive. In being self-reflective, Chertok avoids the kind of solipsistic ruminations that often characterize memoirs. He is both proud of his country's accomplishments and willing to admit failings with honesty. For example, Chertok juxtaposes accounts of the famous aviation exploits of Soviet pilots in the 1930s, especially those to the Arctic, with the much darker costs of the Great Terror in the late 1930s when Stalin's vicious purges decimated the Soviet aviation industry.

---

7. *Izvestiya* correspondent Boris Konovalov prepared these publications, which had the general title "U Sovetskikh raketnykh triumfov bylo nemetskoye nachalo" (Soviets Rocket Triumphs Had German Origins). See *Izvestiya,* March 4, 1992, p. 5; March 5, 1992, p. 5; March 6, 1992, p. 5; March 7, 1992, p. 5; and March 9, 1992, p. 3. Konovalov also published a sixth article on the German contribution to American rocketry. See "U amerikanskikh raketnykh triumfov takzhe bylo nemetskoye nachalo" (American Rocket Triumphs Also Had German Origins), *Izvestiya,* March 10, 1992, p. 7. Konovalov later synthesized the five original articles into a longer work that included the reminiscences of other participants in the German mission such as Vladimir Barmin and Vasiliy Mishin. See Boris Konovalov, *Tayna Sovetskogo raketnogo oruzhiya* (Secrets of Soviet Rocket Armaments) (Moscow: ZEVS, 1992).

8. *Rakety i lyudi* (Rockets and People) (Moscow: Mashinostroyeniye, 1994); *Rakety i lyudi: Fili Podlipki Tyuratam* (Rockets and People: Fili Podlipki Tyuratam) (Moscow: Mashinostroyeniye, 1996); *Rakety i lyudi: goryachiye dni kholodnoy voyny* (Rockets and People: Hot Days of the Cold War) (Moscow: Mashinostroyeniye, 1997); *Rakety i lyudi: lunnaya gonka* (Rockets and People: The Moon Race) (Moscow: Mashinostroyeniye, 1999). All four volumes were subsequently translated and published in Germany.

Chertok's descriptive powers are particularly evident in describing the chaotic nature of the Soviet mission to recover and collect rocketry equipment in Germany after World War II. Interspersed with his contemporary diary entries, his language conveys the combination of joy, confusion, and often anti-climax that the end of the war presaged for Soviet representatives in Germany. In one breath, Chertok and his team are looking for hidden caches of German matériel in an underground mine, while in another they are face to face with the deadly consequences of a soldier who had raped a young German woman (chapter 22).[9] There are many such seemingly incongruous anecdotes during Chertok's time in Germany, from the experience of visiting the Nazi slave labor camp at Dora soon after liberation in 1945, to the deportation of hundreds of German scientists to the USSR in 1946. Chertok's massive work is of great consequence for another reason—he cogently provides context. Since the breakup of the Soviet Union in 1991, many participants have openly written about their experiences, but few have successfully placed Soviet space achievements in the broader context of the history of Soviet science, the history of the Soviet military-industrial complex, or indeed Soviet history in general.[10] The volumes of memoirs compiled by the Russian State Archive of Scientific-Technical Documentation in the early 1990s under the series, *Dorogi v kosmos* (Roads to Space), provided an undeniably rich and in-depth view of the origins of the Soviet space program, but they were, for the most part, personal narratives, i.e., fish-eye views of the world around them.[11] Chertok's memoirs are a rare exception in that they strive to locate the Soviet missile and space program in the fabric of broader social, political, industrial, and scientific developments in the former Soviet Union.

This combination—Chertok's participation in the most important Soviet space achievements, his capacity to lucidly communicate them to the reader, and

---

9. For the problem of rape in occupied Germany after the war, see Norman M. Naimark, *The Russians in Germany: A History of the Soviet Zone of Occupation, 1945–1949* (Cambridge, MA: The Belknap Press of Harvard University Press, 1995), pp. 69–140.

10. For the two most important histories of the Soviet military-industrial complex, see N. S. Simonov, *Voyenno-promyshlennyy kompleks SSSR v 1920-1950-ye gody: tempy ekonomicheskogo rosta, struktura, organizatsiya proizvodstva i upravleniye* (The Military-Industrial Complex of the USSR in the 1920s to 1950s: Rate of Economic Growth, Structure, Organization of Production and Control) (Moscow: ROSSPEN, 1996); and I. V. Bystrova, *Voyenno-promyshlennyy kompleks sssr v gody kholodnoy voyny (vtoraya polovina 40-kh – nachalo 60-kh godov)* [The Military-Industrial Complex of the USSR in the Years of the Cold War (The Late 1940s to the Early 1960s)] (Moscow: IRI RAN, 2000). For a history in English that builds on these seminal works and complements them with original research, see John Barber and Mark Harrison, eds., *The Soviet Defence-Industry Complex from Stalin to Khrushchev* (Houndmills, UK: Macmillan Press, 2000).

11. Yu. A. Mozzhorin et al., eds., *Dorogi v kosmos: Vospominaniya veteranov raketno-kosmicheskoy tekhniki i kosmonavtiki, tom I i II* (Roads to Space: Recollections of Veterans of Rocket-Space Technology and Cosmonautics: Volumes I and II) (Moscow: MAI, 1992) and Yu. A. Mozzhorin et al., eds., *Nachalo kosmicheskoy ery: vospominaniya veteranov raketno-kosmicheskoy tekhniki i kosmonavtiki: vypusk vtoroy* (The Beginning of the Space Era: Recollections of Veterans of Rocket-Space Technology and Cosmonautics: Second Issue) (Moscow: RNITsKD, 1994). For a poorly translated and edited English version of the series, see John Rhea, ed., *Roads to Space: An Oral History of the Soviet Space Program* (New York: Aviation Week Group, 1995).

his skill in providing a broader social context—make this work, in my opinion, one of the most important memoirs written by a veteran of the Soviet space program. The series will also be an important contribution to the history of Soviet science and technology.[12]

In reading Academician Chertok's recollections, we should not lose sight of the fact that these chapters, although full of history, have their particular perspective. In conveying to us the complex vista of the Soviet space program, he has given us one man's memories of a huge undertaking. Other participants of these very same events will remember things differently. Soviet space history, like any discipline of history, exists as a continuous process of revision and restatement. Few historians in the twenty-first century would claim to be completely objective.[13] Memoirists would make even less of a claim to the "truth." In his introduction, Chertok acknowledges this, saying, "I . . . must warn the reader that in no way do I have pretensions to the laurels of a scholarly historian. Correspondingly, my books are not examples of strict historical research. In any memoirs, narrative and thought are inevitably subjective." Chertok ably illustrates, however, that avoiding the pursuit of scholarly history does not necessarily lessen the relevance of his story, especially because it represents the opinion of an influential member of the postwar scientific and technical intelligentsia in the Soviet Union.

Some, for example, might not share Chertok's strong belief in the power of scientists and engineers to solve social problems, a view that influenced many who sought to transform the Soviet Union with modern science after the Russian Revolution in 1917. Historians of Soviet science such as Loren Graham have argued that narrowly technocratic views of social development cost the Soviet Union dearly.[14] Technological hubris was, of course, not unique to the Soviet scientific community, but absent democratic processes of accountability, many huge Soviet government projects—such as the construction of the Great Dnepr Dam and the great Siberian railway in the 1970s and 1980s—ended up as costly failures with many adverse social and environmental repercussions. Whether one agrees or disagrees with Chertok's views, they are important to understand because they represent the ideas of a generation who passionately believed in the power of science to eliminate the ills of society. As such, his memoirs add an important

---

12. For key works on the history of Soviet science and technology, see Kendall E. Bailes, *Technology and Society under Lenin and Stalin: Origins of the Soviet Technical Intelligentsia, 1917–1941* (Princeton, NJ: Princeton University Press, 1978); Loren R. Graham, *Science in Russia and the Soviet Union: A Short History* (Cambridge: Cambridge University Press, 1993); and Nikolai Krementsov, *Stalinist Science* (Princeton, NJ: Princeton University Press, 1997).

13. For the American historical discipline's relationship to the changing standards of objectivity, see Peter Novick, *That Noble Dream: The 'Objectivity' Question and the American Historical Profession* (Cambridge, UK: Cambridge University Press, 1988).

14. For technological hubris, see for example, Loren Graham, *The Ghost of the Executed Engineer: Technology and the Fall of the Soviet Union* (Cambridge, MA: Harvard University Press, 1993).

dimension to understanding the *mentalité* of the Soviets' drive to become a modern, industrialized state in the twentieth century.

Chertok's memoirs are part of the second generation of publications on Soviet space history, one that eclipsed the (heavily censored) first generation published during the Communist era. Memoirs constituted a large part of the second generation. In the 1990s, when it was finally possible to write candidly about Soviet space history, a wave of personal recollections flooded the market. Not only Boris Chertok, but also such luminaries as Vasiliy Mishin, Kerim Kerimov, Boris Gubanov, Yuriy Mozzhorin, Konstantin Feoktistov, Vyacheslav Filin, and others finally published their reminiscences.[15] Official organizational histories and journalistic accounts complemented these memoirs, written by individuals with access to secret archival documents. Yaroslav Golovanov's magisterial *Korolev: Fakty i Mify* (Korolev: Facts and Myths), as well as key institutional works from the Energiya corporation and the Russian Military Space Forces, added richly to the canon.[16] The diaries of Air Force General Nikolay Kamanin from the 1960s to the early 1970s, published in four volumes in the late 1990s, also gave scholars a candid look at the vicissitudes of the Soviet human spaceflight program.[17]

The flood of works in Russian allowed Westerners to publish the first works in English. Memoirs—for example, from Sergey Khrushchev and Roald Sagdeev—appeared in their English translations. James Harford published his 1997 biography of Sergey Korolev based upon extensive interviews with veterans of the Soviet space program.[18] My own book, *Challenge to Apollo: The Soviet Union and the Space*

---

15. V. M. Filin, *Vospominaniya o lunnom korablye* (Recollections on the Lunar Ship) (Moscow: Kultura, 1992); Kerim Kerimov, *Dorogi v kosmos (zapiski predsedatelya Gosudarstvennoy komissii)* [Roads to Space (Notes of the Chairman of the State Commission)] (Baku: Azerbaijan, 1995); V. M. Filin, *Put k 'Energii'* (Path to Energiya) (Moscow: Izdatelskiy Dom 'GRAAL',' 1996); V. P. Mishin, *Ot sozdaniya ballisticheskikh raket k raketno-kosmicheskomu mashinostroyeniyu* (From the Creation of the Ballistic Rocket to Rocket-Space Machine Building) (Moscow: Informatsionno-izdatelskiy tsentr 'Inform-Znaniye,' 1998); B. I. Gubanov, *Triumf i tragediya 'energii': razmyshleniya glavnogo konstruktora* (The Triumph and Tragedy of Energiya: The Reflections of a Chief Designer) (Nizhniy novgorod: NIER, four volumes in 1998–2000); Konstantin Feoktistov, *Trayektoriya zhizni: mezhdu vchera i zavtra* (Life's Trajectory: Between Yesterday and Tomorrow) (Moscow: Vagrius, 2000); N. A. Anifimov, ed., *Tak eto bylo—Memuary Yu. A. Mozzhorin: Mozzhorin v vospominaniyakh sovremennikov* (How it Was—Memoirs of Yu. A. Mozzhorin: Mozzhorin in the Recollections of his Contemporaries) (Moscow: ZAO 'Mezhdunarodnaya programma obrazovaniya, 2000).

16. Yaroslav Golovanov, *Korolev: fakty i mify* (Korolev: Facts and Myths) (Moscow: Nauka, 1994); Yu. P. Semenov, ed., *Raketno-Kosmicheskaya Korporatsiya "Energiya" imeni S. P. Koroleva* (Energiya Rocket-Space Corporation Named After S. P. Korolev) (Korolev: RKK Energiya, 1996); V. V. Favorskiy and I. V. Meshcheryakov, eds., *Voyenno-kosmicheskiye sily (voyenno-istoricheskiy trud): kniga I* [Military-Space Forces (A Military-Historical Work): Book I] (Moscow: VKS, 1997). Subsequent volumes were published in 1998 and 2001.

17. The first published volume was N. P. Kamanin, *Skrytiy kosmos: kniga pervaya, 1960–1963gg.* (Hidden Space: Book One, 1960-1963) (Moscow: Infortekst IF, 1995). Subsequent volumes covering 1964–1966, 1967–1968, and 1969–1978 were published in 1997, 1999, and 2001 respectively.

18. Sergei N. Khrushchev, *Nikita Khrushchev and the Creation of a Superpower* (University Park, PA: The Pennsylvania State University Press, 2000); Roald Z. Sagdeev, *The Making of a Soviet Scientist: My Adventures in Nuclear Fusion and Space From Stalin to Star Wars* (New York: John Wiley & Sons, 1993); James Harford, *Korolev: How One Man Masterminded the Soviet Drive to Beat America to the Moon* (New York: John Wiley & Sons, 1997).

Race, 1945–1974, was an early attempt to synthesize the wealth of information and narrate a complete history of the early Soviet human spaceflight program.[19] Steven Zaloga provided an indispensable counterpoint to these space histories in *The Kremlin's Nuclear Sword: The Rise and Fall of Russia's Strategic Nuclear Forces, 1945–2000*, which reconstructed the story of the Soviet efforts to develop strategic weapons.[20]

With any new field of history that is bursting with information based primarily on recollection and interviews, there are naturally many contradictions and inconsistencies. For example, even on such a seemingly trivial issue as the name of the earliest institute in Soviet-occupied Germany, "Institute RABE," there is no firm agreement on the reason it was given this title. Chertok's recollections contradict the recollection of another Soviet veteran, Georgiy Dyadin.[21] In another case, many veterans have claimed that artillery general Lev Gaydukov's meeting with Stalin in 1945 was a key turning point in the early Soviet missile program; Stalin apparently entrusted Gaydukov with the responsibility to choose an industrial sector to assign the development of long-range rockets (chapter 23). Lists of visitors to Stalin's office during that period—declassified only very recently—do not, however, show that Gaydukov ever met with Stalin in 1945.[22] Similarly, many Russian sources note that the "Second Main Directorate" of the USSR Council of Ministers managed Soviet missile development in the early 1950s, when in fact, this body actually supervised uranium procurement for the A-bomb project.[23] In many cases, memoirs provide different and contradictory information on the very same event (different dates, designations, locations, people involved, etc.).

Academician Chertok's wonderful memoirs point to a solution to these discrepancies: a "third generation" of Soviet space history, one that builds on the rich trove of the first and second generations, but is primarily based on *documentary* evidence. During the Soviet era, historians could not write history based on documents since they could not obtain access to state and design bureau archives. As the Soviet Union began to fall apart, historians such as Georgiy Vetrov began to take the first steps in document-based history. Vetrov, a former engineer at

---

19. Asif A. Siddiqi, *Challenge to Apollo: The Soviet Union and the Space Race, 1945–1974* (Washington, D.C.: NASA SP-2000-4408, 2000). The book was republished as a two-volume work as *Sputnik and the Soviet Space Challenge* (Gainesville, FL: University Press of Florida, 2003) and *The Soviet Space Race with Apollo* (Gainesville, FL: University Press of Florida, 2003).

20. Steven J. Zaloga, *The Kremlin's Nuclear Sword: The Rise and Fall of Russia's Strategic Nuclear Forces, 1945–2000* (Washington, DC: Smithsonian Institution Press, 2002).

21. G. V. Dyadin, D. N. Filippovykh, and V. I. Ivkin, *Pamyatnyye starty* (Memorable Launches) (Moscow: TsIPK, 2001), p. 69.

22. A. V. Korotkov, A. D. Chernev, and A. A. Chernobayev, "Alfavitnyi ukazatel posetitelei kremlevskogo kabineta I. V. Stalina" ("Alphabetical List of Visitors to the Kremlin Office of I. V. Stalin"), *Istoricheskii arkhiv* no. 4 (1998): p. 50.

23. Vladislav Zubok and Constantine Pleshakov, *Inside the Kremlin's Cold War: From Stalin to Khrushchev* (Cambridge, MA: Harvard University Press), p. 172; Golovanov, *Korolev*, p. 454. For the correct citation on the Second Main Directorate, established on December 27, 1949, see Simonov, *Voyenno-promyshlennyy komples sssr*, pp. 225-226.

Korolev's design bureau, eventually compiled and published two extraordinary collections of primary documents relating to Korolev's legacy.[24] Now that all the state archives in Moscow—such as the State Archive of the Russian Federation (GARF), the Russian State Archive of the Economy (RGAE), and the Archive of the Russian Academy of Sciences (ARAN)—are open to researchers, more results of this "third generation" are beginning to appear. German historians such as Matthias Uhl and Cristoph Mick and those in the United States such as myself have been fortunate to work in Russian archives.[25] For example, we no longer have to guess about the government's decision to approve development of the Soyuz spacecraft, we can see the original VPK decree issued on 4 December 1963.[26] Similarly, instead of speculating about the famous decree of 3 August 1964 that committed the Soviet Union to compete with the American Apollo program, we can study the actual government document issued on that date.[27] Academician Chertok deserves much credit for opening the doors for future historians, since his memoirs have guided many to look even deeper.

BECAUSE OF THE IMPORTANCE of Academician Chertok's memoirs, I did not hesitate when Acting Chief of the NASA History Division Stephen Garber invited me to serve as project editor for the English-language version. Jesco von Puttkamer, a veteran of the Huntsville team founded by Wernher von Braun, served as the guiding spirit behind the entire project. He was instrumental in setting up the arrangements for cooperation between the two parties; without his passion and enthusiasm for bringing Chertok's writings to a broader audience, this endeavor might not have gone beyond conception. Once the project was initiated, I was excited to learn that Academician Chertok would be providing entirely *new* chap-

---

24. M.V. Keldysh, ed., *Tvorcheskoye naslediye Akademika Sergeya Pavlovicha Koroleva: izbrannyye trudy i dokumenty* (The Creative Legacy of Sergey Pavlovich Korolev: Selected Works and Documents) (Moscow: Nauka, 1980); G. S. Vetrov and B.V. Raushenbakh, eds., *S. P. Korolev i ego delo: svet i teni v istorii kosmonavtiki: izbrannyye trudy i dokumenty* (S. P. Korolev and His Cause: Shadow and Light in the History of Cosmonautics) (Moscow: Nauka, 1998). For two other published collections of primary documents, see V. S. Avduyevskiy and T. M. Eneyev, eds. *M. V. Keldysh: izbrannyye trudy: raketnaya tekhnika i kosmonavtika* (M.V. Keldysh: Selected Works: Rocket Technology and Cosmonautics) (Moscow: Nauka, 1988); B. V. Raushenbakh, ed., *Materialy po istorii kosmicheskogo korablya 'vostok': k 30-letiyu pervogo poleta cheloveka v kosmicheskoye prostranstvo* (Materials on the History of the 'Vostok' Space Ship: On the 30th Anniversary of the First Flight of a Human in Space) (Moscow: Nauka, 1991).

25. Matthias Uhl, *Stalins V-2: Der Technolgietransfer der deutschen Fernlenkwaffentechnik in die UdSSR und der Aufbau der sowjetischen Raketenindustrie 1945 bis 1959* (Bonn, Germany: Bernard & Graefe-Verlag, 2001); Christoph Mick, *Forschen für Stalin: Deutsche Fachleute in der sowjetischen Rüstungsindustrie 1945-1958* (Munich: R. Oldenbourg, 2000); Asif A. Siddiqi, "The Rockets' Red Glare: Spaceflight and the Russian Imagination, 1857–1957," Ph.D. dissertation, Carnegie Mellon University, 2004.

26. "O sozdaniia kompleksa 'Soyuz' " (On the Creation of the Soyuz Complex), December 4, 1963, RGAE, f. 298, op. 1, d. 3495, ll. 167-292.

27. "Tsentralnyy komitet KPSS i Sovet ministrov SSSR, postanovleniye" (Central Committee KPSS and SSSR Council of Ministers Decree), August 3, 1964, RGAE, f. 29, op. 1, d. 3441, ll. 299-300. For an English-language summary, see Asif A. Siddiqi, "A Secret Uncovered: The Soviet Decision to Land Cosmonauts on the Moon," *Spaceflight* 46 (2004): pp. 205-213.

ters for most of the four volumes, updated and corrected from the original Russian-language editions. In that sense, these English-language versions are the most updated and final versions of Chertok's memoirs.

As editor, my work was not to translate, a job that was very capably done by a team at the award-winning TechTrans International, Inc. (TTI) based in Houston, Texas. At TTI, Documents Control Manager Delila Rollins and Elena Sukholutsky expertly and capably supervised the very large project. Cynthia Reiser, Laurel Nolen, and Lydia Bryans worked on the actual translations with skill, insight, and good humor. With the translations in hand, my job was first and foremost to ensure that the English language version was as faithful to Chertok's vision as possible. At the same time, I also had to account for stylistic considerations for English-language readers who might be put off by literal translations. The process involved communicating directly with Chertok in many cases, and with his permission, taking liberties to restructure paragraphs and chapters to convey his original spirit. I also provided many explanatory footnotes to elucidate points that might not be evident to readers unversed in the intricacies of Russian history.

Many at NASA Headquarters contributed to publication of the memoirs. Steve Garber at the NASA History Office managed the project from beginning to end. I personally owe a dept of gratitude to Steve for his insightful comments throughout the editorial process. We must also thank Steven J. Dick, the current NASA Chief Historian, as well as Nadine J. Andreassen, William P. Barry, Todd McIntyre, and Claire Rojstaczer for all their terrific help. In the Printing and Design Office, Wes Horne expertly copyedited this book, Paul Clements skillfully laid it out, and Jeffrey McLean carefully saw it through the printing process.

I would also like to thank Dmitry Pieson in Moscow for graciously assisting in my communications with Academician Chertok, and Dr. Matthias Uhl for images of German rockets.

Please note that all footnotes in this volume are mine unless specifically noted as "author's footnotes."

A note about the division of material in the volumes. Because of significant additions and corrections, Academician Chertok has altered somewhat the distribution of materials. The English language edition follows a more-or-less sequential narrative storyline rather than one that goes back-and-forth in time. In the first English volume, he describes his childhood, his formative years as an engineer at the aviation Plant No. 22 in Fili, his experiences during World War II, and the mission to Germany in 1945–46 to study captured German missile technology.

In the second volume, he continues the story with his return to the Soviet Union, the reproduction of a Soviet version of the German V-2 and the development of a domestic Soviet rocket industry at the famed NII-88 institute in the Moscow suburb of Podlipki (now called Korolev). He describes the development of the world's first intercontinental ballistic missile, the R-7; the launch of Sputnik; and the first generation probes sent to the Moon.

In the third volume, he describes the historical launch of the first cosmonaut, Yuriy Gagarin. He also discusses several different aspects of the burgeoning Soviet missile and space programs of the early 1960s, including the development of early ICBMs, reconnaissance satellites, the Cuban missile crisis, the first Soviet communications satellite Molniya-1, the early spectacular missions of the Vostok and Voskhod programs, the dramatic Luna program to land a probe on the Moon, and Sergey Korolev's last days. He then continues into chapters about the early development of the Soyuz spacecraft, with an in-depth discussion of the tragic mission of Vladimir Komarov.

The fourth and final volume is mostly devoted to the Soviet project to send cosmonauts to the Moon in the 1960s, covering all aspects of the development of the giant N-1 rocket. The last portion of this volume covers the origins of the Salyut and Mir space station programs, ending with a fascinating description of the massive Energiya-Buran project, developed as a countermeasure to the American Space Shuttle.

IT WAS MY GREAT FORTUNE to meet with Academician Chertok in the summer of 2003. During the meeting, Chertok, a sprightly ninety-one years old, spoke passionately and emphatically about his life's work and remained justifiably proud of the achievements of the Russian space program. As I left the meeting, I was reminded of something that Chertok had said in one of his first public interviews in 1987. In describing the contradictions of Sergey Korolev's personality, Chertok had noted: "This realist, this calculating, [and] farsighted individual was, in his soul, an incorrigible romantic."[28] Such a description would also be an apt encapsulation of the contradictions of the entire Soviet drive to explore space, one which was characterized by equal amounts of hard-headed realism and romantic idealism. Academician Boris Yevseyevich Chertok has communicated that idea very capably in his memoirs, and it is my hope that we have managed to do justice to his own vision by bringing that story to an English-speaking audience.

ASIF A. SIDDIQI
Series Editor
July 2004

28. Konovalov, "Ryvok k zvezdam."

Foreword

After years of being beaten to the punch by our Soviet counterparts during the space race, those of us flying in the Gemini program wondered why the Soviets did not seem to be responding to the string of Gemini successes in the mid-1960s. Aleksei Leonov, my counterpart in 1975 as the Russian commander of the Apollo-Soyuz Test Project (ASTP), had shaken up our early plans for Gemini by conducting the first spacewalk, what we refer to as extravehicular activity (EVA), in March 1965. But after Aleksei's EVA, the Soviet space program was curiously inactive. We flew increasingly long-duration missions. We perfected rendezvous and docking, and practiced the EVA skills we would need for the Apollo missions to the Moon. We were sure the Russians were still in the race to the Moon—but they weren't doing what they needed to do to get there. The entire world knew our intentions, but there no longer appeared to be a Soviet effort to upstage our missions. What was going on?

After the Leonov EVA, there were no Soviet manned missions for over two years. In April 1967, the Soviets unveiled a new spacecraft called Soyuz. Despite the long preparations and all of the past Soviet successes, the first Soyuz mission was a disaster. It lasted only seventeen orbits, and cosmonaut Vladimir Komarov was killed when his Soyuz descent module malfunctioned during re-entry and hit the ground at over 100 miles per hour. The Soviets were so secretive about their space program at the time that they even rejected a U.S. offer to send a representative to Komarov's funeral. They told us the ceremony was "private." This was the harsh and curious reality of the early days of space exploration.

Our relationship with our former Soviet competitors has changed fundamentally in the last forty years. I was fortunate enough to be a part of this change—from commanding our ASTP mission to advising on the International Space Station now being built in orbit. In recent years I have learned that the Soviets really did want to beat us to the Moon in the 1960s. In fact, they had several programs designed to upstage Apollo, but their space and missile programs were starved for cash and torn by competition among their leaders. One of the biggest setbacks to their space program was the death of Sergei Korolev in January 1966. Korolev was the leader of OKB-1, the design bureau responsible for virtually all of the Soviet successes in the early space race. Without Korolev's leadership and ability to get things done in the Soviet bureaucracy, the brilliant folks who worked on the Russian space program were unable to respond effectively to the Gemini and Apollo programs.

Since the late 1980s, the answers to the questions we had in the 1960s have trickled out in the Russian press, in books, and in frank conversations with the participants. I've heard many stories about the bravery, success, and tragedy that our Russian colleagues faced in their space program. The Russians are great storytellers, and many of the tales about their space program are riveting. But Boris Chertok is one of the greatest storytellers of them all. And what a story he has to tell! Chertok played a part in virtually every major event in the development of the Soviet and Russian space programs. As a former deputy to Korolev at OKB-1, Chertok has an insider's perspective on the space race. He has continued to work for the same organization, now known as Rocket Space Corporation Energia, throughout his long and interesting life. In the memoirs translated here, Chertok tells his stories with compassion, humor, and an unflinching eye for the facts. This is far more than the memoir of an interesting life. Chertok has put the great sweep of twentieth century Russian history, and the role of the space program in that history, into perspective. He has pulled together an incredibly detailed narrative with a unique Russian perspective that is written in a delightfully easy style to read. The translators and editors of this English-language version of Chertok's memoirs have done a fantastic job of capturing the tone and nuance of great Russian storytelling. For anyone who has ever wondered, like me, just what was going on in the Soviet space program, this memoir will provide an invaluable and enjoyable insight.

LT. GEN. THOMAS P. STAFFORD, USAF (RET.)
Gemini VI
Gemini IX
Apollo X
Apollo-Soyuz Test Project
September 2004

Preface to the English Language Edition

IN 2001, I ACCEPTED NASA'S OFFER to translate my four-volume memoir, *Rockets and People*, into English for publication in the United States. By then I had accumulated a large number of critical remarks and requests from the readers of the Russian edition. In addition, after three Russian editions had come out, I myself came to the conclusion that in the new edition I must make additions and changes that make it easier for the American reader to understand the history of Soviet cosmonautics. As a result, this new English-language edition is far from being a word-for-word translation of the Russian edition. I changed the total number of chapters and their arrangement among the volumes (to more strictly adhere to chronology) and took into consideration some of my readers' criticisms as far as the need to add information and make clarifications to make it easier to understand complex events.

As a result of these changes, the description of the flight of Yuriy Gagarin in the English-language version has been moved from volume 2 to volume 3. Correspondingly, part of the material in volume 3 of the Russian edition has been moved to volume 4 in the English version.

Making additions, changes, and revisions to the text proved to be much more difficult for me than doing a rewrite. By the way, this is true not only of printed works. The history of aerospace technology abounds with cases where more effort went into modifying and changing rockets or spacecraft after they had been put into service than on the development of the prototypes.

To begin creating an improved four-volume edition at my age is a risky undertaking. Throughout 2003 and 2004, the texts of the first two volumes of the new edition were handed over to NASA. I still hope to finish working on the new edition of volumes three and four in 2005. Huntsville veteran Jesco von Puttkamer has rendered me great moral support.

Over the course of e-mail correspondence and personal meetings, he has convinced me that the work on "Project Chertok" has been met with enthusiasm in NASA's historical research department.

I express my sincere gratitude to all those at NASA Headquarters who are assisting in the publication of my memoirs.

I am particularly grateful to Asif A. Siddiqi, who has agreed to be my editor. His erudition, command of the Russian language, and profound knowledge of the history of Soviet aviation and cosmonautics are a guarantee against possible errors.

Mikhail Turchin has rendered invaluable assistance to "Project Chertok." He transcribes my manuscript notes into electronic copy, keeps a list of the individuals mentioned in each chapter, scans photographs, and handles the transmission of all the information to NASA. He also edits all the material and gives valuable advice on the structure of the books.

I am sincerely grateful to the veterans of cosmonautics whose valuable comments have provided a very strong stimulus for working on the new edition of my memoirs.

Boris Chertok
Moscow
October 2004

A Few Notes about Transliteration and Translation

THE RUSSIAN LANGUAGE IS WRITTEN using the Cyrillic alphabet, which consists of 33 letters. While some of the sounds that these letters symbolize have equivalents in the English language, many have no equivalent, and two of the letters have no sound of their own, but instead "soften" or "harden" the preceding letter. Because of the lack of direct correlation, a number of systems for transliterating Russian (i.e., rendering words using the Latin alphabet), have been devised, all of them different.

Russian Alphabet	Pronunciation	US Board on Geographic Names	Library of Congress
А, а	ă	a	a
Б, б	b	b	b
В, в	v	v	v
Г, г	g	g	g
Д, д	d	d	d
Е, е	yĕ	ye* / e	e
Ё, ё	yō	yë* / ë	ë
Ж, ж	zh	zh	zh
З, з	z	z	z
И, и	ē	i	i
Й, й	shortened ē	y	ĭ
К, к	k	k	k
Л, л	l	l	l
М, м	m	m	m
Н, н	n	n	n
О, о	o	o	o
П, п	p	p	p
Р, р	r	r	r
С, с	s	s	s
Т, т	t	t	t
У, у	ū	u	u
Ф, ф	f	f	f
Х, х	kh	kh	kh
Ц, ц	ts	ts	ts
Ч, ч	ch	ch	ch
Ш, ш	sh	sh	sh
Щ, щ	shch	shch	shch
ъ	(hard sign)	″	″
ы	guttural ē	y	y
ь	(soft sign)	′	′
Э, э	ĕ	e	ĭ
Ю, ю	yū	yu	iu
Я, я	yă	ya	ia

* Initially and after vowels

For this series, Editor Asif Siddiqi selected a modification of the U.S. Board on Geographic Names system, also known as the University of Chicago system, as he felt it better suited for a memoir such as Chertok's, where the intricacies of the Russian language are less important than accessibility to the reader. The modifications are as follows:

- the Russian letters "ь" and "ъ" are not transliterated, in order to make reading easier;
- Russian letter "ё" is denoted by the English "e" (or "ye" initally and after vowels)—hence, the transliteration "Korolev", though it is pronounced "Korolyōv".

The reader may find some familiar names to be rendered in an unfamiliar way. This occurs when a name has become known under its phonetic spelling, such as "Yuri" versus the transliterated "Yuriy," or under a different transliteration system, such as "Baikonur" (LoC) versus "Baykonur" (USBGN).

In translating *Rakety i lyudi,* we on the TTI team strove to find the balance between faithfulness to the original text and clear, idiomatic English. For issues of technical nomenclature, we consulted with Asif Siddiqi to determine the standards for this series. The cultural references, linguistic nuances, and "old sayings" Chertok uses in his memoirs required a different approach from the technical passages. They cannot be translated literally: the favorite saying of Flight Mechanic Nikolay Godovikov (chapter 7) would mean nothing to an English speaker if given as, "There was a ball, there is no ball," but makes perfect sense when translated as, "Now you see it, now you don't." The jargon used by aircraft engineers and rocket engine developers in the 1930s and 1940s posed yet another challenge. At times, we had to do linguistic detective work to come up with a translation that conveyed both the idea and the "flavor" of the original. Puns and plays on words are explained in footnotes. *Rakety i lyudi* has been a very interesting project, and we have enjoyed the challenge of bringing Chertok's voice to the English-speaking world.

TTI TRANSLATION TEAM
Houston, TX
October 2004

List of Abbreviations

BAO	Aerodrome Maintenance Battalion
BON	Special Purpose Brigade
ChK or *Cheka*	Extraordinary Commission for the Struggle with Counter-Revolution and Sabotage
DB-A	Academy Long-Range Bomber
DVL	German Aviation Research Institute
Elektrozavod	Electrical Factory
ESBR	electric bomb release
FON	Special Purpose Faculty
FZU	Factory Educational Institution
GAU	Main Artillery Directorate
GDL	Gas Dynamics Laboratory
GIRD	Group for the Study of Reactive Motion
GKChP	State Committee on the State of Emergency
GKO	State Defense Committee
Glavaviaprom	Main Directorate of the Aviation Industry
Gosplan	State Planning Commission
GTD	gas turbine engine
GTO	Ready for Labor and Defense
GTsP	State Central Firing Range
GULAG	Main Directorate of Corrective Labor Camps
GURVO	Main Directorate of Reactive Armaments
ISS	International Space Station
KB	Design Bureau
KGB	Committee for State Security
KOSTR	Design Department for Construction
LII	Flight-Research Institute
LIS	flight-testing station
MAI	Moscow Aviation Institute
MAP	Ministry of Aviation Industry
MEI	Moscow Power Institute
MEP	Ministry of Electronics Industry

MGU	Moscow State University
MM	Ministry of Machine Building
MOM	Ministry of General Machine Building
MOP	Ministry of Defense Industry
MPSS	Ministry of Communications Equipment Industry
MRP	Ministry of Radio Industry
MSM	Ministry of Medium Machine Building
MSP	Ministry of Shipbuilding Industry
MVTU	Bauman Moscow Higher Technical Institute
Narkomvoyenmor	People's Commissar of Military and Naval Affairs
NEP	New Economic Policy
NII	Scientific-Research Institute
NISO	Scientific Institute for Aircraft Equipment
NII SKA	Scientific-Research Institute for Communications of the Red Army
NII TP	Scientific-Research Institute of Thermal Processes
NKVD	People's Commissariat of Internal Affairs
NPO	Scientific-Production Association
OGPU	United State Political Directorate
OBO	Electrical Equipment Department
ODON	Separate Special Purpose Division
OKB	Experimental-Design Bureau
ORM	Experimental Rocket Motor
OS	final assembly shop
OSO	Special Equipment Department
OTK	Department of Technical Control
PPZh	field camp wives
PUAZO	anti-aircraft fire-control equipment
RD	reactive engine
RD	long-range record
Revvoyensovet	Revolutionary Military Council
RKKA	Workers' and Peasants' Red Army (Red Army)
RL	radio communications link
RNII	Reactive Scientific-Research Institute
RSB	air-to-air transceiver station
RSDRP	Russian Social Democratic Workers' Party
SKB	Special Design Bureau
SMERSH	Death to Spies
SON	fire control radars
Spetskom	Special Committee
SPU	aircraft intercom system
SVA	Soviet Military Administration
SVAG	Soviet Military Administration in Germany

TEKhNO	Department of Technological Preparations
TRD	turbojet engine
TsAGI	Central Aero-Hydrodynamics Institute
TsIK	Central Executive Committee
TsKB	Central Design Bureau
TsSKB	Central Specialized Design Bureau
VEI	All-Union Electrical Institute
VKP(b)	All-Union Communist Party (of Bolsheviks)
VPK	Military-Industrial Commission
VRD	jet engine
VTsIK	All-Russian Central Executive Committee
ZhRD	liquid propellant rocket engine
ZIKh	M.V. Khrunichev Factory
ZIS	I.V. Stalin Automobile Factory

Chapter 1

Introduction: A Debt to My Generation

On 1 March 2002, I turned ninety. On that occasion, many people not only congratulated me and wished me health and prosperity, but also insisted that I continue my literary work on the history of rocket-space science and technology.[1] I was eighty years old when I had the audacity to think that I possessed not only waning engineering capabilities, but also literary skills sufficient to tell about "the times and about myself."

I began to work in this field in the hope that Fate's goodwill would allow my idea to be realized. Due to my literary inexperience, I assumed that memoirs on the establishment and development of aviation and, subsequently, rocket-space technology and the people who created it could be limited to a single book of no more than five hundred pages.

However, it turns out that when one is producing a literary work aspiring to historical authenticity, one's plans for the size and the deadlines fall through, just as rocket-space systems aspiring to the highest degree of reliability exceed their budgets and fail to meet their deadlines. And the expenses grow, proportional to the failure to meet deadlines and the increase in reliability.

Instead of the original idea of a single book, my memoirs and musings took up four volumes, and together with the publishing house I spent six years instead of the planned two! Only the fact that the literary work was a success, which neither the publishing house nor I expected, validated it.

The Moscow publishing house Mashinostroyeniye had already published three editions of books combined under the single title *Rockets and People*. The Elbe-Dnepr publishing house translated and published these books for the German reader. Unfortunately, in the process of reissuing and translating these books, it was not possible to make changes and additions for reasons similar to the series production of technologically complex systems. As a rule, improvements are not made to works of artistic literature. But for a historical memoir, the author has

---

1. The phrase "rocket-space technology," though unfamiliar in Western vernacular, is commonly used by Russians to denote a complete system of elements that include a particular spacecraft, the launch vehicles used to put them into space, and the various subsystems involved. A comparable term in Western English would be "space technology."

From author's archives.

Academician Boris Yevseyevich Chertok speaking with news correspondents in 1992 at the site of the former Institute RABE in Bleicherode, Germany.

the right to make corrections if he is convinced, on his own or with the assistance of his readers, that they are necessary.

I am not a historian, but an engineer who participated directly in the creation of rocket-space technology from its first timid steps to the triumphant achievements of the second half of the twentieth century. Work on the leading edge of the scientific-technical front transformed creatively thinking engineers into noted figures upon whom various academic degrees and titles were conferred. As a rule, the mass media, including the foreign media, refer to such persons simply as "scientists." I, however, must warn the reader that in no way do I pretend to wear the laurels of a scholarly historian. Correspondingly, my books are not examples of strict historical research. In any memoir, narrative and thought are inevitably subjective. When describing historical events and individuals who have become widely known, the author is in danger of exaggerating his involvement and role. Obviously my memoirs are no exception. But this is simply unavoidable, primarily because one remembers what one was involved with in the past. For me, work on these memoirs was not an amateur project or hobby, but a debt owed to comrades who have departed this life. At the Russian Academy of Sciences, I served for many years, on a voluntary basis, as deputy to Academician Boris Viktorovich Rauschenbach—Chairman of the Commission for the Development of the Scientific Legacy of the Pioneers of Space Exploration. After Rauschenbach's death, I was named chairman of this commission. Work in this field not only keeps me busy, but also brings me satisfaction. At times I feel somewhat like a time machine, acting as an absolutely necessary link between times. There is nothing at the present and in the future that does not depend on the past. Consequently, my product has consumer value, for it reveals systemic links between the past and the present, and may help to predict the future.

I lived through eighty-eight years in the twentieth century. In the three thousand year history of human civilization, this brief segment in time will be noted as a period of scientific-technical revolution and of breakthroughs into realms of the macro and micro world that were previously inaccessible to humanity. The scientific-technical revolutions of the twentieth century were intertwined with social revolutions, many local wars, two hot World Wars, and one forty-year Cold War.

In September 2000, while presenting a paper in Kaluga at the Thirty-fifth Annual Lectures dedicated to the development of the scientific legacy and ideas of Konstantin Eduardovich Tsiolkovskiy, I noted that of primary importance for the theory of rocket technology were the works of Ivan Vasilyevich Meshcherskiy on the movement of a body with variable mass and Tsiolkovskiy's paper, published in 1903, the essence of which is expressed in the formula:

$$V_K = W \ln (1 + M_0/M_K).$$

Here V_K is the maximum flight speed of a rocket, whose engine ejects the gas of the spent fuel at a rate of W, while M_0 and M_K are the initial and final mass of the rocket respectively. During the first decades of the twentieth century, a narrow circle of lone enthusiasts who dreamed of interplanetary flight were captivated by the works of Tsiolkovskiy and by analogous, independent works published somewhat later by Hermann Oberth in Germany and Robert Goddard in the United States. In 1905, the German journal *Annalen der Physik* (Annals of Physics) first published Albert Einstein's special theory of relativity containing the formula:

$$E = mc^2$$

From this formula, now known to schoolchildren, it followed that mass is an enormous, "frozen" quantity of energy.

In spite of their revolutionary importance for science, the appearance of these two new, very simple formulas did not lead to any revolution in the consciousness of the world's scientific community during the first decade of the twentieth century. It was only forty years later that humanity realized that technological systems using these fundamental principles of rocket-space and nuclear energy threatened its very existence.

Examples of discoveries by lone scientists that did not initially cause excitement but later would stun humanity are not isolated. During the second half of the twentieth century, the formulas of Tsiolkovskiy and Einstein entered high school textbooks, and strategic nuclear missile armaments came to determine the political climate on planet Earth.

The peoples of the former Soviet Union enriched civilization with scientific-technical achievements that have held a deserved place among the principal victories of science and technology in the twentieth century. In the process of working on my memoirs, I regretfully became convinced of how many gaps there are in the history of the gigantic technological systems created in the Soviet Union after the Second World War. Previously, such gaps were justified by a totalitarian regime of secrecy. Currently, however, it is ideological collapse that threatens the objective recounting of the history of domestic science and technology. The consignment to oblivion of the history of our science and technology is motivated by the fact that its origins date back to the Stalin epoch or to the period of the "Brezhnev stagnation."[2]

The most striking achievements of nuclear, rocket, space, and radar technology were the results of single-minded actions by Soviet scientists and engineers. A colossal amount of creative work by the organizers of industry and the scientific-technical intelligentsia of Russia, Ukraine, Belarus, Kazakhstan, Armenia, Georgia, Azerbaijan—and to one extent or another all the republics of the former Soviet Union—was invested in the creation of these systems. The alienation of the people from the history of their science and technology cannot be justified by any ideological considerations.

I am part of the generation that suffered irredeemable losses, to whose lot in the twentieth century fell the most arduous of tests. From childhood, a sense of duty was inculcated in this generation—a duty to the people, to the Motherland, to our parents, to future generations, and even to all of humanity. I am convinced that, for my contemporaries and me, this sense of duty was very steadfast. This was one of the most powerful stimuli for the creation of these memoirs. To a great extent, the people about whom I am reminiscing acted out of a sense of duty. I have outlived many of them and will be in debt to them if I do not write about the civic and scientific feats that they accomplished.

---

2. The "Brezhnev Stagnation" refers to the period in the 1970s characterized by economic stagnation, the suppression of political and artistic dissent, and the growth of a massive inefficient bureaucracy.

Chapter 2

On the Times and My Contemporaries

This chapter should not be considered part of my memoirs, recollections, or reflections, but rather excerpts from the history of rocket-space technology and state politics that I deemed necessary to cite at the very beginning of my memoirs in order to move the reader into the "frame of reference" that will facilitate his familiarization with the subsequent content of my works.

Rocket-space technology was not created in a vacuum. It is worth remembering that during the Second World War the Soviet Union produced more airplanes and artillery systems than our enemy, fascist Germany. At the end of the Second World War the Soviet Union possessed enormous scientific-technical potential and defense industry production capacity. After the victory over Germany, American and Soviet engineers and scientists studied Germany's developments in the field of rocket technology. Each of these countries profited from the captured materials in their own way, and this had a decisive role at the beginning of the postwar phase of the development of rocket technology. However, all of the subsequent achievements of our cosmonautics are the result of the activity of our own scientists, engineers, and workers.

I will attempt to briefly describe the foundation upon which cosmonautics was erected and the role of individuals in the history of this field of science and technology.

In the history of our rocket-space technology, a decisive role belongs to Academician S. P. Korolev and the Council of Chief Designers created under his leadership—a body that was unprecedented in the history of world science.

The initial members of the Council were: Sergey Pavlovich Korolev, Chief Designer of the rocket system as a whole; Valentin Petrovich Glushko, Chief Designer of rocket engines; Nikolay Alekseyevich Pilyugin, Chief Designer of the autonomous guidance system; Mikhail Sergeyevich Ryazanskiy, Chief Designer of radio navigation and radio guidance systems; Vladimir Pavlovich Barmin, Chief Designer of ground fueling, transport, and launch equipment; and Viktor Ivanovich Kuznetsov, Chief Designer of gyroscopic command instruments.

The role played by Mstislav Vsevolodovich Keldysh was also very great. He was considered the "chief theoretician of cosmonautics" and was actually the organizer of a school of mathematics that solved many practical problems of rocket dynamics.

As the sphere of its activity expanded, the Council was enriched with new names, and in the ensuing years of the space age its ranks have included Aleksey Mikhaylovich Isayev, Semyon Ariyevich Kosberg, Aleksey Fedorovich Bogomolov, Andronik Gevondovich Iosifyan, Yuriy Sergeyevich Bykov, Armen Sergeyevich Mnatsakanyan, Nikolay Stepanovich Lidorenko, Fedor Dmitriyevich Tkachev, Semyon Mikhaylovich Alekseyev, Vladimir Aleksandrovich Khrustalev, Gay Ilyich Severin, and Aleksandr Dmitriyevich Konopatov.

All of these people recognized Korolev as the leader, director, and commander of Soviet cosmonautics. Each of the individuals listed above had the official title of Chief Designer. Each became the founder of his own school, developing his own special field of emphasis. The ideas that emerged in the organizations of these Chief Designers could only have been realized using the scientific potential of the entire nation, with the assistance of powerful industry. Hundreds of factories and industrial, academic, military, and higher education scientific institutions were drawn into the Council of Chiefs' ideological sphere of influence.

Ministers and government officials who were directly involved in rocket-space matters did not oppose the authority of the Council of Chiefs. Sometimes, they themselves participated in its work. Disposed with real economic and political power, the ruling echelons of the Soviet state by and large supported the Council's technical proposals.

Later Councils of Chiefs were created using the Korolev Council as a model. These were headed by Mikhail Kuzmich Yangel, Vladimir Nikolayevich Chelomey, Aleksandr Davidovich Nadiradze, Viktor Petrovich Makeyev, Dmitiriy Ilyich Kozlov, Georgiy Nikolayevich Babakin, and Mikhail Fedorovich Reshetnev.

The interconnectedness of these Councils was unavoidable. The Chief Designers—members of the first Korolev Council enriched by the experience of producing the first rocket systems—began the development of rockets for other Chief Designers and entered into new Councils. Glushko created engines for Korolev as well as for Chelomey and Yangel; Isayev created engines for Korolev and Makeyev; Pilyugin created guidance systems for Korolev and then for Yangel, Chelomey, and Nadiradze; Barmin created launch systems for Korolev, Yangel, and Chelomey's rockets. The development of Kuznetsov's gyroscopic systems proved to be the most universal. They were used on the majority of Soviet rockets and many spacecraft. It was only a matter of time before the command and measurement complex equipped with the radio systems of Ryazanskiy, Bogomolov, and Mnatsakanyan became the standard for everyone.

Similar technocratic structures also existed in the nuclear industry (under the leadership of Igor Vasilyevich Kurchatov) and in the field of radar (under the leadership of Aksel Ivanovich Berg, Aleksandr Andreyevich Raspletin, Grigoriy Vasilyevich Kisunko, and Boris Vasilyevich Bunkin). Their sphere of scientific-technical activity included production, scientific institutions, and military organizations.

Long before the Councils of Chief Designers, which directed the production of rocket-space technology, the aviation industry had set up its own system of chief and general designers. The name of the chief designer was given to all of the aircraft created under his leadership as something akin to a trademark. There were no democratic Councils in aviation similar to the Korolev Council until they became involved with the development of air-defense and ABM missiles.

By the end of the 1970s, the Soviet Union had the strongest technocratic elite in the world. While remaining outwardly devoted to the politics of the Communist state, the leaders of this elite did not shy away from criticizing among themselves the obvious shortcomings of the political system, the continuing offenses of the Cold War, and the persecutions of "dissident" individuals that flared up from time to time. However, the technocracy undertook no actions to exert political pressure on the powers that be. The persecution of Andrey Sakharov serves as a typical example of this. It seems to me that this case demonstrated a lack of skill in organizing politically that is characteristic of the intelligentsia in general, and the Russian intelligentsia in particular.

I HAVE TRIED TO REMEMBER the outstanding, unique individuals whom I have worked with and encountered. They were all different, and it is impossible to impose a certain standard on a Soviet scientist or Chief Designer. For all the diversity of their characters, work styles, and the thematic directions of their activity, they were characterized by common traits that distinguished their creative work substantially from the established notions about great scientists of the past. This was true not only for the people mentioned above, but also for other figures of Soviet science and technology who were involved in the scientific military-industrial complex. Perhaps a list of these general traits will, to a certain degree, serve as a response to the question as to why, despite possessing colossal potential strength, these individuals never tried to obtain real power in the country. Allow me to present my formulation of these general traits.

1. Technical creative work was a vocation, the meaning of life. Pure science was viewed not as an end in itself, but as a means for attaining technical results, and in some cases, results in the interests of state politics.
2. Individual scientific-technical creative work was combined with organizational activity and with the search for the most fruitful work methods for the teams that each Chief Designer headed. To a greater or lesser degree each one tried to be an *organizer* of science.
3. Chief Designers, scientists, and leading specialists were personally responsible to the state for the final results of their creative activity. The greatest scientists of the past never had such a degree of responsibility. Maxwell, Einstein, Rutherford, Mendeleyev, Tsiolkovskiy, Zhukovskiy, Oberth, the Curies, and others, the names of which are firmly ensconced in the history of science, were also born to create. They accomplished scientific feats, but they did not have state structures standing over them to monitor their scientific activity and demand compulsory scientific results within strictly regulated deadlines.

4. During the Second World War, science was militarized in all the warring countries. The development of new armaments—nuclear, missile, aviation, and radar—required the participation of the most eminent scientists. The warring states did not spare the means to create new types of weaponry, but demanded practical results within the shortest possible time. In the postwar years scientists were not released from this sort of military service. The circumstances of the Cold War were no less tense than the war years for all those involved in the military-industrial complex. The Iron Curtain compelled us to look in an original and independent way for a solution to the problems that had evolved. The possibility of blind imitation, mimicry, and borrowing was virtually excluded. This taught scientists and engineers to rely only on the intellectual potential of their own country.
5. Collectivism in scientific work proved to be a completely necessary means of conducting research and achieving final practical results. The most outstanding results were achieved at the interfaces of various sciences and branches of technology. These successes were the result of close collaboration among scientists from various fields of knowledge. Outstanding accomplishments were achieved only by those scientific schools whose leaders, from the very beginning, could boldly attract other talented people to their creative efforts who were capable of working under the specific conditions required for the creation of large systems.
6. Each scientist recognized himself as a member of a gigantic technocratic system closely associated with the state and with the ideology of a socialist society. Everyone was a true patriot, thinking of nothing except honest service to the Motherland. Their general worldviews differed only in the details. The general requirement for the higher echelons could be reduced to the expression, "Help, but don't interfere!"

In the twentieth century, the strongest stimulus for the development of rocket technology was its military use. Among all the nations of the world, the first leader in this field was Nazi Germany; and then, after World War II, the Soviet Union and the United States. The names of scientists who were pioneers in rocket and space technology and the founders of national schools in these fields are well known. In recent years their activities have been subjected to serious study, and this has been reported widely in scientific, historical, and even artistic literature; in movies; and through various other mass media. The development of rocket-space technology, however, was determined not only by the activity of scientists, the leaders of design and science schools, and the heroic feats of cosmonauts and astronauts, but also to a great extent by state policies.

During the Cold War, rocket-space technology became one of the determining factors in the politics of the leading nations; the struggle for world power was expressed through strategic nuclear capabilities and supremacy in space.

Theoretically, the government of any nation is a certain abstract entity standing over society. It is obliged to safeguard the economic and social structure of society, depending on the sovereign will of the people, and to be the sole source of power. However, the interests of the state, its policies, and its forms of power are determined by specific people—the political leaders that are in power, who rely on the state apparatus, which is not faceless either, but consists of specific people who are obliged to ensure the realization of strategic, economic, and social doctrines.

When studying the role of the State and its leaders in the history of cosmonautics and rocket technology in the Soviet Union and Russia, it is advisable to take the year 1933 as the starting point. For the rocket technology and cosmonautics of our country, this year is marked by an event that had very significant historical consequences—the state decree concerning the creation of the world's first Reactive Scientific-Research Institute (RNII).[1]

The initiator of this decision was Mikhail Nikolayevich Tukhachevskiy, the Deputy People's Commissar of Military and Naval Affairs (*Narkomvoyenmor*) and Deputy Chairman of the Revolutionary Military Council (*Revvoyensovet*), who was responsible for arming the Red Army. Cosmonautics is indebted to him for bringing together under one roof the theoreticians and practical engineers Yuriy Aleksandrovich Pobedonostsev and Mikhail Klavdiyevich Tikhonravov; future Academicians Korolev, Glushko, and Rauschenbach; enthusiasts of solid-fuel rocket projectiles, the future *Katyushas* or "guards mortars", Ivan Terentyevich Kleymenov and Georgiy Erikhovich Langemak; and many others. At RNII, they conducted a wide-ranging research program and developed various ballistic and cruise missiles with various types of engines.

The history of RNII is instructive in the sense that during the first years of the country's industrialization the state stimulated the broad-scale organization of work for this new, promising field of research. Tukhachevskiy, being a prominent military leader in the state system, understood that rocket building had to be supported first and foremost by leading-edge technology and modern industry, rather than by the enthusiasm of lone individuals dreaming of interplanetary flights. Therefore, in late 1933, by resolution of the Council of Labor and Defense, RNII was transferred to the jurisdiction of the People's Commissariat for Heavy Industry, which was headed by Sergo Ordzhonikidze. However, four years after the creation of the RNII, the totalitarian state executed the initiator and patron of this field of research, Marshal Tukhachevskiy, and after him the leadership of RNII: Director Kleymenov and Chief Engineer Langemak. In the midst of the mass repressions, Ordzhonikidze ended his life by committing suicide. In 1938, the state took punitive measures against Glushko and Korolev.

1. RNII—*Reaktivnyy nauchno-issledovatelskiy institut.* The word "reaktivnyy" (literally meaning "reactive") in Russian is commonly used to denote "jet propulsion" such as in the phrase "jet propulsion engine." However, strictly speaking, the word "reactive" encompasses not only air-breathing jet engines but also rocket engines which carry all of their own propellants. It was not uncommon in the 1930s and 1940s for the word "reactive" to denote both jet and rocket propulsion. In the particular case of the RNII, "reactive" was meant primarily to represent the development of what we now call rocket engines.

The institute's work was disrupted for a prolonged period of time. In the ensuing period up until 1944, no serious work was performed in the Soviet Union on long-range rockets as weapons, and much less on rockets as a means for penetrating into outer space.

In January 1937, RNII was transferred to the People's Commissariat of the Defense Industry and renamed Scientific-Research Institute 3 (NII-3).[2] Two years later, NII-3 was transferred to the People's Commissariat of Ammunitions, and in July 1942 it was directly subordinated to the USSR Council of People's Commissars and renamed the State Institute of Reactive Technology. In February 1944, the institute was transferred to the People's Commissariat of Aviation Industry and named Scientific-Research Institute 1 (NII-1). Twenty-one years later, NII-1 was transferred to the newly created Ministry of General Machine Building and given the name Scientific-Research Institute for Thermal Processes (NII TP).[3] Finally, at the end of the twentieth century, NII TP became part of *Rosaviakosmos* in 1992, and in 1993 was renamed the M.V. Keldysh Research Center.

The Soviet State, which was the first in the world to stimulate the practical development of cosmonautics, actually inhibited the development of large rocket technology for six years beginning in 1937. The behavior of leaders in a totalitarian state does not always lend itself to explanation from the standpoint of common sense.

IT WAS WINSTON CHURCHILL who stimulated the new approach of actively working on pure rocket technology in the USSR. In July 1944, he appealed to Stalin with the request that English specialists be allowed to inspect the German rocket firing range that Red Army troops were about to capture in the territory of Poland. Our troops had the opportunity to capture the Germans' most secret weapons, about which the English intelligence service knew more than ours. We could not allow that, and our specialists received an order to inspect everything that could be inspected by the attacking troops before the English would be allowed in.

I will write in greater detail in other chapters about the activity of the Germans in Poland and subsequent events regarding the Anglo-Soviet searches for German rockets.

Here, I would like to note that Stalin entrusted the inspection activity to the People's Commissar of Aviation Industry, Aleksey Ivanovich Shakhurin, and this responsibility was in turn placed on NII-1—the former RNII, which was subordinate to Shakhurin. Aviation generals and aviation scientists were among the leadership of NII-1 at that time, and it would seem that it was then that the prospects for seizing a new field of research unfolded before the leadership of the aviation industry. Stalin himself instructed Shakhurin to do this, rather than Vannikov,

---

2. NII-3—*Nauchno-issledovatelskiy institut* 3.
3. NII TP—*Nauchno-issledovatelskiy institut teplovykh protsessov.*

another powerful People's Commissar (for Ammunitions) or Ustinov (the People's Commissar for Armaments).

After the most difficult war in our history, our government needed at the very least to catch its breath, but the Cold War did not allow it. It was unrealistic to begin the development and construction of large rocket systems in a country that was starving and mutilated by war. The decision to reproduce German rocket technology on German territory—using Russian specialists assisted by German rocket specialists—was risky, but it turned out to be a very good one. Active work by Soviet specialists in the field of large rocket technology was transferred to Germany for two years (from May 1945 through January 1947). Working in Germany, we reconstructed the history of German rocket technology and correspondingly the role of the Nazi totalitarian regime in the organization of super-large-scale programs for the production of long-range ballistic missiles.

On 4 May 1945, the troops of Marshal Rokossovskiy occupied, virtually without opposition, the area of the German rocket scientific research center in Peenemünde. The reconstruction of the Germans' activity in Peenemünde was handled not by Aviation General Petr Ivanovich Fedorov, as might be expected, but by Artillery Major General Andrey Illarionovich Sokolov.

Sokolov was one of the first distinguished figures in the history of our rocket-space technology. During the most difficult years of the war, he was the person authorized to represent the State Defense Committee (GKO) during the introduction of the *Katyusha* to factories in the Urals and its subsequent production there.[4] *Katyusha* was the name the army gave to combat vehicles mounted with multiple solid fuel rocket launchers. This type of rocket armament was developed as early as 1937 at RNII. One of the primary authors of this development, RNII Chief Engineer Langemak, was executed.

Vehicle-mounted rocket launchers were not accepted as weaponry until 1941. It is very likely that this was the only Soviet weapon that stunned the Germans during the first months of the war. Serial production had to be set up almost from scratch, and troop units had to be organized for the effective mass use of the new multiple rocket launcher systems for salvo fire. At Stalin's initiative, all *Katyushas* designated for the front were combined into a new military branch: the Guards Mortar Unit of the Supreme Command Headquarters. At the end of 1944, Andrey Sokolov carried out his mission to organize the production of and military acceptance of the "Guards mortars" in the Urals and was appointed Chief of Armaments and Deputy Commander of the Guards Mortar Units.

The Guards Mortar Units had command of the *Katyushas*, which were purely a tactical rocket weapon, but the Germans' example suggested that there was no time to waste. They had to seize the initiative to produce strategic rocket

4. The GKO—*Gosudarstvennyy Komitet Oborony* (State Defense Committee)—headed by Stalin, was the principal state organization that oversaw the Soviet Union's wartime activities during World War II.

weaponry. Sokolov was instructed to head a State Commission to inspect Peenemünde. In Peenemünde, he was not only the representative of the Army Command, but also the plenipotentiary of the Central Committee of the Communist Party. He did not wait for instructions from the government, but seized the initiative and enacted his own decisions. Thus, in the very first months after the collapse of Nazi Germany, small unguided rocket projectiles, which were then called RS's, led to active work on the creation of large guided missiles—long-range ballistic missiles. Upon returning from Germany, Sokolov took a management position as chief of rocket armaments in the Main Artillery Directorate of the Ministry of Defense. Later at NII-4, his activity was of decisive importance in the organization of the command measurement complex. After the catastrophe of 24 October 1960, in which the Commander-in-Chief of the Strategic Rocket Forces was killed (Chief Artillery Marshal Mitrofan Ivanovich Nedelin), Sokolov was appointed chairman of the State Commission for flight-design tests for the first Yangel rocket: the R-16. Sokolov rehabilitated Yangel and his rocket. This was of decisive importance for the fate of Yangel himself as well as his design bureau.

Another figure whose work between 1945 and 1946 had a lasting impact on our cosmonautics was General Lev Mikhaylovich Gaydukov. Gaydukov was a member of the military council of the Guards Mortar Units; in other words, he was a military commissar—a representative of the Central Committee of the Communist Party. Having become familiar with the "partisan" organization and the Institute RABE that Isayev and myself had founded in July 1945 in the village of Bleicherode, he understood that rapid governmental support was needed for this initiative from the very top. During August and September 1945, Gaydukov was gearing up for frenetic activity by using his personal connections with Central Committee members and two Deputy Chairmen of the Council of Ministers, Vyacheslav Aleksandrovich Malyshev and Nikolay Aleksandrovich Voznesenskiy. Bypassing the all-powerful Lavrentiy Beriya, he was received by Stalin. Gaydukov reported on the work being conducted in Germany to restore German rocket technology and asked Stalin to allow the temporary posting to Germany of known specialists in rocket technology, the former *zeki* who worked in the so-called Kazanskaya *sharaga*.[5] Gaydukov's list included Korolev, Glushko, Sevruk, and another twenty former "enemies of the people." After returning to Germany, Gaydukov headed the Institute Nordhausen, which pursued several different projects. Korolev was appointed Chief Engineer.

If it had not been for General Gaydukov's exceptional energy and bold decisionmaking, it is possible that many names—including Korolev, Glushko, Pilyugin, Mishin, Chertok, and Voskresenskiy—would not be listed today among the pioneers of Russian cosmonautics.

---

5. *Zek* (prisoner) and *Sharaga* (prison work camp) were slang terms commonly used during the Soviet era.

The little *Katyushas* provided our great rocket-space technology with more than just two generals. Future chiefs of firing ranges and cosmodromes Vasiliy Ivanovich Voznyuk and Aleksey Ivanovich Nesterenko came to rocket technology from posts where they had commanded combat Guards Mortar Units.

The future chief of Scientific-Research Institute 88 (NII-88) and future first deputy to the Minister of General Machine Building, Georgiy Aleksandrovich Tyulin, was chief of staff of a *Katyusha* regiment commanded by General Aleksandr Fedorovich Tveretskiy. At the end of the war, Tveretskiy was deputy commander of a Guards Mortar Unit group. Such a position existed because the Guards Mortar Units were subordinate to the Supreme Command Headquarters. But in 1945, Tveretskiy was instructed to form the first "special assignment brigade," which began the combat launching of A-4 rockets in 1947 and R-1 rockets in 1948 at the firing range in Kapustin Yar.

Here, I would like to return again to Gaydukov's feat. Reporting to Stalin, he requested that someone from among the ministers of the defense industries be instructed to further develop and produce rocket technology. Stalin did not make hasty decisions and proposed that Gaydukov himself talk with the ministers and then prepare the appropriate resolution. Gaydukov met with Boris Vannikov, who announced that he had quite enough responsibility with the production of the atomic bomb and that it was absolutely no use talking to him about rockets. The Minister of Aviation Industry, Aleksey Shakhurin, was preoccupied with the production of jet aircraft. For him, too, the troubles of unmanned rockets seemed excessive. Armaments Minister Dmitriy Ustinov thought about it, but before making a decision sent his first deputy, Vasiliy Mikhaylovich Ryabikov, to Germany to examine everything on site.

Our rocket-space technology was obviously lucky in terms of great Chief Designers. But we were no less fortunate when it came to talented, brilliant organizers of state industry. One of the results of the Second World War was not the waning, but rather the substantial acceleration of science-intensive technologies. The push was beginning for the use of new physical principles in the creation of weaponry. The statesmen of the victorious powers devoted particular attention to fundamental scientific research. It was statesmen rather than scientists who had the primary responsibility for developing strategic doctrines to achieve national and international military security through the effective use of pure and applied science. Ustinov was one of the Soviet statesmen who met this challenge. Ustinov and Chief of the Main Artillery Directorate (GAU) Marshal Nikolay Dmitryevich Yakovlev drew up a memorandum for Stalin with proposals for organizing rocket technology work in occupied Germany and the Soviet Union. This memorandum, dated 17 April 1946, was signed by Beriya, Malenkov, Bulganin, Voznesenskiy, Ustinov, and Yakovlev. Of these six, Voznesenskiy, an outstanding economist and organizer of the national economy during the very difficult war years and during the period of transition to peacetime life, was executed in 1950 on Stalin's orders. In 1953, Beriya was tried and executed. For a certain period of time, Malenkov formally directed rocket affairs, but of the six signers, the actual organizer of the Soviet rocket industry was Ustinov. Ustinov and Ryabikov were among the main authors of the historic Council of Ministers resolution No. 1017-419ss, dated 13 May 1946 and signed by Stalin under the stamp Top Secret (Special File).

This decree laid the foundation for the creation of the country's entire rocket industry infrastructure, from the very top state agencies to the scientific research, design, and production organizations and enterprises to the military units that tested and used the rocket armaments.

The resolution created a higher state agency for rocket technology—the "Special Committee"—under the Soviet Council of Ministers chaired by Georgiy Maksimilianovich Malenkov. Not one of the Chief Designers was appointed to the Committee. They simply were not there yet in terms of power and influence. They had only begun to grow accustomed to one another in Germany at the RABE, Nordhausen, and Berlin institutes. Dmitriy Fedorovich Ustinov, Minister of Armaments, and Ivan Savelyevich Zubovich, who was relieved of his duties as Deputy Minister of the Electric Industry, were appointed deputy Chairmen of the Special Committee. The Committee was given the responsibility for the development of the new field and invested with very broad powers. The resolution of 13 May 1946 made projects concerning rocket technology the most important state mission, and it was imperative that their execution be given top priority. Georgiy Nikolayevich Pashkov, who directed the department for the development and production of armaments in Gosplan, determined the distribution of work between the ministries.[6] The Ministry of Armaments—with Minister D. F. Ustinov—was appointed the head ministry for the development and production of rocket projectiles with liquid-propellant engines. The Seventh Main Directorate, headed by Sergey Ivanovich Vetoshkin, was formed within the ministry. In the city of Kaliningrad in the Moscow region, the State Scientific-Research Institute No. 88 of the Ministry of Armaments was created using the infrastructure of artillery factory No. 88.

In the USSR Armed Forces Ministry (the former People's Commissariat of Defense and the future Ministry of Defense), the Directorate of Reactive Armaments was created within the GAU, and a corresponding directorate was created in the Navy. The military also created the NII-4 of the GAU and the State Central Firing Range (GTsP) to serve all ministries involved with missiles.[7]

A resolution defined the responsibilities of the other ministries (with the exception of the Ministry of Armaments) for the creation of reactive armaments. The Ministry of Aviation Industry was entrusted with developing and manufacturing

6. Gosplan—*Gosudarstvennaya planovaya komissiya* (State Planning Committee)—founded in 1921 by the Council of People's Commissars, was in charge of managing allocations for the Soviet economy.

7. GTsP—*Gosudarstvennyy tsentralnyy polygon.*

liquid-propellant rocket engines and conducting aerodynamic research. The Ministry of Electrical Industry was responsible for ground-based and onboard equipment for guidance systems, radio equipment, radar tracking stations, and ground testing and electrical engineering equipment. The Ministry of Shipbuilding Industry was entrusted with gyroscopic stabilization instruments, shipboard launch systems, and homing warheads for use against submarine targets. The Ministry of Machine and Instrument Building was instructed to develop launch equipment, fueling units, compressors, pumps, all of the lifting and transport equipment, and all the systems comprising the ground equipment. The Ministry of Chemical Industry was tasked with research and production of liquid propellants (fuels and oxidizers), catalysts, and plastic articles, paint and lacquer coatings, and industrial rubber articles. The Ministry of Agricultural Machine Building (the former Ministry of Ammunition) was tasked with developing explosive devices, filling warheads with explosives, and manufacturing pyrotechnic compounds. Main Directorates for Rocket Technology were created in all of these ministries. Additionally, the Ministry for Higher and Secondary Special Education was ordered to organize the training of specialists in reactive technology.

The Ministry of Armed Forces was assumed to be the final customer for every type of combat rockets. At the same time, the Ministry's specialists were obliged to participate in their development, the organization of flight tests, the creation of firing-range tables, standard documentation for troop operation, and the formation of special troop units.

The resolution of 13 May 1946 can serve as an example of the precise definition of strategic priorities at the highest state level: "Work for the development of reactive technology shall be considered the most important State mission and all ministries and organizations shall be obliged to fulfill reactive technology assignments as top-priority."[8] Over the course of many years, very important resolutions concerning rocket technology signed by Khrushchev and then Brezhnev borrowed this Stalinist turn of phrase: ". . . shall be considered the most important State mission."

Within three months after the resolution, all of the pertinent ministries had determined the organizations and enterprises responsible for work on rocket technology and the directors of the new enterprises had been appointed.

In August 1946, Lev Robertovich Gonor, Major General of the technical engineering service and one of the first Heroes of Socialist Labor, was appointed director of the head Scientific-Research Institute No. 88 (NII-88).[9] He had spent the war, from the first day to the last, as director of artillery factories. In 1942, he heroically defended the Barrikady factory in Stalingrad, and was appointed direc-

---

8. The entire text of this important government decree was declassified and first published in I. D. Sergeyev, ed., *Khronika osnovnykh sobytiy istorii raketnykh voysk strategicheskogo naznacheniya* (Chronicle of the Primary Events in the History of the Strategic Rocket Forces) (Moscow: TsIPK, 1994), pp. 227-234.

9. "Hero of Socialist Labor" was one of the highest awards bestowed upon civilians during the Soviet era.

tor of NII-88 after being released from his post as director of the Bolshevik factory in Leningrad.

During the first postwar years, the apparatus of the Communist Party Central Committee and the KGB began to zealously follow the anti-Semitic moods of Stalin. Gonor was not only a Jew, but a member of the Soviet Jewish Anti-Fascist Committee. His appointment to the post of director of the main scientific-research institute could not have passed without the approval of Stalin and Beriya. In January 1947, I arrived from Germany as chief of the guidance department and deputy to the chief engineer of the new head institute NII-88. I had been appointed on Ustinov's orders. During a confidential conversation Gonor warned me: "We were appointed to leadership posts in the new field at Ustinov's insistence. Over in Germany you did not sense that here, anti-Semitism has been inculcated into the consciousness of the officials of the state apparatus by secret instructions from above. I worked under Ustinov for many years and I could never accuse him of that syndrome. But even in his own ministry he will not be able to oppose that policy for long."

Gonor was right. In 1950, Ustinov sent him far from Moscow to Krasnoyarsk, to be director of an artillery factory. But in January 1953, during the infamous "Doctors' Plot," Gonor was arrested.[10] Ustinov could not save him. What saved him was Stalin's death.

In terms of their historical significance in the field of armaments—during the war and during the postwar organization of work on a broad spectrum of rocket-space technology—I would equate the accomplishments of Dmitriy Ustinov with the feats of Marshal Zhukov.

The Special Committee on Reactive Technology under the USSR Council of Ministers, which was created by the resolution of 13 May 1946, was soon reorganized into Special Committee No. 2 under the Council of Ministers. In all, there were three Special Committees: No. 1 was the supreme state agency on nuclear technology and No. 3 was the agency for radar and air defense issues.

In February 1951, Special Committee No. 2 was reorganized into the Main Directorate for Rocket Technology, under the USSR Council of Ministers. In 1953, it was temporarily combined with the Council of Ministers Main Directorate for the nuclear program. In 1955, a new state agency was formed on the basis of these organizations—the sole Special Committee of the USSR Council of Ministers—for the three most important strategic and most science-intensive programs: nuclear, rocket, and radar. This union, however, did not last long.

Vasiliy Mikhaylovich Ryabikov was the de facto head of these committees from 1951 through 1957. In the system of all the Special Committees, Vyacheslav

10. The "Doctors' Plot" was orchestrated by Stalin in 1953 to blame nine doctors, six of them Jewish, for planning to poison the Soviet leadership. Their arrest was a pretext for the future persecution of Jews in the Soviet Union. Fortunately, after Stalin's death in March 1953, all the accused were released.

Aleksandrovich Malyshev and Boris Lvovich Vannikov were the directors of the nuclear program, Dmitriy Fedorovich Ustinov was director of the rocket program, and Valeriy Dmitryevich Kalmykov was director of the radar program. All of these directors and the personnel of the state apparatus of the Special Committees and ministries went through a rigorous school of leadership and industrial organization during the years of World War II, and the post-victory euphoria did not dull their very acute sense of responsibility to the State. They exercised their state power in the most advanced sectors of the scientific technical front, with the intensity of the Cold War increasing each year.

During the Second World War, the Soviet Union suffered enormous material and human losses on a scale that cannot be compared with that of other countries. During the postwar years, the most urgent need was to restore the destroyed cities and towns, to provide the people with the basic conditions for work and to expand production for the entire domestic economy. In this situation, the development of science-intensive nuclear, rocket-space, and radio electronic technology diverted significant resources from many industries that were vitally necessary to the national economy.

During the first decade after victory, the country's material, intellectual, and spiritual resources were expended neither for the sake of political prestige nor to prove the superiority of socialism, but because the very real threat of nuclear aggression on the part of the United States hung over our country. The political leaders of the United States, not without reason, feared the expansion of the Soviet Union's hegemony and the weakening of American positions in Europe and Asia. Striving to eliminate the threat of nuclear aggression and World War III, the Soviet Union achieved parity with the United States in the basic types of strategic armaments. The experience of mobilizing industry during the war was used by prominent state leaders to organize work for the new fields of research using the method of "progressive-mobilization" economics. It is difficult to name the actual authors of this new economic system. It was, undoubtedly, a collective creation that relied on confidence in the heroic labor of the people, on the country's scientific-technical potential, and on the active political support of the Party's Central Committee.

One of the authors of "mobilization economics" was no doubt the outstanding statesman Academician Nikolay Aleksandrovich Voznesenskiy. During the pre-war and war years, he was Chairman of the USSR's Gosplan, a member of the State Defense Committee, and a candidate member and then full member of the Politburo. His book, *War Economy of the USSR during the Period of the Patriotic War* (1947) confirmed the nickname secretly given to him of "economic dictator." But he was a progressive dictator. He unwaveringly supported Gaydukov's rocket initiatives, and then those of Ryabikov and Ustinov in 1945, and he was one of the authors of the note addressed to Stalin concerning rocket issues in April 1946, and of the carefully formulated text of the aforementioned resolution of 13 May 1946. Voznesenskiy was too noticeable and progressive an individual in the higher leadership of the country. In 1950, he was executed for the fabricated "Leningrad Affair."[11] This was a heavy blow for our economy.

THE STORY OF THE FAMED R-7 MISSILE, which was the basis for the Soyuz-U launch vehicle known to everyone in the world, serves as one example of the rapid solution of the most difficult scientific-technical problems using progressive "mobilization economics."

In 1954, under the directorship of S. P. Korolev, who already had experience producing intermediate-range missiles, the development of the first intercontinental missile was begun. Tests were prepared for the R-5M missile, which would carry a nuclear warhead with four to five times the yield of the one dropped on Hiroshima by the Americans. It was natural that such a nuclear device, which had already been perfected, be the proposed "payload" for the warhead of the first-ever intercontinental missile. By design, a missile with such a warhead had a launch mass of 170 metric tons and a range of 8,000 kilometers. The range was specified by governmental decree as early as 1953.

At that time it was believed that such a nuclear charge was completely sufficient to have a sobering effect on the American "hawks." But on 12 August 1953, near Semipalatinsk, tests on a thermonuclear (hydrogen) bomb were successfully conducted. One of the directors of these tests was Vyacheslav Aleksandrovich Malyshev, a member of the Central Committee of the All-Union Communist Party (of Bolsheviks) [VKP(b)].[12] At that time he was Minister of Medium Machine-Building and subsequently deputy chairman of the Council of Ministers.

When it became clear that, in principle, the problem of the hydrogen bomb was solved, Malyshev confronted its creators with the task of reducing its mass and dimensions so that it would be acceptable for the missile that Korolev was designing. The effectiveness of the hydrogen bomb was ten times greater than the atom bomb. But no matter how hard they tried at the famed Arzamas-16, the mass and dimensions still greatly exceeded those of the atom bomb.[13] The charge for the atomic warhead had a mass of 1 metric ton, while the hydrogen warhead was around 3.5 metric tons. Nevertheless, on his own initiative (and who was there to ask, as Stalin and Beriya were already gone, and Khrushchev and other members of the Politburo did not know about this business) Malyshev took this data, went directly to Korolev in May 1954, and proposed changing the design of the intercontinental missile so that the atomic warhead was replaced with a thermonuclear warhead—with the compulsory stipulation that the flight range of 8,000 kilome-

11. The "Leningrad Affair" involved a massive purge of powerful Party and government officials in Leningrad in the late 1940s that was orchestrated by Stalin in order to weaken Leningrad as a power base.

12. VKP(b)—*Vsesoyuznaya Kommunisticheskaya partiya (bolshevikov)* [All-Union Communist Party (of Bolsheviks)]—was the full designation of the Communist Party of the Soviet Union during the Lenin and Stalin eras.

13. Arzamas-16 was the closed city where one of the Soviet Union's two major nuclear weapons laboratories was founded.

ters be preserved. The first calculations showed that the mass of the payload had to be increased from 3 metric tons to 5.5 metric tons while the launch mass of the rocket had to be increased by 100 metric tons.

The design had to be completely redone. In June 1954, a government resolution was issued concerning the production of R-7 two-stage ballistic missiles, which satisfied Malyshev's proposal. According to this resolution, the production of this missile was a goal of special importance to the State.

Korolev's designers, and those of allied organizations, "went nuts," to use a slang expression. Serious conflicts arose between Korolev and Barmin.[14] Now a fundamentally new system would have to be created. Thus, the now world-famous launch system for the R-7, the modern Soyuz-U, emerged. The deadlines stipulated previously by the government had to be shifted somewhat, but in accord with Korolev's famous motto "No later than May!", the first launch of the R-7 took place on 15 May 1957.

Since then, two of the first stages of the R-7 have remained the reliable basis for manned and many other space programs up to the present day. The world owes gratitude not only to Korolev and the other creators of the R-7 "packet," but also to Vyacheslav Malyshev, who in 1954, using the State's power, obliged Korolev to redesign the rocket. Unfortunately, Vyacheslav Aleksandrovich himself, the initiator of such a historic turning point in the fate of the first Soviet intercontinental ballistic missile, did not live to see its first launches. The powerful doses of radiation that he received participating in the tests of the first nuclear devices had taken a toll on his health.[15]

AS A RULE UNDER STALIN, the creation of the first scientific-research organizations and the industrial base for missile production did not take place in a vacuum, but at the expense of, and frequently to the detriment of, work in the other defense industry fields. The NII-88 head institute was created on the basis of Artillery Factory No. 88. Production of anti-aircraft guns was transferred to other factories. Factory No. 456 in Khimki, which had mastered the production of transport aircraft even before the war under license to the American firm Douglas, was appropriated from the aviation industry for use as a design bureau and factory for liquid-propellant rocket engines. NII-885 was created on the basis of an electromechanical factory to serve as an institute and factory for guidance systems. The automobile industry was hardest hit. In the city of Dnepropetrovsk, the construction of the largest automobile factory in the country had been completed. It was supposed to produce amphibious vehicles for the infantry, and in the future, all-terrain trucks and tractors. In 1951, the factory was handed over to the Ministry of Armaments for the series production of Korolev's R-1 and R-2 missiles and

14. Academician Vladimir Barmin was the Chief Designer of ground launch complex for the R-7.
15. Malyshev died of cancer in 1957.

their engines. Ustinov personally directed the reconstruction of the factory, the organization of the new production, and its staffing. Dnepropetrovsk Factory No. 586 literally mushroomed. A Council of Ministers resolution, dated 10 April 1954, called for the design department of the factory to be converted into a Special Design Bureau—OKB-586.[16] On 9 July of that year, M. K. Yangel was appointed Chief Designer of OKB-586. In 1991, Yuzhnoye Design Bureau (formerly OKB-586) was named after Academician M. K. Yangel.

The Soviet leader Stalin, and then Khrushchev and Brezhnev, devoted particular attention and patronage to the development of missile, and subsequently, space technology at Dnepropetrovsk. The present-day state-owned Yuzhnoye Design Bureau and Yuzhniy factory enable an independent Ukraine to lay claim to the title of "space power," thanks to Ustinov's initiatives in the early 1950s.

IT IS NECESSARY to pay particular attention to the role of Nikita Sergeyevich Khrushchev in the history of Russian cosmonautics.

There were many contradictions in his activity as General Secretary of the CPSU Central Committee and as head of the government. However, without any doubt one should recognize his unequivocally positive role in the history of cosmonautics. As head of state, he not only made the final decisions concerning the production of the first intercontinental missiles, but also, in spite of objections from the military, personally decided to allocate two R-7s—which were undergoing flight design tests for the Ministry of Defense—for the launch of the first artificial satellite. After this success stunned the world, he demanded that Korolev immediately launch a second satellite. In response to the timid objections of the Minister of Defense, he announced that the political success of space flights was more important to us than ten combat missiles. At that historic stage he was right. In 1959, during a visit to the United States, Khrushchev presented President Eisenhower with the gift of an exact copy of the pennant delivered by an R-7 rocket to the surface of the moon. In 1961, he very happily accepted congratulations from heads of state across the world on the occasion of Yuriy Gagarin's triumphant flight.

The logic of the Cold War required convincing proof about the advantage of the social structure in each opposing side. This political requirement advanced science and technology more than a hundred abstract scholarly dissertations.

A great service performed by Khrushchev was his skillful use of the first practical achievements in cosmonautics to unite society politically and spiritually. However, it is worth remembering that, while sparing no resources for the development of space technology, Khrushchev did not dare break the taboo imposed by the security agencies on divulging the names of the true authors of our space

---

16. OKB stood for both *Osoboye konstruktorskoye byuro* (Special Design Bureau) or *Opytno-konstruktorskoye byuro* (Experimental-Design Bureau). In the case of OKB-586, it was the former.

conquests. Academician Kapitsa wrote that the author and organizer of such a scientific feat as the launch of the first artificial satellite is entirely deserving of the Nobel Prize.[17] There is no doubt that worldwide public opinion would have been positive if the Nobel Committee had awarded this prize to the Chief Designer of the launch vehicle and first satellite. But the name of the Chief Designer was kept secret until his death, and the Nobel Prize is not awarded to anonymous authors.

AFTER THE LAUNCH of the first satellites, the rocket-space industry was forced to rapidly expand its industrial base and attract new production capacity. It needed to further improve the coordination of operations for armaments and military technology production for the conventional and new branches of the armed forces. Obsolete military doctrines needed to be revised. The interaction between industries, and the organization of new and broader cooperation between enterprises required the restructuring of industrial management.

Khrushchev consistently implemented a policy of developing the rocket-space industry to the detriment, above all, of the aviation industry. He believed that if the Soviet Union had intercontinental missiles it did not need heavy long-range bombers, and he believed the number of intermediate-range combat aircraft and ground attack aircraft could also be reduced if we learned how to crank out rockets "like sausages."

One of the first "gifts" to cosmonautics was the handing over of Kuybyshev Aviation Factory No. 1, also called Progress Factory, to the rocket industry. The factory was commissioned to take over the series production of the Korolev R-7 intercontinental missiles. In 1962, Progress Factory also took on space technology. It started its own monopoly in the field of reconnaissance and surveillance spacecraft. Today the Central Specialized Design Bureau (TsSKB) and the Progress factory are a single enterprise.

In 1951, at Fili in Moscow at the base of the huge Aviation Factory No. 23, Experimental-Design Bureau 23 (OKB-23) was formed for the development of heavy bombers. By the late 1950s, OKB-23 Chief Designer Vladimir Mikhaylovich Myasishchev was producing bombers that, in terms of their parameters, surpassed the American B-52; but in spite of his indisputable achievements, he was transferred to the post of director of the Central Aero-Hydrodynamics Institute (TsAGI). The staff of OKB-23 became Branch No. 1 of OKB-52, headed by V. N. Chelomey, while Factory No. 23, named for M.V. Khrunichev, was transferred in its entirety to the production of missiles and Proton launch vehicles. Today the Khrunichev State Rocket-Space Scientific Industrial Center is the largest rocket-space enterprise in Russia and the pride of the Russian Aviation and Space Agency. One can assume that if the Khrunichev factory had remained in the

17. Academician Petr Leonidovich Kapitsa (1894-1984), who won the Nobel Prize in 1978, was one of the pioneers of Soviet nuclear physics.

aviation industry it would have "gone begging" during the years of the devastating market reforms, as did our entire aviation industry, which was once the second most powerful in the world.

The full list of enterprises that were retooled for rocket-space production is very long. I have cited only the most significant examples.

In 1957, at Khrushchev's initiative, the defense ministries were converted into State Committees. The primary scientific-research institutes and experimental-design bureaus remained under their management, but production—primarily series production—was transferred to regional economic councils. The problem of coordinating the work of all the fields of defense technology turned out to be very acute. In this regard, in December 1957, the CPSU Central Committee and the USSR Council of Ministers decided to create the Commission of the Presidium of the USSR Council of Ministers for Military-Industrial Issues. Subsequently, this body was named the State Military-Industrial Commission of the Presidium of the USSR Council of Ministers, and later the State Military-Industrial Commission of the USSR Cabinet of Ministers.[18]

The Military-Industrial Commission was tasked with coordinating the work of the State Committees (i.e., the former ministries) and monitoring operations for the creation and very rapid production cycle of military technology, including rocket and space technology, without regard to the departmental affiliation of the executives. The Commission had the right to make rapid decisions on behalf of the State, but did not have its own money. The Ministry of Finance provided resources to the agencies only per the decisions of the Central Committee and the Council of Ministers. Dmitriy Fedorovich Ustinov was appointed the first Chairman of the Military-Industrial Commission, and simultaneously, deputy Chairman of the USSR Council of Ministers. Thus, Ustinov, until then Minister of Armaments in charge of artillery and rocket-space technology, became the actual boss of the entire military-industrial complex of the USSR. One must say that his experience, his strong-willed and decisive nature, and his (at times) very strict and exacting temperament were in the right place at the right time.

In March 1963, Ustinov was appointed chairman of the All-Russian Council of the National Economy. At a later plenary session of the Central Committee, he was selected to be Secretary of the CPSU Central Committee for Defense Issues and a candidate member of the Politburo. Finally, in 1976, the rocketeers' dream came true—Ustinov was named USSR Minister of Defense, and thereby became one of the most influential members of the Politburo.

After Ustinov, Leonid Vasilyevich Smirnov was appointed chairman of the Military-Industrial Commission. He held that post for 22 years! Smirnov was replaced in 1985 by Yuriy Dmitrievich Maslyukov, formerly of Gosplan. Subsequently the Military-Industrial Commission was headed by Igor Sergeyevich

18. This Commission was more commonly known as the Military-Industrial Commission (VPK).

Belousov, the former Minister of Shipbuilding. During the last phase of this complex period for our defense industry, the Commission was headed once again by Maslyukov. The Commission was liquidated in December 1991 after the collapse of the Soviet Union.

Traditionally, the Military-Industrial Commission oversaw nine ministries, which corresponded to nine key industries: nuclear (MSM[19]); aviation (MAP[20]); rocket-space (MOM[21]); shipbuilding (MSP[22]); radio engineering (MRP[23]); electronics (MEP[24]); defense (MOP[25]); communications equipment (MPSS[26]); and machine-building (for munitions, MM[27]).

According to data that Oleg Dmitryevich Baklanov cites in a yet-to-be-published work, by the late 1970s, the military industry, i.e., what we refer to as the military-industrial complex, was concentrated into 1,770 enterprises under these nine main ministries, of which 450 were scientific research organizations and 250 experimental-design organizations. A total of 10.45 million individuals worked in the industry.

In addition, another approximately 546,000 persons were involved in civilian industries associated with the military-industrial complex (chemical, electrical, textile, automobile, etc.). In all, in spite of the Cold War, no more than approximately 10 percent of the scientific-technical and industrial potential of the USSR was working in the interests of the military-industrial complex. This is approximately 12 million persons, or around 30 million persons when counting family members. This figure does not count the industrial and construction organizations of the Ministry of Defense, which did not formally enter into the "Big Nine" of the military-industrial complex.

From the total number of industry personnel involved in defense production, 33.7 percent worked in aerospace; 20.3 percent worked in radio engineering, electronics, and communications; and 9.1 percent worked in shipbuilding. Meanwhile, the military-industrial complex contributed more than 20 percent of the volume of the gross nation product. Thus, rocket technology and cosmonautics were far from being the only concern of the Commission on Military-Industrial Issues.

IT WOULD BE A SIMPLIFICATION to suggest that the rapid development of rocket-space technology during the period of the progressive mobilization economy was a process devoid of conflict. Not only were there confrontations between design

---

19. MSM—*Ministerstvo srednego mashinostroyeniya* (Ministry of Medium Machine-Building).
20. MAP—*Ministerstvo aviatsionnoy promyshlennosti* (Ministry of Aviation Industry).
21. MOM—*Ministerstvo obshchego mashinostroyeniya* (Ministry of General Machine-Building).
22. MSP—*Ministerstvo sudostroitelnoy promyshlennosti* (Ministry of Shipbuilding Industry).
23. MRP—*Ministerstvo radio promyshlennosti* (Ministry of Radio Industry).
24. MEP—*Ministerstvi elektronnoy promyshlennosti* (Ministry of Electronics Industry).
25. MOP—*Ministerstvo oboronnoy promyshlennosti* (Ministry of Defense Industry).
26. MPSS—*Ministerstvo promyshlennosti sredstv svyazi* (Ministry of Communications Equipment Industry).
27. MM—*Ministerstvo mashinostroyeniya* (Ministry of Machine-Building).

schools, but also embittered debates surrounding the doctrines and strategies of development between the very statesmen endowed with real power. These debates were not antagonistic, since no one was struggling to appropriate or seize state public property for mercenary aims in the interests of some clan.

A great qualitative advancement that resolved many organizational conflicts was the resolution of 1965 under the Brezhnev Politburo concerning the creation of a special Ministry of General Machine-Building (MOM). The name of the new ministry had nothing in common with the actual content of its work; however, to openly announce to the entire world that in the Soviet Union a rocket-space ministry or nuclear industry ministry had been created was considered impermissible. There was not a problem with aviation or radio engineering, but rocket-space was strictly forbidden!

Sergey Aleksandrovich Afanasyev was named Minister of "General Machine-Building." Afanasyev's biographical early life is typical of many other defense industry managers. In 1941, he graduated from the Bauman Moscow Higher Technical Institute (MVTU).[28] During the war he was a foreman and designer at an artillery factory, as well as shop chief and deputy chief mechanic. After the war he was transferred to the Ministry of Armaments, where, beginning in 1955, he headed the Main Technical Directorate. In 1957, he became deputy chairman, and beginning in 1958, chairman of the Leningrad Economic Council. In 1961, he became deputy chairman of the All-Union Council of the National Economy. And then on 2 March 1965, he was appointed MOM minister. He worked in this high-level state post for eighteen years!

Afanasyev left for another ministry—not voluntarily or for reasons of health, but at the will of a member of the Politburo, Minister of Defense Marshal Ustinov. In 1965, unexpectedly for many, Ustinov promoted Afanasyev to the post of minister of the rocket industry and supported him in every possible way. However, he did not forgive him for his opposition in the conflict that we called the "little civil war," which resulted from disagreements between General Designers Yangel and Chelomey on principles of defense doctrine and the development of strategic nuclear missile armaments. The "little civil war" began in 1964 under Khrushchev and did not end until 1976 when Ustinov, having become Minister of Defense, ended it.

Ustinov profoundly felt and understood the importance of science for state security. At one of the meetings on the lunar program, while addressing the president of the USSR Academy of Sciences, Academician Keldysh, he said that science must be the "headquarters" of government. He also suggested this to Afanasyev, who was beginning his ministerial career. The fact that Afanasyev succeeded in convincing Brezhnev to transfer to his new ministry not only all of the main rocket-space design bureaus, scientific-research institutes, and series

28. MVTU—*Moskovskoye vyssheye tekhnicheskoye uchilishche.*

production factories from the other State Committees, but also all of the other allied enterprises as well, should be considered his first personal contribution.

Under Afanasyev, the legendary Council of Chief Designers was placed for the first time under a single ministry. The Chief Designers were transferred as follows: Korolev came from the State Committee on Defense Technology; Glushko came from the State Committee on Aviation Technology; Pilyugin and Ryazanskiy came from the State Committee on Radio Electronic Technology; Barmin came from the State Committee on Machine and Instrument Building; Kuznetsov came from the State Committee on Shipbuilding.

Accordingly, the tens of factories that fulfilled the orders of these Chief Designers as well as other factories were transferred to MOM.

Georgiy Aleksandrovich Tyulin was appointed First Deputy Minister of MOM. He was very familiar with our history. He participated in the work in Germany, was a former chief of NII-4, and former Deputy Chairman of the State Committee on Defense Technology. Tyulin was an organizer/scientist, very much a kindred spirit to Korolev and to all the members of the Council of Chief Designers. He also enjoyed the support of Ustinov, but in his own subsequent work his relationship with Afanasyev did not improve. During the so-called "small civil war" he and Afanasyev were on opposite sides of the front.

MOM was given almost all of the duties that NASA had in the United States in terms of cosmonautics, plus a whole host of other responsibilities, including the production of all onboard and ground equipment; the weighty responsibility, together with the Ministry of Defense, of producing strategic nuclear missile forces; and ensuring the social well-being of all the workers in this branch of industry.

Therefore, Minister Afanasyev and the corresponding deputies had to bear the responsibility for all unmanned and manned flights, for the N1-L3 lunar program, had to speak in front of the Central Committee either for or against Chelomey's alternative proposal concerning the creation of the UR-700 lunar launch vehicle, had to sort out the debate between Minister of Defense Andrey Grechko and Yangel concerning his proposals about the mortar-type launching of strategic missiles, had to have a new mission control center built in order to be ready for the Soyuz-Apollo program, and had to keep track of construction progress at hundreds of other industrial and social facilities from Dnepropetrovsk, Ukraine to Ussuriysk in the Russian Far East.

In and of itself, the epic of the major construction of the rocket-space industry deserves separate historical study. The MOM enterprises, which formed entire towns, created new living conditions in the cities of Korolev, Khimki, Reutov, Peresvet, Yurg, Nizhnyaya Salda, Zheleznogorsk, Novopolotsk, Miass, and many others.

Independent of MOM, because it had its own budget, the Ministry of Defense made a large contribution to the creation of rocket-space "science cities." The cities of Leninsk (today Baykonur), Mirnyy (Plesetsk), Yubileynyy, Krasnoznamensk; the military towns of the command and measurement complex; and finally the grandiose complexes of the engineering and launch sites of the cosmodromes, were all primarily the contribution of the Ministry of Defense builders.

Today, our capital and other large cities are filled with advertisements about the construction and sale of luxury homes for the modern elite of Russia. For "only" $1 million you can acquire a very nice apartment. It would be useful for the modern slanderers of our history to remember that, during the period from 1966 through 1990, the rocket-space industry alone built more than 14 million square meters of residential housing. In terms of modern housing prices that is $1.5 trillion! In addition, MOM alone built schools for the general public to accommodate 59,300 pupils, kindergartens to accommodate 74,000 children, hospitals with 8,550 beds, polyclinics accommodating 19,100 visits per shift, professional institutes accommodating 14,360 students, clubs and cultural centers accommodating 7,400 individuals, sport complexes, swimming pools, dispensaries, pioneer camps, recreational areas, and holiday retreats. Practically all of this was accessible to hundreds of thousands of individuals, creating the necessary psychological frame of mind for the field of cosmonautics and the peoples' belief in tomorrow.

In light of these facts, one must say that Afanasyev was a unique minister. He, his deputies, and the directors of the allied ministries bore personal responsibility for a broad range of scientific, technological, political, and social issues that were critical to the prestige, security, and might of the State. Afanasyev and his deputies worked in close contact with many of the scientific institutions of the Academy of Sciences. Virtually every fundamental decision concerning cosmonautics was made with the participation of the President of the Academy of Sciences.

THE EXPERIENCE OF WORKING in rocket-space technology shows that success occurs where and when each individual anywhere in the enormous and complex system always performs his or her duty. In this sense, it is difficult for history to select leaders in politics, economics, and science.

It would be a mistake to think that statesmen, who possessed an elevated sense of responsibility and who were personally interested in the successes of cosmonautics, always made the optimal decisions. An example of this is the history of our N1-L3 program—the manned expedition to the Moon. The lunar program required the expenditure of great resources and the concentration of design and production capacities under a single painstakingly selected project. Khrushchev personally tried to reconcile the technical disagreements between Chief Designers Korolev and Glushko. He did not succeed. Moreover, he supported Chelomey's alternative proposals. Khrushchev, and after him Brezhnev, as the Secretaries of the Central Committee; Ustinov, who was responsible for rocket-space technology; Minister Afanasyev; the Ministers of Defense; and all the others comprising the "Big Nine" of the military-industrial complex were required to obtain the resources for achieving both military-space "strategic parity" and the peaceful lunar program from a single state budgetary pocket.

In the final analysis, parity was not only achieved, but according to some indicators, we passed the United States in terms of nuclear missile armaments. In any case, the threat of a third world war was removed. The two superpowers began to understand that bad peace was better than mutual annihilation. But the initial economic positions at the start of the Moon race were considerably stronger for the Americans.

In the early 1970s, we won the nuclear missile race, but lost the moon race. Statesmen did not support the designers' proposals to hold on to what had been begun on the lunar program, modify the N-1 launch vehicle, and gain prestigious "revenge" by creating a permanent lunar base by the end of the 1970s. It really would have been possible to achieve this. But every cloud has a silver lining. The successful landing of Americans on the Moon prompted the rapid acceptance of the program for creating a series of Salyut long-duration orbital stations, which served as the scientific-technical basis for the Mir orbital complex. The Mir station, in turn, initiated the beginning of American operations to construct an even larger orbital station.

American officials in charge of NASA were convinced that using Soviet-Russian intellect and the experience gained on Mir, it would be possible to reduce both the costs of the station and the time required for the program's implementation. Thus, the International Space Station (ISS) appeared, the total cost of which, according to predictions, would be significantly higher than the American expenditures on all seven lunar expeditions.[29]

An outstanding achievement of Soviet rocket-space technology was the creation of the Energiya-Buran reusable space system. This time, the initiative for the creation of the system, which is similar in performance to the American Space Shuttle system, originated not from scientists and designers, but from the top, from the State, which for prestige and for political purposes, and also out of fear of an American advantage in a new space system for delivering sudden strikes, spared no resources on this exceptionally complex program.

The Energiya-Buran system was the largest-scale program in the history of domestic cosmonautics. More than 1,200 enterprises and organizations and almost 100 ministries and departments participated in its creation. As a result, using mobilization economics methods, the Soviet state produced the Energiya launch vehicle, which in terms of its capabilities had no rivals in the world, since the Americans had halted work on the Saturn V superheavy launch vehicle after the lunar expeditions. The Energiya superheavy launch vehicle offered cosmonautics the broadest of possibilities. The Buran space carrier vehicle in its single unmanned flight also demonstrated advantages over the American Space Shuttle. The success of the Energiya-Buran program was made possible by the fact that expenditures on the "nuclear missile shield" were substantially reduced after

29. There were seven attempted Apollo lunar landings. One of them, Apollo 13, did not reach the surface.

parity was achieved and treaties had been concluded for the reduction of strategic offensive arms.

In a struggle for personal power, new statesmen—without asking the permission of their people—destroyed the Soviet Union with a swiftness that not even its most ardent enemies could have dreamed. The collapse of the Soviet Union led to the creation of "sovereign" states that were not concerned with the fate of cosmonautics. They were no longer in the mood for Energiya and Buran. In spite of the efforts of the leading enterprises, which had developed proposals for the practical use of the very rich resources on hand from the Energiya-Buran system, all operations were terminated.

RUSSIA IS THE LEGAL HEIR of the space programs and achievements of the Soviet Union. Over the last decade of the twentieth century, Russian cosmonautics virtually had no state support. According to various estimates, the amount of actual funding from the Russian state budget for the rocket-space industry was between 0.05 and 0.01 percent of the amount it received in the mid-1980s. The new Russian governments that came to power and rapidly replaced one another proved to be incompetent and did not want to assess and claim their very rich intellectual and technological legacy. But then they quickly assessed the tremendous value of our natural resources and legalized their plunder with impunity. An enormous safety margin enabled our cosmonautics to survive under conditions of general systemic crisis; economic crisis; and ideological, moral, and spiritual crises.

The new government leaders were in a great hurry to reject previous political and social doctrines, and together they discarded our enormous experience in state planning and the organization of the economy. In exchange, neither a new strategy nor new prospects appeared. A significant portion of the rocket-space enterprises were organizationally combined under the management of the Russian Space Agency (RKA). This saved them from the threat of greed-driven privatization and plunder. The aviation industry did not act fast enough to carry out a similar self-organization at the State level. Under the threat of the total destruction of the domestic aviation industry, the latest Russian government handed over the surviving portion of this industry to the space agency. The RKA was converted into the Russian Aviation and Space Agency (RAKA or *Rosaviakosmos*).[30]

Modern cosmonautics has acquired vital military strategic importance. The military doctrines of the United States for the twenty-first century stipulate fundamentally new methods for conducting future "local" wars. Space navigation, space "omniscience," communications, and data transmission and control play a defining role in them. During the times of the Soviet Union, our military-space and missile forces, in terms of their potential capabilities, nearly matched similar American forces.

30. In 2004, Rosaviakosmos became the Federal Space Agency.

Chapter 3

Between Two Aerodromes

Almost every type of literary memoir proves that we are products of our childhood. Normally, childhood is defined as ranging from infancy to the early school years. I am not about to break with literary tradition, and I will confirm this thesis with an example of my own biography.

My passport and all manner of questionnaires that I have filled out over the course of at least eighty years claim that I was born 1 March 1912, in the city of Lódz, Poland.

The "initial conditions" imposed by an individual's parents and childhood environment determine the fate of the majority of people. Initial conditions are the critical factor in solving any problem in the scientific disciplines. There are no super-powerful computer systems that modern science or the science of the foreseeable future can use to describe the life of a human being. Literary memoirs solve this problem with an essential simplification of reality and the initial conditions. I will begin the description of my initial conditions, as in classic memoirs, with my parents.

When I arrived in this world, my father was forty-two and my mother was thirty-two. I was an only child. My mother, Sofiya Borisovna Yavchunovskaya, was the third daughter in a wealthy Jewish family in the city of Gomel.[1] Her parents, my grandmother and grandfather, had five daughters. I never met any of my grandparents. My aunts and uncles, under various circumstances and for various reasons, unanimously confirmed that of the five Yavchunovskiy daughters, Sofiya, my future mother, was the most beautiful and capable.

Finding worthy suitors for five daughters who met the strict requirements of wealthy and conservative parents was not easy. No one doubted that the beautiful Sofiya would be left wanting. Quite unexpectedly, however, Sofiya Yavchunovskaya was expelled from her last year of preparatory school for belonging to an illegal revolutionary organization. To the horror of her family and admirers, she expressed no remorse for her beliefs, and instead turned into a professional revolutionary. To be independent, she completed doctor's assistant and midwifery courses in St.

---

1. Gomel is located in the southeastern portion of present-day Belarus.

Petersburg, and returned home an active member of the Russian Social Democratic Workers' Party (RSDRP).[2]

After the crushing defeat of the revolutionary movement of 1905, she was threatened with arrest, trial, and—in the best-case scenario—exile to Siberia. Almost all of her comrades-in-arms had been repressed, and she had gone into hiding, using the connections of her venerable parents and two elder sisters, who had married representatives of the hated bourgeois class. Her only salvation was to flee Russia. Relatives quickly found a respectable but quite poor teacher named Yevsey Menaseyevich Chertok, whom they finally talked into immediate marriage and emigration.

In 1907, teacher Yevsey Chertok and midwife Sofiya Chertok lived peacefully for some time in Poland. There, the elder sister of Sofiya's mother, who had married a Lódz textile manufacturer, watched over them. However, Poland was not outside the reach of the czarist secret police and my future parents emigrated farther west.

In Germany, Father learned bookkeeping and perfected the German language, and in France my future mama completed her education in medicine and mastered French so that she could study the history of the French Revolution in the original texts. In 1911, my parents returned to the Russian Empire and began to work in Lódz. My father worked as a bookkeeper in the textile industry. Mother worked as a nurse and finally started to dream of having a baby. She ended any activism, but remained a sympathizer of the Mensheviks and the Jewish social democratic party Bund.[3] In March 1912, Mother completely switched her energy for absorbing revolutionary activity over to caring for my health and upbringing.

At the very beginning of the First World War, my parents decided that Poland was an unsafe place for their only son, and they moved to Moscow. And so, in 1914, I became a Muscovite. I remember the red brick, four-story building on Olkhovskaya Street near Razgulyay. We lived there until the autumn of 1917.

The structure of human memory is amazing. I do not remember important telephone numbers and often I am not capable of recalling events from the previous week. It is my "random-access memory" that cannot perform those operations. But my long-term memory still remembers the broad street filled with jubilant, shouting, and singing demonstrators with a multitude of red flags. My father is holding me tightly by the hand and the whole time saying, "Look, we're having a revolution." This was the February Revolution of 1917.

I vaguely remember the platform at some train station. For some reason, soldiers were on the roofs of the freight cars pushing about with bags, and old women were shouting things I couldn't understand. Mother offered something

---

2. The Russian Social Democratic Workers' Party—*Rossiyskaya sotsial-demokraticheskaya rabochaya partiya* (RSDRP)—was one of the most important political groups who opposed Tsarist rule. The party was established in 1898, and in 1903 split into two factions, the Bolsheviks and the Mensheviks.

3. The Bund was one of the earliest Jewish socialist political parties and helped to galvanize various Jewish groups in Russia.

From the author's archives.

The author at the age of two in Lódz, Poland—1914.

shiny to some man, and he measured out something for her into a white bag—maybe flour, or groats.

I was often sick. A significant portion of Father's small income went for consultations with luminaries on childhood diseases. But it was not I who became seriously ill, but Father. He became severely lame, and I remember the diagnosis was "water on the knee." It was very difficult for him to get to work in the center of Moscow. Uncle Moisey, the husband of my mother's elder sister, came to his aid. He offered both of my parents work and an apartment on the far outskirts of Moscow—beyond the Presnenskaya Gate. There, at the textile factory, they needed a good bookkeeper and wanted to set up some kind of medical attendant's office, preferably with a midwife, because there were no maternity homes in the surrounding areas. We received a two-room apartment in a one-story wooden house. The same building housed the factory office, and they set up the medical nook in the hallway.

There were no city conveniences like running water, a sewer system, or central heating, but to make up for that, our windows overlooked an apple orchard. Clean air, nature, and produce—supplied by the closest villages, Mnevniki and Shelepikha—were far more available here than in Razgulyay.

Our factory-owned house stood right on the bank of the Moscow River, which at that time was very clean. To the south of the factory were the vegetable gardens of the peasants from the village Shelepikha, and the potato and rye fields of the villages Mnevniki and Khoroshevo extended from the northern side.

I am eternally indebted to my parents. In the first place, I am grateful to them for selecting the place where we lived. They chose wisely. In order to show how the geography and social environment of the former outskirts of Moscow influenced my fate, I shall excavate some not yet faded fragments from my memory. I have selected something that, to my way of thinking, should be of interest to everyone who is interested in the unique history of Russia and Moscow.

Boys today have hardly taken their first steps before they have gained a knowledge of the different models of automobiles. For me, knowledge of transport issues began with the temperaments and nicknames of the horses that were the basis of all the factory's transport services—trips to Moscow, trips to the villages for potatoes and vegetables, and the ambulance service delivering the gravely ill to the famed Soldatenkovskaya (today Botkinskaya) hospital. The factory committee and the governing body of the Nizhnekhodynskaya textile factory, where my parents lived and worked, provided horses during the most difficult and hungry years to bring joy to the children of the workers.

From the author's archives.

The author, now a young Muscovite, photographed in 1918 at the age of six.

On New Year's Eve in 1919, these horses brought us, dozens of children, in sleighs to the first New Year's party in Russian history in the Hall of Columns of the Noblemen's Club. This was my first visit to the Hall of Columns of the future House of Unions. I have since been in that most popular old Muscovite hall innumerable times. Many visits to the Hall of Columns have been completely erased from my memory. But some of the events associated with it have been stamped in my memory forever.

I remember the New Year's celebration of 1919 in the Hall of Columns clearly after eighty years. Pieces of real white bread with jam have stuck in my memory just as clearly as the enormous fir tree that stunned my childhood imagination with its vast assortment of ornaments, the radiance of the electric chandeliers, and the music. Indeed, at home, the primary source of light was kerosene lamps. The factory, located ten kilometers from the center of the capital, did not have transmission lines from the Moscow electric grid until 1922.

For people of the younger generation, it is difficult to imagine that in today's prestigious region of Serebryanyy Bor, in Khoroshevo-Mnevniki, and along the entire Khoroshevskoye Highway people lived and worked without enjoying such basic achievements of civilization as gas, electricity, telephones, refrigerators, running water, and so on. At the factory, all of the machines were driven by a single diesel engine with a complex system of multi-stage belt transmissions. This same diesel engine also provided the factory village with light for a couple of hours in the evening.

The factory workers and their children, with whom I quickly became friends, determined the social microclimate. The Civil War was still going on, and we, of course, played "Reds and Whites," rather than "Cowboys and Indians". Nobody wanted to be White. Frequently, when visiting the workers' dormitories, which were called "bedrooms," I heard conversations about the imminent victory of "our side." Our side was the Red Army, and there were no doubts as to the just cause of the proletariat. At home, the *Sotsialisticheskiy vestnik* (*Socialist Herald*) sometimes appeared. It was an underground Menshevik newspaper that was somehow delivered to Mother. She carefully hid it, but that was the very thing that made me curious.

In all, 300 men worked at the factory, which had its own Communist party cell, its own factory trade union committee, and even a Pioneer unit.[4]

I recall that the Pioneers put on a play at the Red Army club at the Presnya station about the French Revolution. My father chuckled, "What do you know about this revolution!" But Mother encouraged my social exploits in every way possible. She was the first to tell me about the times of Robespierre, Marat, and Danton, and she explained what the Bastille and the guillotine were.

The geographic location contributed to my being a pretty good swimmer by the time I reached the age of seven, and soon my comrades and I took a liking to river voyages. We rowed against the rapid current to the mysterious Studenyy ravine. Archeology students often visited this ravine, which was really cold even on hot days. They filled their knapsacks with ancient fossils and gladly enlightened us curious natives about "who was who" hundreds of thousands of years ago. At the bottom of the ravine was a spring that produced water that was crystal clear and, as they used to say, even had healing properties.

In 1932, these places were completely wiped out by the construction of the Karamyshevskaya Dam of the Moscow-Volga Canal. To begin with, they filled in the ravine. In its place they built a camp surrounded by barbed wire for the prisoners who built the canal. Now a major transportation line runs through these sites. As we got older, we rowed upriver to Fili, and then to Kuntsev. Our dream was always to reach Krylatskoye. On its high bank, like a beacon, rose a white bell tower. It was very difficult to overcome the rapid current with our oars.

I began to read at a young age. During our journeys along the Moscow River we made rest stops on the willow-thicketed banks. My traveling companions made a campfire, roasted potatoes, and demanded one of my celebrated readings of the literature that I used to snatch from my parents' library for such occasions.

The Adventures of Tom Sawyer and *Huckleberry Finn* met with the greatest success, while Harriet Beecher Stowe's *Uncle Tom's Cabin* gave rise to aggressive moods: "Hey—after the revolution in Europe, we'll deal with the American slaveholders!"

By the time we were ten, my comrades and I got to know the outlying areas, on bicycles made by the *Duks* factory. The *Duks* factory, the future "No. 1" aviation factory, and subsequently the Progress Factory, produced surprisingly durable road bikes in those years. Skilled workers could afford to buy such a bicycle for their sons.

---

4. The *Pionery* (Pioneers) was a nationwide organization that existed during the Soviet era to inculcate socialist and communist values among youth.

The area surrounding Moscow was amazingly pure ecologically. One could judge this simply by the fact that our table was never without fish caught in the Moscow River or the factory pond, which was formed by a dam on the Khodynka River. During those hungry years of rationing cards, this was a great help.

The Khoroshevskoye Highway, which connected Krasnaya Presnya with Khoroshevskiy Serebryanyy Bor, passed a kilometer from the factory. This highway was paved with cobblestones and therefore considered to be an all-weather thoroughfare. At the age of seven, some friends of the same age and I ran off to the highway to watch the few cars that drove by. We secretly hoped we would see them blow up the powder depots located across the highway next to the Khodynskaya radio station towers. I got into a lot of trouble for such absences, since the Khoroshevskiy arsenals instilled fear in everyone living in the vicinity. There was talk that enemy counterrevolutionaries would sooner or later blow them up and then our factory and every living thing in it would be obliterated.

The interesting thing was that the arsenals really did blow up. It was the summer of 1920 and I had asked for permission once again to "go see how the radio station operates." Along the way, I saw smoke and flame of an extraordinary height dancing over the powder depots. People were running toward me shouting: "The depots are burning!" I was a pretty good runner and I took to my heels as fast as my eight-year-old legs could carry me toward the village of Shelepikha. Behind me it was already rumbling and thundering. The people running beside me were yelling that we needed to take cover behind the water tower—there was a very sturdy embankment there. When I reached the riverbank, I ran at full speed to the stone water tower building looming up ahead. Suddenly flames leapt up in front of me, and a hot wind full of sand and clumps of earth hit me in the face.

It seemed as if I had fallen into some sort of pit. Someone's strong arm pulled me out and did not let go of me. Now I tried to break into a run, drawn in by that arm. Only when I heard the cry, "Don't try to break loose, I'm still not going to let go of you!" did I recognize Vera, the young worker from the dye shop who often came to see my mother. In the rushing stream of people we reached the railroad bridge by Fili. The guard let everyone across, and finding ourselves on the other Fili bank, we were able to catch our breath. Back where I had run from, a mushroom-shaped column shot high into the air. Out of the cloud, hot clumps of some unknown material were flying in various directions. I remember being given water to drink and eating from the mess-tins of our Red Army soldiers. That night we slept in their tents.

The next day we were allowed to return. Under Vera's protection, I arrived home. My mother had been convinced that I had died or lay wounded near the depots. She had rushed to the fire, and according to eyewitness accounts, remained alive only because Red Army soldiers stopped her and held her in a shelter until the most dangerous phase of the explosions was over. Then they helped her in fruitless searches, and finally despairing, they drove her home.

A line had already formed there of people with minor injuries who needed bandages. Amazingly, no one had been killed or seriously wounded among all our acquaintances.

Our neighbor—the factory foreman, a former artilleryman—had taken cover during the explosions with his entire family in the cellar. When the most terrifying part was over, he also attempted to search for my remains. When he saw me whole and unharmed, with my mother almost beside herself, he offered to give me a thrashing as a warning. He often used this method on his own five children. This time his offer was not accepted. But soon another occasion presented the opportunity for such an educational lesson.

The immediate destruction and fires from the depots in the area blowing up were surprisingly small. However, thousands of artillery shells of various calibers, hand grenades, and boxes of cartridges were scattered over a radius of more than three kilometers. Red Army units mobilized to collect this ammunition that was so hazardous for the populace and so essential to the army, but they could not manage to gather and neutralize it all. No one could clearly explain why such a large amount of artillery shells had been stored at the depots since they had been in catastrophically short supply during the czarist army's war with the Germans. The Red Army was also on hunger rations and was compelled to make use of booty captured in battles with the Whites.

Sappers, who had blown up whole piles of ordnance in the Studenyy ravine, enlightened the curious little boys as to "what was what" as far as ammunition goes. I decided that the war was continuing, and that in any case, it wouldn't do any harm to have my own arsenal. Unbeknownst to the grownups, my comrades and I stashed a couple dozen 3- and 6-inch unexploded shells and bottle-necked and fragmentation hand grenades under the porch of our house, which was at the same time my parents' place of business. Fortunately, the grenades had no fuses. We learned how to get the explosive out of the ammunition and were delighted by the lighting effects when we threw it into the campfires that we made on the riverbank. While putting away gardening tools under the porch, a grownup discovered our arsenal. The sappers were called immediately, and the factory cell of the Communist Party demanded an investigation into the possible counterrevolutionary plot to seize power in our area, or at least, at the factory and in the village of Shelepikha. I confessed our secret plan to my parents: we had been preparing a gift for the Red Army for the crushing defeat of the White Poles. The plot scenario faded, but in this case, the neighbor insisted that his educational method be applied. For the first and last time in my life, my father gave me a thrashing, using an ordinary office ruler as the implement for the corporal punishment.

The edge of the antenna field of the Khodynskaya radio station, the largest in Russia, lay just 2 kilometers from the annihilated powder depots. The steel and wood towers, which were over 100 meters high, stood at intervals of 100 meters. Strung between them, on garlands of insulators, hung the sausage-shaped antennas. A wire net for counterbalance was suspended above the ground. Barbed wire surrounded the entire area, and it was considered closed to the public. However, the factory trade union committee organized tours for workers and school children to the radio station.

At the station, I got my first glimpse into the work of the powerful wireless transmitter sending its dots and dashes into the ether in a series of blinding sparks. Next, they showed us what they called the accumulator hall. "Twelve thousand of these glass jars produce 24,000 volts. Any contact with them is lethal," explained our tour guide. It was frightening, mysterious, and terribly interesting. In another hall, the high-frequency generators were humming. Here, for the first time, I saw the famed machines of Professor Vologdin.[5] I later met the Professor himself when I was a student. I went on at least three tours of the Khodynskaya radio station, trying to understand why it was possible to hear it thousands of *versts* away.[6] Perhaps living in such close proximity influenced my subsequent passion for electricity and radio engineering.

My cousin Mikhail Solomonovich Volfson often came to visit us. He was six years my senior and knew how to hold interesting conversations about the wonders of technology. He introduced me to the literary genres of adventure and science fiction, which often resulted in conflicts with my parents. As soon as the opportunity arose, I set aside Turgenev's *Sportsman's Sketches* and became absorbed in *Aelita*, *Captain Grant's Children*, or one of the books from James Fenimore Cooper's famous Indian series.[7] Somehow Father took me to Moscow, and for the first time I found myself at a real movie. It was the Ars movie house on Tverskaya Street. On the screen I saw the Martian beauty Aelita and was completely stunned. Here was my calling in life. It was possible to receive secret signals via radio from Mars: "Anta, Odeli, Uta!" This meant: "Where are you, son of Earth?"

And so I became mad about radio engineering. This enthusiasm went in tandem with an enthusiasm for airplanes. About five kilometers to the east of our house was the notorious Khodynka. There, on the day of the coronation of Tsar Nicholas II, hundreds of people died in a stampede for free vodka. In the twentieth century, they filled in the multitude of pits and ditches and it became the Central Airfield of the Republic. I used to love to go to the airfield with my friends, and sometimes alone. After making myself comfortable in the sweet-smelling grass, I'd watch the takeoffs and landings of the airplanes, which looked a lot like bookcases whose shelves were tied together with string.

---

5. Corresponding Member of the Academy of Sciences Valentin Petrovich Vologdin (1881-1953) was a famous Russian scientist in the field of radio engineering.

6. "Verst" is an archaic Russian unit for distance. One verst is equal to slightly less than 1 kilometer.

7. *Aelita*, written by Aleksey Nikolayevich Tolstoy (1883-1945), was a science fiction novel originally published in 1923. It tells the story of a Soviet expedition to Mars. In 1924, director Yakov Aleksandrovich Protazanov (1881-1945) produced a famous film of the same name based on the novel. It was the first science fiction film made during the Soviet era that depicted spaceflight.

Soon my knowledge of the construction and layout of airplanes became more substantial. The same Vera who dragged me away from the firestorm of the burning powder depots married a flight mechanic who worked at Khodynka. With his assistance, I began to study the various types of aircraft. Whenever I saw an airplane in the sky, I had to make a public announcement of its name. And there were a multitude of them, single- and two-engine, biplanes, monoplanes, even triplanes, and all of them were foreign: Junkers, De Havilland, Avro, Fokker, Dornier, Sopwith, Vickers, Newport. "Just wait," the flight mechanic assured us, "we will have ours." Soon our airplanes appeared at Khodynka. They were very similar to the De Havillands. These were the first domestic reconnaissance aircraft, the R-1 and R-2.

During this time of the New Economic Policy (NEP) in 1923, aviation had already arrived at our factory.[8] The meadow on the opposite side of the Moscow River became the airfield of the Junkers subsidiary factory. The Soviet government offered the German company Junkers the vacant buildings of the *Russko-Baltiyskiy* (Russo-Baltic) factory in the forest tract of Fili. German designer Hugo Junkers was the first to begin construction of all-metal military aircraft. The Treaty of Versailles forbade the production of military aircraft on the actual territory of Germany. It was therefore at this factory in Soviet Russia where the technology was created for the manufacture of all-metal aircraft. Until then, throughout the world, airplanes were constructed using wood and textile technology.

Swimming across the river, we had the opportunity to walk right up to the Junkers standing at the edge of the forest. There was no strict security. In return for doing small chores and helping the flight mechanics, we young boys were allowed to look at the airplanes and even touch them with our hands. The Fili plant had been set up to produce single-engine, two-seater Ju-20 and Ju-21 reconnaissance planes. The Junkers aircraft had the later-classic, cantilevered monoplane design made completely out of corrugated duralumin.[9] Some of the airplanes were assembled on floats rather than on wheeled landing gear. The seaplanes were hauled down to the Moscow River on special carriages. The seaplanes' takeoffs and landings on the river, which we had imagined to be our home territory, disrupted the peaceful coexistence of fishermen, Red Army soldiers bathing their horses in the river, and guests arriving from Moscow for relaxation and boat outings.

A year or two later, two- and three-engine Junkers appeared at the airfield. These airplanes also flew in the winter after the landing gear wheels were replaced with skis.

---

8. The New Economic Policy (NEP) was a program initiated by Lenin in 1921 that called for a mixed socialist-capitalist economy and significant concessions to the peasantry in Russia. The program was seen as a temporary response to the ravages of revolution, Civil War, and War Communism.

9. Duralumin, introduced widely in the aviation industry in the 1930s, was an aluminum-based alloy that was stronger and more resistant to corrosion than aluminum.

In 1923, yet another event took place that gave me reason to consider myself a person fully steeped in aviation and announce to my chums that in the future I would select a flying career. In Moscow, the First Agricultural and Crafts Exhibition opened in the area that today is Gorky Park. The exhibition was a big event in the life of the a nation making the transition from a system of War Communism to a New Economic Policy—NEP, which allowed and even encouraged capitalistic enterprise in small manufacturing and commerce. Our Nizhnekhodynskaya factory was transferred from the state sector for lease by a private stockholder company. Cloth wares from the factory, expensive for those times, were presented at the exhibition. On one of these occasions my father took me along. He was occupied with business matters, and I roamed around the exhibition the entire day studying the tents of northern reindeer herders, the *yurts* of Central Asian nomads, and the new show houses of average peasants.[10] The housing was presented together with the proprietors, live camels, reindeer, fattened horses, and various and sundry fowl. At the same time, agricultural produce from every geographical area of the country was on sale.

But for me, the most interesting sight was the passenger Junkers. On the bank of the Moscow River there was a small line of respectable citizens, who for a price unknown to me, were seated four at a time in the seaplane. It took off, circled over Moscow, and five minutes later taxied back up to the dock.

At the end of the day, after hunting for my father, I evidently made him feel so sorry for me, that after speaking with someone who obviously had connections, he led me to the line for the Junkers. The rest was like a fairy tale. For the first time in my life, I flew! Afterwards, this first "commercial" flight gave me the gumption to look down on all those who had never flown.

On 21 January 1924, Lenin died. In spite of the regular publication of bulletins concerning Lenin's grave illness, the news of his death was perceived as a great misfortune in our family and among the factory workers. The grief was sincere. I remember my mother's words: "Now everything could be lost." Father was cautious and asked her not to speak unnecessarily, especially with her many patients. During the days of mourning, the factory almost came to a standstill. In spite of the intense cold—the temperature fell to -25° C—the majority of the workers went to stand in line at the House of Trade Unions to say goodbye to Lenin.

After some domestic dissension, Mother announced that she must go to say goodbye to the great man, and she took me and other children whose parents gave their permission. Bundled up beyond recognition, eight or ten boys led on foot by my mother headed for House of Trade Unions. I remember that we often warmed ourselves by campfires that the Red Army soldiers kept burning along the length of the entire line. We entered the Hall of Columns after spending six hours in the subzero cold. I recall that someone said, "Let the children come closer." And so, for

---

10. Yurts were domed tents used by nomadic groups in central Asia.

the second time in my life I found myself in the Hall of Columns. This time it was nothing like the New Year's celebration.

The people moved slowly, trying to get a better look at Lenin lying in the red coffin. My friend Pashka Lebedev said rather loudly, "Looks like in the portrait." In response, someone immediately gave him a whack in the back of the head. My mother leaned down and whispered to me, "Look, there are Krupskaya, Bukharin, Zinovyev, and Dzerzhinskiy!"[11] But we had already been gently nudged out the door, and once again found ourselves outside in the freezing cold. The ten-kilometer march back home, now in the dark, was an arduous trial, but everyone returned without frostbite.

Two days later, a newspaper devoted to Lenin's mourning was posted on the wall of the factory. It contained my drawing "Lenin in the coffin" and a description of our march to the Hall of Columns.

After Lenin's death, the Central Committee of the Communist Party announced the "Lenin enrollment." All upright workers were called to join the Party in order to compensate for the loss that it had suffered from the death of Lenin. The factory's Party cell proposed that my mother join the Party. "Sofiya Borisovna selflessly works without regard to the time spent caring for the workers' health," announced the secretary of the Party organization. A family council was convened to discuss this issue. One of my uncles, the husband of Mother's younger sister Berta Borisovna, had been a member of the Party since 1917. He told Mama that she should not be hasty. There was to be a "purge" in the Party and she might be expelled as a former Menshevik. Being purged from the party was far worse than simply not being a member. And so, my mother, who before my birth was an active member of the Russian Social Democratic Party, remained a non-party woman.

---

11. Nadezhda Konstantinovna Krupskaya (1869-1939) was a leading Russian revolutionary who married Lenin in 1898. Nikolay Nikolayevich Bukharin (1888-1938) was an influential Bolshevik, Marxist theoretician, and economist. Grigoriy Yevseyevich Zinovyev (1883-1936) was a revolutionary, Bolshevik, and a central figure in the Communist Party leadership. Feliks Edmundovich Dzerzhinskiy (1877-1926) was a member of the Bolshevik Party's Central Committee and also headed the Soviet Union's first security police agency, the dreaded ChK (*Cheka*).

Chapter 4

School in the Twenties

In the 1920s, nine-year schools provided comprehensive education. The first four years were called the first level or primary school. The next three years completed the compulsory seven-year education. After completing seven years of school you could enter a technical school, go to work, or continue to study for another two years. For the last two years, the eighth and ninth grades, each school had its own emphasis—a specialization enabling graduates to obtain a certificate conferring on them one profession or another. Factory educational institutions (FZU) also provided comprehensive education.[1] I dreamed of getting into a radio and electrical engineering FZU, or if worst came to worst, an aeronautical engineering FZU. But there was nothing of the sort in our immediate vicinity.

In autumn 1924, I went straight into the fifth grade of the nine-year school, having passed the exams for the first level, which I got through thanks to my parents' efforts.

Father exhibited a great deal of patience and spent long hours trying to secure in my memory the fundamentals of Russian grammar and syntax. I persistently failed to understand the difference between the genitive and accusative cases, and I hated penmanship, but gladly solved arithmetic problems. I was more enthusiastic about the fundamentals of physics and chemistry than the prescribed reading of Turgenev's *Sportsman's Sketches* and Tolstoy's *Childhood*. My mother kept track of my literary training. I remember how exasperated she was when she found me in the garden reading a volume of Pushkin's selected works edited by Valeriy Bryusov and published in 1919. It turned out that the rest of the famous poem "The blush of dawn covered the east . . . " was quite involved and not at all what one would find in a primer; and many of Pushkin's epigrams were absolutely inappropriate for my age. The book was returned to the airplane mechanic who lived at our factory with the aforementioned Vera. He was surprised. "Why hide this from children? Everyone knows that Pushkin was a hooligan."

Another of Mama's passions was her desire to teach me French. To this day I regret that during my childhood, under the influence of my neighborhood friends,

1. FZU—*Fabrichno-zavodskoye uchebnoye zavedeniye*.

I showed no zeal for studying French, and later none for English. As far as German was concerned, my father laid the foundation, combining it with mathematics and Russian lessons.

I was accepted into School No. 70 of the Krasnaya Presnya region. It was located on Sadovo-Kudrinskaya Street.[2] Until 1918 that building had been occupied by a girls' preparatory school. The teachers of the lower grades, having shown their loyalty to the Soviet authority, remained in their positions. New teachers came from the former technical school that had been located next door. The girls' preparatory school and boys' technical school, the teachers told us, had been schools for the privileged in that area of Moscow. Both buildings, which were built in the late 1800s, were distinguished by the architectural monumentalism of the classical Russian Empire style and the grand scale of the interior spaces. The extremely wide hallways, spacious classrooms, excellently appointed offices, sumptuous library, and large assembly hall were all now given to workers' children. Incidentally, the class in which I was placed had considerably more children of the intelligentsia, white-collar workers, and the new NEP bourgeoisie than Krasnaya Presnya workers' children. Bordering the school building was a park with ancient linden trees. The park contained sports fields and even an equestrian school.

Sadovo-Kudrinskaya Street was very garden-like, as was the case with the entire Garden Ring of that era. Linden trees separated all of the buildings from the roadway, the width of which was mostly taken up by the streetcar tracks of Ring Line B. The street noise did not interfere with our activities in the least when we had our windows wide open during pleasant weather. The large territory of the schoolyard bordered on the sprawling Moscow Zoo. Getting a little ahead of myself, I will add that soon thereafter the country's first planetarium was built between our school and the School of Higher Marxism, which had taken over the former technical school building. After the war, the school building was transferred into a scientific-research institute for biophysics that handled the problems involved with preserving Lenin's body.

In 1984, while enlightening my grandson Boris, I felt like boasting a bit about my old school. Driving up in fine fashion in my Volga to the old familiar entrance, I tried to gain entry with him into the building. My attempt was cut-off right at the entrance. I had to change our itinerary and instead showed my grandson Patriarchs' Ponds, which in the winter months in the 1920s was one of Moscow's best skating rinks.

"After skating to exhaustion I would accompany one of my classmates to her home on Krasina Street," I explained to my grandson.

"Let's drive there. Show me," he proposed unexpectedly.

---

2. This is a northwestern segment of the *Sadovoye Koltso* or Garden Ring. Now the middle of Moscow's three ring roads, the Garden Ring marked the city limits of pre-Stalinist Moscow.

We rolled out onto the Garden Ring. Turning at Mayakovskaya Square, we arrived at Krasina Street. I drove very slowly here, convinced that the arch leading to the courtyard and the familiar building itself would be long gone. But to my surprise, everything was in place.

"OK, you saw her to her house, and then what?"

"Then I went home. I walked to Kudrinskaya Square, then I waited about thirty minutes for the bus. For five kopecks I rode the bus to the Silicate Factory on Khoroshevskiy Highway, and then I walked—or rather, since it was freezing cold, I ran the two kilometers home."

"Can we drive there?"

"Let's see."

And we set out on the reconstructed Khoroshevskiy Highway to look for the road to the former Nizhne-Khodynskaya textile factory. When I got lost on unfamiliar streets, it occurred to me that it was a lot simpler sixty years ago without an automobile. Now I had to make out the numbered Silikatnyy thoroughfares jammed with trucks. With difficulty I turned onto Karamyshevskaya Riverside Drive and finally stumbled onto a building site, beyond which I surmised where our little wooden house on the riverbank had been located in those days.

"Every day except holidays, I used to make my way from here to the school that they wouldn't allow us into today. And in the evening I went back, sometimes on foot if I missed the last bus. If I walked, I went as the crow flies, not along Khoreshevskiy, but through Krasnaya Presnya, along the Zvenigorodskiy Highway, and past the Vagankovskiy cemetery. It took me an hour to walk home from school. Judging by a modern map of Moscow, it's impossible to drive that route today.

"And look here. Down there is the river where we used to swim to our heart's content in the summer. On the opposite bank there was a sandy beach and over there, where those factory buildings are standing, was a marvelous meadow. This was the Junkers airfield and then later, Factory No. 22."

"And you flew from this tiny, little airfield?"

"I flew, though of course not as a pilot, but as a technician and later as an engineer. It's too bad I can't show you the airplanes that I flew on. Not a single museum has them."

From the not-so-distant '80s, let us return to the school days of the 1920s. As I explained to my grandson, I usually made my way to school on Leyland buses, which began making runs from Theater Square to Serebryanyy Bor in 1924. This was the first bus line in Moscow. I was only given money for the trip and for a six-kopeck loaf of French bread. Pupils received free breakfasts until the seventh grade, so the French bread took the place of lunch. Officially, the school had a liberal arts and library science emphasis; but the mathematics, physics, and chemistry teachers had just as much class time as the humanities teachers, and moreover, they showed initiative in organizing study groups for their subjects.

The physics teacher organized a radio study group. Soon the group's activity extended beyond the confines of school—I became a member of the school section of the Central Ham Radio Club, which was located on Nikolskaya Street. There I saw professor Mikhail Aleksandrovich Bonch-Bruyevich in person for the first time, and ham radio operators and engineers who were already well-known to me from radio journals: Shaposhnikov, Pavel Nikolayevich Kuksenko, and Lev Sergeyevich Theremin, inventor of the world's first electronic musical instrument. My next meeting with Kuksenko took place twenty-one years later in the office of the Minister of Armaments. I will write about the events of this meeting in the chapter "Air Defense Missiles."

In 1926, at the ham radio club at No. 3 Nikolskaya Street, Lev Sergeyevich Theremin demonstrated the first electronic musical instrument in the world, the thereminvox—the Voice of Theremin. This concert sparked enormous interest not only among radio aficionados, but also among professional musicians. The audience was captivated by the elegant thirty-year-old engineer who literally pulled sounds from the air. The wooden box had two antennas: one in the form of a loop, the second a rod. Using light, fluid hand movements Theremin changed the pitch and volume of the sound. The music emanating from out of nowhere was reminiscent first of a violin, then a flute, then a cello.

The leader of our section told us that Lev Sergeyevich first demonstrated his instrument in 1921 for the Eighth All-Russian Electrical Engineering Convention, and then again at the Kremlin for Lenin himself. Not long thereafter we were distressed to learn that there would be no Theremin concerts in Moscow for the time being—he was going abroad. I forgot about Theremin for a long time.

In 1928, I subscribed to the Technical Encyclopedia. This was a costly publication, but my parents, encouraging my passion for engineering, spared no expense. The twenty-six volumes condensed a colossal mass of technical knowledge encompassing the enormous realm of the applied science and practical technology of the time. In the last volume I discovered a description and electrical diagram of the thereminvox. It turned out that the box contained a circuit assembled from eleven electronic tubes. No mention was made about the fate of Theremin himself after he emigrated from the USSR.

Sixty-five years after the concert on Nikolskaya Street, I once again heard the sounds of the thereminvox and saw—it seemed incredible—Theremin in person. The meeting took place in the apartment of Natalya Sergeyevna Koroleva, the daughter of Sergey Pavlovich Korolev. In honor of her father's birthday, Natasha gathered his relatives and colleagues. She used to track down and invite people to these events who had known Korolev long before he became Chief Designer. Only after the gathering at Natasha Koroleva's home did I find out a bit about the astounding fate of Theremin. A talented writer or journalist could simply describe Theremin's active, creative life and—without making anything up—produce a bestseller.

Theremin emigrated to the United States from the USSR not against the wishes of the government, but with its consent. With its assistance, from 1929 through 1938, he organized and headed a company in New York that produced electronic musical instruments. The company thrived, and Theremin himself, while remaining a Soviet citizen, became an American millionaire. In America he got married, but in 1938 was called back to Moscow and immediately arrested. Taking into consideration his contributions, a special conference sentenced him to eight years in a labor camp. He was sent not to Kolyma, but to Central Design Bureau

Archives of Natalya Sergeyevna Korolev. Photo by N.A. Syromyatin.

B. Ye. Chertok enjoying a lesson on the thereminvox with its creator, L.S. Theremin, at the apartment of N.S. Koroleva, Moscow, 1986.

29 (TsKB-29), a *sharaga* headed by "enemy of the people" Andrey Tupolev.[3] There he was tasked with developing the guidance system for an unmanned radio-controlled aircraft. Convicted prisoner Sergey Korolev was a consultant on the aircraft's design.

In 1941, Theremin was evacuated to Omsk along with all of the workers at TsKB-29. By then, however, the leaders from Lavrentiy Beriya's department had found a more urgent subject for Theremin's talent: they transferred him to a *sharaga* to develop equipment for secret communications, eavesdropping, encoding, and voice recognition. Aleksandr Solzhenitsyn describes the activity of this institution

---

3. TsKB-29—*Tsentralnoye konstruktorskoye byuro 29*. Kolyma was the location of the notorious GULAG camp in eastern Siberia.

in detail in his novel *First Circle*. Theremin continued to invent until the last days of his ninety-seven-year life!

In and of itself, the meeting with the ninety-three-year-old Theremin was utterly fantastic. He arrived at Natasha's home on 12 January 1986 with his thereminvox and proposed that we try out our musical abilities. In 1926, we young radio enthusiasts had not been allowed to touch Theremin's marvelous wooden box. Sixty years later, Lev Sergeyeich Theremin himself patiently taught me to play the instrument he had invented in 1920!

LET US RETURN FROM THIS DIGRESSION to the 1920s. My parents could not provide me with enough funds to acquire new, expensive radio parts. We had only enough money for new shoes and clothes (I was growing rapidly) and for new textbooks—and I still needed skates and skis! Father gave me money for literature on radios separately. I bought all three of the popular radio journals that were published at that time: *Radio Enthusiast*, *Radio for Everyone*, and *Radio News*.[4] In order to read serious literature, I headed after school to the reading room at the Lenin Library, which was located in the famous Pashkov House, and from force of habit still often called the Rumyantsev Library. I often sat there late into the evening over the journal *Wireless Telegraph and Telephone*, John Moorcroft's *Electronic Tubes*, and novelties of radio engineering literature.[5] Sometimes I did not have enough knowledge to read such works, especially when they involved higher mathematics!

Beginning in 1923, book fairs were held on Tverskoy Boulevard once a year. There you could acquire the latest literature at lower prices. When I became a schoolboy, my parents looked over the literature syllabus for the next three years, checked through our home library, and drew up a list of the Russian classics we lacked. After providing me with money, they instructed me to go to Tverskoy Boulevard and buy the most inexpensive publications on the list. Imagine their anger when I revealed that, instead of *Hero of Our Times, Rudin*, and collected poems of Nekrasov, Blok, and Bryusov, I had bought six small books from a series on theoretical physics published in Berlin.[6] During this domestic scandal my older cousin advised my father to hide the expensive editions of *War and Peace* and *Anna Karenina* and eight leather-bound volumes of Gogol, lest I exchange them for amateur radio literature. During the wartime relocations we were unable to keep the unique edition of Tolstoy, but of the eight volumes of Gogol, only one disappeared. Five of the small volumes of the 1923 edition of *Theoretical Physics* are still in my library, intact to this day.

---

4. In Russian, the titles were *Radiolyubitel* (Radio Enthusiast), *Radio vsyem* (Radio for Everyone), and *Novosti radio* (Radio News).

5. The Russian title was *Telegrafiya i telefoniya bez provodov* (Wireless Telegraph and Telephone).

6. Nikolay Alekseyevich Nekrasov (1821-1877), Aleksandr Aleksandrovich Blok (1880-1921), and Valeriy Yakovlevich Bryusov (1873-1924) were famous Russian poets of the 19th century. The latter two were part of the Russian symbolist movement.

A year later, the domestic scandal recurred. Father noticed the absence of the three-volume *Life of Animals* by Alfred Brehm given to me as a birthday present. I confessed that I had sold the books in order to buy brand new microtubes so that I could assemble a two-tube receiver. I had urgently needed to make a receiver that was better than the one that my classmate Sergey Losyakov had devised.

I remember my school years with contentment. Learning was interesting; I had consuming new interests and a circle of good comrades. The boys were into chess, physics, and radio engineering. But there were study groups where girls also thrived. Our chemistry teacher made the potential effects of the chemical industry on our economy sound so fascinating that when he organized a chemistry study group, almost the entire female contingent of our class joined it. I excelled in chemistry and was also involved with this study group for two years. The skills that I gained during that time proved extremely useful to me later.

Another thing we were crazy about was the military. The daughter of one of the prominent military chiefs of the Moscow Military District studied at my school. Her father sponsored a military training program for the school. Small-arms groups were formed. We fired small caliber weapons once a week at the military shooting range; and then—at the Khamovnicheskoye range—we fired real "three-line" rifles.[7] At a marksmanship competition, I won what was for those times an expensive prize—a chess set. The military training course ended with mastery of the Maxim machine gun and field firing. During the course no time was wasted on drill training, regimentation, or studying the guard duty manual.

During our last two years, an emphasis was placed on library science. Having studied the classification of literature in theory, we were expected to put this into practice in large libraries. I found myself at the military library of the Central Airfield—once again on the renowned Khodynka.

The liberal arts emphasis required us to read the Russian literary classics and study art history and the history of the revolutionary movement. "Izo" was what we called the young architect who spent a great deal of time strolling around Moscow with us explaining different architectural styles. We went with him to old churches, the Tretyakov Gallery, and the Museum of Fine Arts. We could look at a building and say without error, "That's Empire," or "That's Baroque . . . Modern . . . Russian Classicism," and so on. The school had the resources to organize field trips. In the spring of 1928, I visited Leningrad for the first time.

Not only was the reading material in the nine grades completely free of charge for school children, but so were the expensive long-distance field trips. All eight classes of our school spent a week getting to know Leningrad. First we went to the tsars' palaces and saw the actual "Bronze Horseman."[8] In Kronstadt we visited the

---

7. 'Three-line' refers to a subunit of an infantry battalion.

8. The "Bronze Horseman" is a statue of Peter the Great in St. Petersburg commonly referred to by this name and the subject of a poem by Aleksandr Sergeyevich Pushkin (1799-1837).

cruiser *Oktyabrskaya Revolyutsiya.* We shivered with cold in the chambers of the Peter and Paul Fortress and admired the works of the Italian and French architects. On the verge of exhaustion, we made our way to the site of Pushkin's duel on the Chernaya River and then in the evening, out of fatigue, we fell asleep at the ballet in the former imperial Mariynskiy Theater. Leningrad won us over with its grandeur and its history. My love for this city has been preserved from those school years to this day.

The scope of the humanities we studied in school was not great, but I often recall with gratitude the teachers who, at the risk of losing their job, digressed from the instructions and literally sowed "reason, kindness, and the eternal."[9] We studied neither Russian nor world history. These subjects simply were not offered. Instead we had two years of social science, during which we studied the history of Communist ideas from Thomas More to Lenin, as well as the worldwide revolutionary movement. Our clever social sciences teacher conducted lessons so that, along with the history of the French Revolution and the Paris Commune, we became familiar with the history of the European peoples from Ancient Rome to World War I; and while studying the Decembrist movement and 1905 Revolution in detail we were forced to investigate the history of Russia.[10]

Many years later I understood what tremendous educational impact there is in direct contact with living history, with genuine works of art and architecture. I understood that this was because I received a greater dose of the humanities in a Soviet nine-year school in the late 1920s than my sons did in the postwar eleven-year schools, or my grandson sixty years later! True, they had television, modern cinema, and extensive home libraries at their disposal. However, it is one thing to feel for yourself the dampness of the Alekseyevskiy ravelin and quite another to look at the Peter and Paul Fortress in the comfort of your home on television.[11]

Of course, during those five years of school, as far as I can remember, I never had enough time. On winter evenings you still had to manage to go skating. Patriarchs' Ponds, Iskra on the Presnya, and the Young Pioneers' Stadium were three of the rinks where we would meet in our free time. We did more than just skate there. Rinks were venues for rendezvous and declarations of love. In those strict, puritanical times it was considered inappropriate for a young man of fourteen or fifteen to walk arm in arm with a young woman. But while skating, you could put your arm around a girl's waist, whirl around with her on the ice to the point of utter exhaustion, and then accompany her home without the least fear of reproach. Such was the code of honor at skating rinks. Skiing competitions were held at Petrovskiy Park. The section

---

9. The line is an excerpt from Nekrasov's poem "To the Sowers."

10. The Decembrist movement refers to one of the first attempts to overthrow the Tsarist regime. The failed coup took place in December 1825.

11. The Alekseyevskiy ravelin is the exterior fortification of Peter and Paul Fortress built in 1733 and named in honor of Tsar Aleksey Mikhaylovich (1629-1676). A "secret house" was built in 1797 at the ravelin and was used until 1884 as a prison.

of what is now Leningradskiy Prospect between the Dynamo and Aeroport metro stations was an excellent route. On Sundays it was filled with Moscow's cross-country skiing elite. The rare cabs and automobiles were no obstacle.

Schoolboys did not shy away from politics, particularly the noisy campaign over the struggle with Trotskyism.

It was the autumn of 1927. At that time, a fierce battle was going on between the Stalinists and the Trotskyite opposition. On 7 November, returning from a demonstration on the occasion of the tenth anniversary of the October Revolution, we witnessed an attempt by Trotsky supporters to make a speech on Mokhovaya Street. They hung out a portrait of Trotsky on the building of the Central Executive Committee (TsIK), where the Kalinin reception room was located.[12] Zinoviyev was giving a speech from the fourth floor balcony. Suddenly, soldiers appeared on the balcony and began to tear down the portrait of Trotsky with long poles. The crowd below broke into a rage. You could not tell who outnumbered whom, the Trotsky supporters or their opponents. A column of Trotskyite university students emerged from the gates of Moscow State University (MGU) singing the "Internationale."[13] On the street a brawl broke out; in the melee it was difficult to determine who was fighting for whom.

The next day our class, 7A, had a lively discussion about ways to fight Trotskyites entrenched in the school. During recess, shouting "beat the Trotskyites," we fought with the neighboring class, 7B. Their defenses were ready. The following slogan was inscribed on the blackboard: "Shoot the *kulak*, the NEPman, and the bureaucrat!"[14] We were met with shouts of "Opportunists, traitors!" After a slight brawl we agreed to conduct our discussions in a more civilized manner.

BY THE TIME WE HAD REACHED the sixth grade, groups had been formed according to interests. I struck up a friendship with Sergey Losyakov and Slava Kutovoy that would last for many years. It all started with each of us trying to prove his superiority in mastering the art of ham radio and his erudition in those fields of science not included in our school program. Sergey surpassed Slava and me in his mastery of Einstein's theory of relativity and Freud's psychoanalysis. To talk freely and solder circuits we usually met at Losyakov's two-room apartment on the broad Kondratyevskiy Lane, near Belorusskiy train station. Sergey was the only son of a single mother. His father had abandoned the family when Sergey was six years old. His mother worked in a railroad accounts department. She managed to make ends meet, though with difficulty, and was completely devoted to her son. Of the three of us, only Sergey had his own separate room.

---

12. TsIK—*Tsentralnyy Ispolnitelnyy Komitet.*

13. MGU—*Moskovskiy Gosudarstvennyi Universitet.*

14. *Kulak* was a derisive term used to refer to wealthy peasants after the Revolution. "NEPmen" were those traders and businessmen who benefited most from the mixed socialist-capitalist policies of the NEP era in the 1920s. The term, like *kulak*, had strong pejorative connotations when used by those who supported the Bolsheviks.

When we gathered at Slava's house on Protochnyy Lane, the close quarters of his radio nook prevented us from tinkering with a soldering iron, much less debating issues of world order. Slava's father was a Ukrainian man of letters with a vividly expressed nationalist bias. He was eager to show his son's comrades how much we were missing by not studying Ukrainian literature. Slava's mother was a literature teacher, also with a pro-Ukrainian bias. Slava's younger brother Igor fluently spoke the languages of both fatherlands, and under coercion had also mastered French and German. When we met at the Kutovoys' house, instead of fiddling with the latest single-tube regenerative receiver, we listened to fascinating stories about the works of Taras Shevchenko.[15] With obvious satisfaction and inspiration, Slava's father read to us—in Ukrainian—the poetry of Shevchenko and other Ukrainian poets who were unknown to Sergey and myself.

After finishing the seventh grade, Slava transferred to a different school. There, after the eighth and ninth grades, the students acquired the specialty of electrician. Sergey, Slava, and I remained friends. During our subsequent years as married engineers and then respectable scientists and family men, we went on vacations together, celebrated anniversaries, and yes, tried unsuccessfully to teach one another some sense. Both Sergey Losyakov and Slava Kutovoy acquired their Ph.D.'s in technical sciences. Until his final days, Sergey was a professor in the Department of Radio-Controlled Devices at the Moscow Institute of Radio Engineering and Electronics. His boyhood passion never changed.

I saw Slava Kutovoy off to his final journey in the summer of 2001. I was accompanied by scientists from the Institute of Engines, where until the last days of his life he was the most highly respected specialist on the injection, atomization, and combustion of fuel in diesel engines for tanks.

After this sad digression, the time machine again returns me to the 1920s.

In the eighth and ninth grades, during the lessons on various subjects, a practice was established whereby during the lessons on various subjects the teacher would seat the boys and the girls in pairs, underscoring the fact that, unlike tsarist times when there were separate schools for boys and girls, schools under Soviet authority were "unified and for the workers."

Mathematics teacher Mariya Nikolayevna Yakhontova was in charge of our class. She fostered in us a love for mathematics and strove to arrange the pupils in pairs so that the young people were indifferent to each other and all their attention was devoted to mathematics. During mathematics lessons I sat with the stuck-up, taciturn Nadya Sukhotskaya, who never asked for prompting, and in every way possible demonstrated her complete indifference not only to me, but to all boys in general. In German class, I sat next to the class monitor, Zhenya Taratuta. She was from a family of pre-revolutionary intellectuals. She had spoken German since childhood, honestly fought for discipline during lessons, and fearlessly defended

15. Taras Hryhorovich Shevchenko (1814-1961) was a famous Ukrainian poet and playwright.

the honor of our class during intra-school competitions. Having her as a neighbor worked to my advantage in mastering German. Zhenya Taratuta did not finish school. At the beginning of her last academic year, fulminant tuberculosis took the life of the most advanced and—in the opinion of our teachers—the most promising young woman of our class.

Freedom reigned during literature lessons. The tall, handsome, bearded and dark-haired Aleksandr Aleksandrovich Malinovskiy seated his pupils according to "ladies' choice." He read us his own poems, which were far from the prescribed syllabus, all the while attentively noting how the female contingent of the class reacted to them.

When it was announced that we were free to sit anywhere, Zoya Sudnik, generally recognized as the prettiest girl in the class, sat next to me. She was Polish. Our observant and sharp-tongued teacher did not pass up the opportunity to say to me: "Chertok, you now have the responsibility for the literary success of the beautiful Pole." We had already read Gogol's *Taras Bulba* in the seventh grade, so I wasn't shaken.[16]

I said, "Andrey's fate does not threaten me. The beautiful Pole and I have already mastered range rifle firing together and Taras Bulba doesn't have a pass to the Khamovnicheskoye range."

I never had enough time to read all the literature in the syllabus. Aleksandr Aleksandrovich knew this and did not miss his chance to put me in my place.

"Your charming neighbor will not enhance the fighting efficiency of our Red Army with her rifle. But Zoya has gotten through three volumes of *War and Peace* ahead of schedule and I hope that she will help you to get a decent grade in my class."

Soon our literature teacher got his chance to check whether his instructions regarding *War and Peace* had been fulfilled.

In the winter of 1928, in search of earnings and glory, I decided on a venture. I had gotten an idea for a fundamentally new radio receiver from the German journal *Radio für Alle* (Radio for Everyone), but having neither the time nor the means to manufacture a receiver that really worked, I decided to develop the circuitry and design on paper only. The description of the two-tube reflex receiver that I invented was very detailed. I attached its external view to a separate sheet in accordance with all the canons of design of that era. Completing the detailed article, I cited a list of European radio stations that anyone who reproduced my circuitry could successfully receive. I sent all of this to the editorial staff of the journal *Radio for Everyone.*

On a sunny spring morning on my way to school, I passed a newspaper kiosk on Kudrinskaya Square and saw my drawing in living color on the cover of *Radio*

16. *Taras Bulba* is the story of a Cossack, Taras Bulba, and his sons Ostap and Andrey. Andrey betrays the Cossacks for the love of a beautiful Polish woman and Taras kills him.

for Everyone. Shelling out every kopeck I had, I bought two copies of the journal. In class, it wasn't so much the content of the article as the very fact of its appearance that caused a sensation. It became the object of admiration, jests, and practical jokes. During our first class, mathematics, we were supposed to be called up to the blackboard to be grilled on the binomial theorem. When Mariya Nikolayevna entered the room and began to look around at her inexplicably excited class, Lev Nirenburg, the class wit, stood up and announced:

"Mariya Nikolayevna! We have a request. Rather than test our knowledge of the great Newton's binomial theorem today, let's listen to the report of a scientist who, though not yet great, is one of our own."

Nirenburg strode to the teacher's desk and placed a copy of *Radio for Everyone* in front of the perplexed and formidable grande dame of our class, whom we secretly called Maryasha.[17] To our surprise, she began thoughtfully to peruse the journal.

"Chertok! To the blackboard!"

The class fell silent. The command performance by the hero of the day, in spite of the collective appeal, did not bode well. Chertok would start swimming in Newton's theorem, and in the best-case scenario Maryasha would give him a "C" and tell him that the article would get him no special treatment when it came to evaluating his knowledge of the subject at hand.

"You will draw a diagram of your reflex receiver without referring to the journal, explaining to all of us its operating principle and why you call it a reflex receiver!"

This was so unexpected that I stood speechless for a minute in front of the stony silence of the class.

Gradually, the same inspiration came over me that I had experienced when I had thought up and drawn the diagram of the receiver at home. I spoke for about thirty minutes. To everyone's surprise, Maryasha began to pose questions about the electronic tube specifications and the variable capacitor design. When the bell rang she explained, "That's it. Today during second period I am not going to call anyone to the board—I'll waste time again on examples of the binomial theorem. Next Tuesday there will be a test for everyone. And a week from now, if the weather is good, we will take a field trip to Barvikha!" The class applauded, but during the break it was Nirenburg who was showered with gratitude for his successful initiative, not me.

After two hours of mathematics the schedule called for literature class. As soon as the teacher had taken his seat, my neighbor Zoya, inspired by Nirenburg's example, leaped up and approached our teacher with my journal. Without a trace of a smile, she loudly pronounced, "Aleksandr Aleksandrovich, instead of my report on 'The Image of Natasha Rostovaya', I propose that we listen to the new writer from our class. Here is his article." The class fell silent. Our teacher, after leafing through

---

17. Russian diminutive form connoting familiarity

the journal, returned it, commenting, "I see that today it is going to be difficult for you to transport yourselves back to the time of *War and Peace*. I excuse you, Zoya, from your report, but a week from today you and the new writer will present a report together on 'Natasha and Prince Andrey.' Meanwhile, today I will deviate from the syllabus and tell you about Russian Symbolists."

The class broke into applause. We already knew about our teacher's penchant for the Symbolists; he was a bit of a poet himself. For two hours, holding our breath, we listened to what was for those times a forbidden lecture about the poetry of Balmont, Belyy, early Blok, and Bryusov.[18]

With their initiatives, Nirenburg and Zoya had eclipsed my glory for the time being. But the following week, the school newspaper *Iskra* (Spark) was posted on the wall. In the traditional "Comic Quiz" section the following question appeared: "How do you use radio tube reflexes to change a "D" to an "A" in mathematics and literature?"

Three months later I received an honorarium of sixty rubles, which for those times was not much at all, but it was my first paycheck in the field of science. This article was the crown of my radio engineering activities during those years.

In the spring of 1929, having completed nine grades, we were triumphantly given certificates attesting to the successful completion of comprehensive school. We were released into a life where each of us was faced with selecting our own path. We of course all dreamed of going straight into an institute of higher learning. The resources that School No. 70 had given me did not go to waste.

Forty years later, after meeting with several former classmates, we calculated that our class alone had produced four doctors of science, five candidates of science, and three or four production managers. Only three girls became professional librarians; the others had obtained higher pedagogical, civil engineering, or literary education. One of our classmates even graduated from the conservatory as a pianist. Everyone whom we recalled from the graduating class of 1929 had sooner or later obtained higher education. Three did not return from the front during World War II.

My entry into school coincided with the end of the Russian State's general systemic crisis (1914–1923). Industrial production during the crisis period was one-fifth of the production level for 1913, and agricultural production two-thirds. Six years later, when we finished school, industrial production had already surpassed the 1913 level. The country's integrity was restored and the backward peasant nation had begun moving headlong toward achieving the status of a great industrial power.

A universal, burgeoning passion for technology and precise science had already begun in those years. Seven to ten individuals competed for each opening in the

---

18. Konstantin Dmitriyevich Balmont (1867–1942) and Andrey Belyy (1880–1934; his real name was Boris Nikolayevich Bugayev) were famous Russian Symbolist poets.

technical institutes of higher learning. Entrance exams were not the only barrier to admission. In addition to the usual admissions board, a sort of screening board was at work. Such boards were supposed to ensure that the vast majority of those admitted were workers who had served at least three years in an industrial internship, members of a trade union, or children of pure proletarian lineage. Next priority was given to peasants, and white-collar workers and their children were admitted to the remaining positions.

According to the "social lineage" chart, I was the son of a white-collar worker and had virtually no hope of being accepted the first time around. Nevertheless, I applied to enter the MVTU school of electrical engineering. Out of naiveté, assuming that my radio engineering works might play some role, I wrote about them in detail in my autobiographical essay, citing the three inventor's certificates I had already received and my journal publication. I passed the exams, but naturally did not fit into the social lineage chart. A member of the acceptance board, who was specially designated to interview applicants, explained this to me quite frankly: "Work about three years and come back. We'll accept you as a worker, but not as the son of a white-collar worker."

Of all the working-class specializations, the most attractive to me was electrician. The Krasnaya Presnya silicate factory was the closest to our house. Equipped with imported processing equipment, it began to produce white silicate brick. I was accepted at the factory as an electrician on a probationary basis. Finding myself placed under the charge of a stern senior electrician, a Latvian who spoke Russian poorly, I probably would have been taken on as an apprentice. But suddenly it turned out that the factory had acquired a German power shovel with electric drives. The company sent a German fitter to assemble the machine and put it into operation. The German didn't know Russian and he needed an assistant who understood him. My reserves of German vocabulary from my school days proved to be sufficient for this man to say *sehr gut* (very well), and after working with him for a month, I received a high evaluation as a "foreign specialist." Thus, I skipped over the humiliating (in terms of self-esteem and income) apprentice level and became a fourth-class electrician on a seven-class scale.

The work of an electrician at a brick factory proved to be anything but easy. Silicate brick was manufactured from a mixture of sand and unslaked lime. The sand was added using the German power shovel. It chomped into a mountain of sand that in the winter had served as the favorite and sole venue for alpine skiing. Carts loaded with sand were raised by a cable drive along a trestle to high towers where the sand was mixed with the lime. The electrician's duties included, among other things, splicing the steel cable, which frequently broke. To this day I remember how many times I cursed that cable when I had to splice the frayed steel wires with bare hands in an icy winter wind, while following all the safety regulations. The wires pricked my numbed, disobedient fingers until they bled.

Troubleshooting a failure in the electric drives of the ball mills that pulverized the lime was even less pleasant. The acrid cloud of lime dust made it impossible to breathe without a respirator, which prevented you from closely examining the tangle of wires in the distribution boxes. One false move and the shock made you remember the rubber gloves and galoshes that were mandatory for such work. But no galoshes were available and the thick gloves prevented you from adjusting the delicate mechanisms of the switches manufactured by the Swedish company Asea. I learned to talk back at the comments of our excessively faultfinding foreman, who was always drunk by the end of the workday. As punishment, under the ridicule of the other electricians, I was sent to run power lines on the poles in the workers' settlements and to do the dirtiest electrical work of all: repair the lighting network in the dormitories. However, after a year in the school of hard knocks I became a full member of the workers' collective, a member of the trade union, and proved that I was fully capable of sustaining myself.

Soon, my comrades from my former boyhood expeditions began to campaign for me. Their parents were quitting work at the textile factory (they had set up a boat crossing over the Moscow River) and were now working in Fili at Factory No. 22. They had found jobs for their own sons there too. They promised to pull strings for me.

More and more I gazed longingly at the opposite bank of the river. There at the airfield factory, along a forest clearing, stood a row of new, two-engine TB-1 bombers. In the autumn of 1929, a TB-1 named "Land of the Soviets" completed a Moscow-New York flight. The entire country followed this flight with tremendous excitement. The names of the pilots, S. A. Shestakov and F. Ye. Bolotov, the navigator, B. F. Sterligov, and the flight engineer, D. V. Fufayev, appeared regularly in the newspapers from August through October.

In the autumn of 1930, the brick factory authorities conferred on me the title of "jobhopper" when I left that factory and was accepted at Factory No. 22, named after the tenth anniversary of the October Revolution.

Chapter 5

Factory No. 22

I worked at Factory No. 22 for eight years. In August 1930, I was hired to work in the electrical shop of the equipment department (OBO) as a class-four electrician.[1] In August 1938, I was already the chief of a design brigade for airborne weapons and special equipment. For those times, this was a senior engineering position. But I still did not have a higher education diploma, so I left in September to complete my studies at an institute.

I had managed to get through four years at the Moscow Power Engineering Institute (MEI) while continuing my work at the factory. The fifth year, the last one, required daily attendance. I promised Boris Nikolayevich Tarasevich, the factory director, that I would return in a year and do my degree thesis on the cutting-edge subject of a new high-speed dive-bomber. I was guaranteed work in the factory design bureau with a substantial raise in pay, and I was even promised an apartment. However, I did not return to Factory No. 22 a year later.

Thirty years would go by before I would pass through that same entrance gate to the territory formerly occupied by Factory No. 22, which is now the M.V. Khrunichev Factory (ZIKh).[2] Since then, I have often gone to ZIKh on space-related business, and each time it reminds me of my days there as a young man.

After that brief orientation, I will begin with the history of Factory No. 22, which I did not yet know when I went to work there.

THE AIRCRAFT FACTORY IN FILI traces its roots back to 1923. Everything started with the unfinished buildings of the Russian-Baltic carriage works that lay forgotten in a forest tract. In 1923, the German company Junkers started to develop these buildings. According to the Treaty of Versailles, Germany did not have the right to build combat aircraft on its own soil.

At that time, Junkers was the only one in the world that had mastered the technology for building all-metal aircraft. Hugo Junkers had already developed and was building all-metal aircraft out of corrugated Duralumin before World War I, but

1. OBO—*Elektrotekh Otdela Oborudovaniya.*
2. ZIKh—*Zavod imeni M.V. Khrunicheva* (ZIKh), or literally, "Factory named after M.V. Khrunichev."

they did not manage to see action. According to the contract with the Soviet government, the Germans were obliged to arrange for the construction not only of all-metal combat airplanes, but also aircraft engines. However, engine production simply never started. This was one of the pretexts for breaking off the relationship.

In 1926, the Soviet government abrogated the concessionary contract with the Germans. The few Soviet specialists remaining at the factory were faced with the task of using German experience and mastering the series production of the first domestic all-metal airplanes developed at the Central Aerohydrodynamic Institute (TsAGI) by Andrey Nikolayevich Tupolev. These were two-seat reconnaissance ANT-3 (R-3) aircraft and ANT-5 (I-4) fighters.

With difficulty, Fedor Malakhov, the factory's first director, assembled a group of forty designers and production engineers who began to rework the Tupolev aircraft drawings for series production.

The factory equipment and buildings were in a sorry state. With the onset of winter, the workers built campfires right in the shops to keep warm because the heating still did not work. The workers lived in neighboring villages. Muscovites rode to work on the train and slogged from the Fili station through kilometers of mud. The streetcar line from Dorogomilov to the factory was not built until 1929. Forty-four years old, with light brown hair and a powerful physique, director Malakhov rallied his work force under the slogan "Working People, Build an Air Fleet!"

At the initiative of Deputy Chairman of the Revolutionary Military Council (*Revvoyensovet*) Mikhail Tukhachevskiy and Air Force Commander Petr Baranov, the decision was made to build a gigantic aircraft factory at Fili. The factory was subordinate to People's Commissar for Military and Naval Affairs Kliment Voroshilov. They assigned it the number "22" and named it after the tenth anniversary of the October Revolution. Construction of new production buildings, hangars, dormitories, and multi-story apartment blocks was underway on a grand scale.

In the summer of 1928, Tupolev R-3s and I-4s began to take off from the factory airfield instead of Junkers. They were flown by the first factory test pilots, Yakov Moiseyev and Petr Lozovskiy. The same year, the factory design bureau, which was named KOSTR (Design Department for Construction), also began to release drawings of the TB-1 (ANT-4) aircraft for series production.[3] TsAGI's own experimental design factory had built only two of them.

At that time the TB-1 was the best all-metal, heavy, two-engine bomber in the world. Its low-wing cantilever monoplane design subsequently became the basis for the development of long-range heavy bombers in the USSR and elsewhere. Boeing engineers do not hide the fact that their first Flying Fortresses and Superfortresses trace their genealogy to the TB-1 design. Americans first saw the TB-1 airplane dubbed "Land of the Soviets" (*Strany sovetov*) on 30 October 1929 in New York.

---

3. KOSTR—*Konstruktorskiy Otdel Stroitelstva.*

The famous flight tour of the unarmed series-produced bomber—from Moscow to the Russian Far East to Japan to the United States—lasted sixty-nine days. This was the first flight in history from Russia to America. The Americans enthusiastically greeted the aircraft's crew: commander S. A. Shestakov, co-pilot F. Ye. Bolotov, navigator S. A. Sterligov, and flight mechanic D.V. Fufayev.

In all, 212 TB-1 airplanes were built. This was good training for our aircraft industry in terms of the organization and technology required for the production of all-metal airplanes. I saw this aircraft in 1930 when it returned from the United States for repairs at the factory. The silver surface of the Duralumin skin was completely covered with the scribbled autographs of enthusiastic Americans. An inscription in Russian stuck in my mind: "I, a Russian tsarist gendarme, am thrilled with this feat of my people." The signature was illegible.

The TB-1 was series-produced at Factory No. 22 until the beginning of 1932 and remained standard equipment until 1936. It was used by the Main Directorate of the Northern Sea Routes (*Glavsevmorput*). On 5 March 1934, pilot A. V. Lyapidevskiy flew crewmembers of the stranded *Chelyuskin* expedition from an ice floe to the mainland. He was the first to receive the title, "Hero of the Soviet Union."[4]

When I came to work at the factory, the new Tupolev R-6 was going onto the assembly line. According to its design, this was to have been a long-range reconnaissance aircraft and "air cruiser" for escorting bombers and for air battles. Factory No. 22 released fifty of these aircraft and then turned their production over to another factory.

Production of the first Tupolev ANT-9 passenger airplanes began at Factory No. 22 in 1930. This was a three-engine, nine-seat aircraft with a two-man crew. In contrast to the Junkers and Fokker passenger aircraft of that time, it had a more spacious cabin, comfortable wicker seats, and even a buffet. In 1929, with eight passengers onboard an ANT-9 dubbed "Wings of the Soviets" (*Krylya Sovetov*), Mikhail Mikhaylovich Gromov completed a remarkable series of flights: Moscow to Travemünde, Berlin, Paris, Rome, Marseille, London, Paris, Berlin, and Warsaw back to Moscow—a distance of 9,000 kilometers in 53 flight hours. The factory released more than seventy ANT-9s. Their production was halted in 1932.

The release of hundreds of heavy TB-3 (ANT-6) bombers brought the greatest fame to Factory No. 22. At that time, this aircraft was Tupolev's most outstanding work. The TB-3 was the first four-engine, cantilever monoplane in the world, built as a further development of the TB-1 design. The design of four engines built into cantilevered wings was later used as the model for many aircraft in other countries. The first TB-3s had a speed of 200 kilometers per hour and a range of 1,400 kilo-

---

4. The goal of the *Chelyuskin* expedition was to travel from the Barents Sea through the Arctic Ocean all the way to the Sea of Japan during a single navigational season. The crew was, however, stranded in a heavy ice floe in the Chukchi Sea on 13 February 1934, and was later saved in a daring rescue that was widely publicized in the Soviet Union. The "Hero of the Soviet Union" was a famous national award instituted on 16 April 1934 given to both military and civilian persons for heroism in service to the USSR.

meters. They were capable of carrying 3 metric tons of bombs. Five machine-gun turrets were included to protect the bomber against fighter planes. Reconstruction and new construction was underway at the factory to produce a spacious, lofty final assembly shop; hangars; and painting facilities for the airplanes.

It seems to me that with the torrent of one new aviation achievement after another, much in aviation history has been undeservedly forgotten. In the 1930s, the Soviet Union attempted to create the most powerful strategic military air force, with the TB-3 heavy bomber as its foundation. Over a period of five years, the young aircraft industry produced more than 800 of these aircraft. This took place during a time of peace, which was also a time of hunger; the majority of the country's population was in acute need of foodstuffs and the most fundamental necessities of civilized life.

From the author's archives.

TB-3 (ANT-6) heavy bomber, 1933.

I would like to tell the history of the TB-3 heavy bomber in greater detail for three reasons. First, the mastery of the large-scale series production of an aircraft that, at that time, outstripped the technological level of aircraft construction in Europe and the United States, was an indicator of the effectiveness of "mobilization economics" for a country that lagged behind in its scientific and technical development. The mastering of TB-3 production technology shows that the motto of those years to Catch Up and Pass Up the leading capitalist countries was not empty propaganda, but was based on the Soviet society's creative upsurge. In this regard, the history of the TB-3 anticipates the history of the creation of our domestic nuclear missile shield. There was a thirty-five-year break between these two technological and economic feats of the Soviet people, but the historical analogy is instructive.

Second, it was not the fate of the TB-3 heavy bomber to serve the global military-political assignment that the nuclear missile forces fulfilled. By the beginning of World War II, the airplane was hopelessly obsolete. However, it brought fame to the Soviet Union with its conquest of the North Pole, just as twenty years later the modified R-7 intercontinental missile stunned the world with the first satellites and first manned spaceflight.

Third, and finally, the TB-3 played a decisive role in my own personal life.

The history of the TB-3 goes back to 1925. The Leningrad Special Technical Bureau developed heavy, remote-controlled naval torpedoes that were supposed to be dropped from an aircraft and then controlled by radio. TsAGI, which was effectively directed by Andrey Tupolev, received the assignment to develop the aircraft.[5]

To begin with, they tried to use a TB-1 equipped with floats instead of wheeled landing gear as the torpedo launcher. However, the full-scale, radio-controlled torpedoes proved to be too heavy for the TB-1, and its flight range did not satisfy the naval command.

The design work on a four-engine modification of the TB-1 at TsAGI was completed in 1929, and as early as 22 December 1930, Mikhail Gromov executed the first flight. During that flight, Gromov just barely avoided an accident on the ground. The throttle levers that the pilot used to control the engines' RPM became deformed and vibrated. During takeoff, the right engines pulled so much harder than the left engines that the airplane turned toward the hangars. With tremendous effort, using the rudder, Gromov successfully avoided a collision and took off. There were also many other problems that were corrected during the test process, which continued until February 1931.

In the early 1930s, during the production of the heavy bombers, the Moscow-based Central Design Bureau (TsKB) in the Khodynka region and the K. A. Kalinin Design Bureau in Kharkov competed with Tupolev's design bureau (KB). The Air Force Command was going to select one of these design bureaus and then make a decision on the design to be used for the large-scale series production of bombers.

However, in those years General Secretary of the Communist Party I. V. Stalin and Peoples' Commissar for Military and Naval Affairs K. Ye. Voroshilov made the final decision. Tupolev and TsAGI had a significant advantage. Factory No. 22, which served as Tupolev's main production facility, had already mastered the technology for the large-scale series production of Tupolev all-metal TB-1 bombers, the R-6 "air cruiser," and the ANT-9 passenger plane. On its own, TsAGI had manufactured only two TB-3 aircraft for state tests. Without wasting time, they needed to hand over the drawings and begin preparation for production at Factory No. 22.

Air Force Commander P. I. Baranov did not hesitate, but now it was up to Stalin. In June 1931, Stalin, Voroshilov, and the other Politburo members arrived at the former Khodynka, which was now the central airfield of the country. The Air Force NII, which conducted the state tests on all combat airplanes, used this airfield. Without its final report, not a single airplane could be released into production. At the airfield, the Politburo members were shown all the aircraft prototypes that were competing to be accepted as armaments. Air Force NII test pilots flew demonstration flights. The highlight of the program was the Tupolev

---

5. Tupolev was technically not chief of TsAGI, but chief of its main design department: the Aviation, Marine Aviation, and Experimental Design Department (AGOS).

TB-3. Stalin climbed into the aircraft. Its dimensions and weaponry made the proper impression, and he wanted to see how this hunk flew.

It turned out that only two pilots, Gromov from the Air Force NII and Volkovoinov from TsAGI, had mastered control of the airplane. But Gromov was ill and Volkovoinov was out of town. After consulting with Air Force NII Chief Turzhanskiy, Air Force Commander Baranov decided to take a risk and entrust the TB-3 flight demonstration to pilots Kozlov and Zalevskiy, who had never before flown that airplane. It was a very big risk; the future fate of the airplane literally hung in the air. However, the new pilots coped with the task brilliantly. After a 40-minute flight, they landed safely on the very edge of the airfield. Stalin was told the reason for the delay in the flight. The length of the flight was explained by the fact that Kozlov and Zalevskiy were learning to control the aircraft to keep from crashing it while landing.

The decision to begin series production was approved and the aircraft was accepted as an armament, though with serious reservations. Flight tests were begun with American Curtiss engines, which were soon replaced with German BMW-VI engines under the condition that our industry would master their licensed variant, the M-17 engine.

I began to work at Factory No. 22 when the first domestic M-17 "flaming engines" were just being mastered. In a popular march, we sang that we had a flaming engine instead of a heart. M-17 engines could literally make this claim. The problem for the engine designers was the torsion vibrations in the crankshafts. These vibrations caused the deterioration of the shafts and fire in the engine, which could cause a fire in the airplane if the crew did not shut down the engine and close the emergency fuel cut-off valve in time. But every cloud has a silver lining. The M-17 engine failures taught pilots how to land with only three and sometimes two engines. Updated M-17s were not introduced until 1932, but they were not powerful enough for the TB-3 in its further modernized state.

In 1933, Aleksandr Mikulin's AM-34 engines were introduced. The TB-3's low-altitude speed increased, but its other parameters did not improve. The aircraft engines required boosting, the introduction of a reduction gear to reduce the rotation of the large propellers, and supercharging to increase speed at altitude. Thus, the AM-34FRN engine was introduced.

The series-produced version of the TB-3 with these engines, with a takeoff mass of 20 metric tons, an eight-man crew, and a bomb payload of 1,000 kilograms, had a range of up to 2,200 kilometers. Its speed at an altitude of 3,000 meters was only 220 kilometers per hour. These specifications might provoke no more than an ironic smile in today's readers, but I would remind the reader that in the early 1930s the military doctrine of air power was under the powerful influence of Italian General Giulio Douhet's theory. Douhet asserted that the emergence of the airplane revolutionized all military strategy, and he placed absolute significance on strategic bombers. Douhet's theory depended on the creation of powerful, independent air forces capable of deciding the outcome of a war through independent actions.

At a lecture for the factory's managerial staff in 1932, I heard an instructor from the Air Force Academy assert that hundreds of TB-3 bombers flying in a solid formation would protect one another against fighters. Dropping thousands of high-explosive and incendiary bombs, they would ensure strategic success. General victory would be guaranteed.

During the production process, in addition to replacing the engines, the TB-3 aircraft underwent other updates that entailed changes to its weapons, bomb payload, and chassis design. TB-3 airplanes set several world records for altitude while carrying large loads. The arctic version of the aircraft became world-renowned after the landing of Ivan Dmitriyevich Papanin's expedition on the North Pole in 1937.[6]

Factory No. 22 produced more than 800 TB-3 bombers between 1932 and 1937. In 1939, the aircraft gained fame in combat at Khalkhin-Gol.[7] This was the last successful use of the TB-3. At the beginning of World War II, German fighter planes annihilated the TB-3 bombers with impunity. TB-3s flying in formation at low altitude, each equipped with five small-caliber machine guns, proved powerless when under attack by maneuverable Messerschmitt-109 fighters armed with rapid-firing guns. Douhet's doctrine also failed to vindicate itself when a large number of American Flying Fortresses appeared on the scene. They delivered sensitive strikes against German cities in the rear, but could not break the German war machine. The infantry forces and tactical air force of the USSR accomplished this goal.

By the way, if we had started World War II rather than the Germans, and in 1937 rather than in 1939, it is possible that 600–700 TB-3s might have saved the world from Nazi aggression. But in 1937 Stalin was too carried away with the struggle against "enemies of the people." This saved Hitler.

BUT LET'S RETURN TO THE HISTORY of the factory. In 1927, Sergey Petrovich Gorbunov, a twenty-five-year-old graduate of the N. Ye. Zhukovskiy Red Army Air Force Academy, was appointed chief of the factory technical bureau—which today would be called the chief production engineer's department. He had already done practical work at the factory when the Germans were there in 1924. During his studies at the Academy he demonstrated brilliant capabilities. Sergey Gorbunov could have become an outstanding aircraft designer. Before he entered the academy, Gorbunov was a *Komsomol* secretary, and then secretary of the Party organization in Zaraysk *uyezd*, Ryazan province.[8] He was a delegate to the Third

---

6. Pananin and three others established the first drifting base near the North Pole in 1937–38.

7. The Khalkin-Gol engagement, also known as the Nomonhan incident, was a major confrontation between Soviet and Japanese forces in Mongolia.

8. *Komsomol*, which stood for *Kommunisticheskiy Soyuz Molodezhi* (Young Communist League of the Soviet Union), was a massive Communist youth organization established during the Soviet era. Most major institutions in Soviet society, such as Factory No. 22, had their own *Komsomol* organizations. Such entities were involved in fostering social and youthful activities that celebrated Communist rule. *Komsomol* organizations also had significant influence over personnel appointments at factories since they were asked to provide character references for workers. *Uyezd* is the Russian word for an administrative unit within a province.

Komsomol Convention and heard Lenin's famous speech.[9] It is possible that his aspirations for organizational work got the upper hand. He was at the factory day and night. He developed plans for capital reconstruction and new construction. At that time there were not many design institutes. Gorbunov, Malakhov, and a small group of enthusiasts designed the giant aircraft factory on their own.

In 1929, Gorbunov became chief engineer, and in 1930, technical director. He introduced what were then new ideas for aircraft construction—line production, closed-cycle shops, and laboratory tests of new technologies. He created an abundantly equipped laboratory for general factory use, along with shop test benches, and he regularly held conferences on quality with representatives from military units.

In 1930, the factory was made subordinate to People's Commissar for Heavy Industry Sergo Ordzhonikidze. At his suggestion, the entire aircraft industry was combined into a single entity called *Aviatrest*. In August 1931, Ordzhonikidze appointed Malakhov chief of *Aviatrest*. At the same time, at age 29, Sergey Gorbunov became director of the largest aircraft factory in Europe. Gorbunov was fated to be the sovereign factory director for only two years. I will take the liberty of saying that those were the best years in the pre-war history of the factory. During those years Gorbunov also managed to meddle with my fate.

In August 1930, before the final staffing had been completed in the personnel department at Factory No. 22, I was interviewed by the chief of the OBO department, who was interested in my work history, general educational training, and knowledge of the fundamentals of electrical engineering. Each applicant had similar interviews with the directors of the production departments and shops. If a group of technical school or FZU graduates was sent to the factory, or applicants transferred from other enterprises, a chief engineer or technical director conducted the interview. The director himself personally interviewed each young specialist who had graduated from an institute of higher learning and wanted to work at the factory.

The day after my interview, I received a photo ID pass, a time card, a cafeteria lunch card, and something especially valuable at that time—a pair of worker's coveralls that fit me perfectly, with many pockets. The chief of the electric shop explained that I would be the duty electrician for the factory for at least three months on a three-shift team.

A duty electrician's work required that he learn the general layout of the factory and the technological processes of each shop. I quickly gained an idea of how airplanes were built. Working the second and night shifts enabled me to attentively observe the production processes in each shop, get to know people, and hear their complaints about equipment or their praise of new gadgets. More than anything, the round-the-clock mechanical shops kept the duty electricians hopping.

Hundreds of new, imported metal-working stands were grouped according to the sort of production operations they performed. An unmanned section of Pitler

---

9. This is a reference to Lenin's speech, "The Tasks of the Komsomol," given on 2 October 1920.

ZIKh museum archives.

Sergey Petrovich Gorbunov, the stalwart young director of Factory No. 22, shown in 1931.

machines turned out almost the entire list of standard hardware: screws, bolts, and nuts.[10] The stands had a multi-step belt drive driven by a common transmission, which in turn was driven by the only electric motor on the entire line. If the electric motor's drive belt broke, a line of five to eight machines shut down. The shop foreman would then storm around the factory in search of the duty electrician, whose job it was to sew together and retighten the belt, start up the transmission for the entire line, check the lubrication of the bearings, and see whether the electric motor was overloaded. The shop foreman also noted in the duty logbook how long the stoppage lasted. If the total stoppage time over the course of a month exceeded the norm, this had a substantial impact on your paycheck.

It was much more pleasant to do business with the milling machines section. The Wanderer milling machines had individual electric drives with their own circuit breakers.[11] Work on the milling machines was considered women's work. The female milling machine operators beat out their male counterparts in terms of the cleanliness of their machines and entire workplace. In terms of productivity, however, especially on the night shift, the women lost. For a respite, the cleverest millers learned how to select an operating mode that would trip the circuit breaker relay. Then a meeting with the duty electrician was unavoidable. A mutual exchange of pleasantries about the accident might end with an instructive admonition, or even an agreement to meet at the factory stadium where we trained to pass the physical fitness tests to receive "Ready for Labor and Defense" (GTO) pins.[12]

It was primarily the old guard of workers who worked on the various models of lathes at the factory. Lathe operators tried to outdo each other and receive orders for the most complex operations. They knew how to correct simple electrical failures themselves if the lathe had an individual drive, and they did not bother the electricians without good reason.

---

10. Pitler was a machine tool manufacturer.

11. By the mid-1930s, the German company Wanderer Werke AG was the largest milling machine manufacturer in Europe.

12. GTO—*Gotov v Trudu i Oborone.*

The fuselage, center wing section, and wing shops thundered with the din of hundreds of pneumatic hammers. The manufacture of a single all-metal airplane required the pounding of hundreds of thousands of rivets. The corrugated exterior skin made of Duralumin was secured to the truss-like structure of longerons, ribs, and frames using covered rivets. The longerons along the length of the wing were riveted using tubes in special building cradles. The covered, "blind" method of riveting required the participation of two persons—a skilled driller/riveter, and an assistant who supported the rivet that could not be seen in the long tube. Next, the riveted structures of the entire center wing section, wing, and fuselage were assembled in the cradles. Riveting entailed such bursts of noise that it was only possible to communicate with gestures. The primary concern of the electricians in these shops was to provide all the cradles with portable lights and make sure that the insulation on the electric drill cords was in good condition. The workers had not been instructed in electrical safety engineering. It was the duty of the electricians to make sure that the 220 volts of tension did not stun the riveters working at the cradles. We were not always successful.

Once during the evening shift, a young fellow came to me in the duty room carrying a young woman in his arms—his riveting assistant. She had been shocked by a damaged electric drill, lost consciousness, and fell from the cradle. For some reason the first aid station was not open and the young man decided that the person responsible for the electrical malfunction was obliged to help. Like all electricians, I had taken an exam on rendering first aid to a shock victim, and according to all the rules, for the first time in my life, I started to revive her using artificial respiration. After about fifteen minutes, the young woman came to, and the strapping young man, overcome with joy, began to hug me. "Hug her, not me," I advised him.

"I can't—she's *really* strict—she'd never forgive me," he objected. Several times as I passed by the center wing section shop, when our shifts coincided, I saw this pair working with great concentration on the building cradle. When the guy saw me, he would smile affably, but the young woman pretended that she didn't notice anything.

The danger of shock dropped as we replaced the electric drills with pneumatic ones. A complete switch to pneumatic drills was finished in a year. The characteristic howling of compressed air escaping from hundreds of drills joined the thunder of the pneumatic hammers.

The supply shops and chassis and frame shops caused a lot less trouble for the duty electricians. Skilled metal workers, tinsmiths, and welders worked in them.

As a believer enters God's temple, so I entered the final assembly shops with a combination of joy and trembling. Here it was clean, quiet, and bright. The aircraft mechanics and assemblers were considered to be the worker aristocracy. They did not call the duty electricians for trivial problems. They had their own aircraft electricians who could also cope with the routine repair of auxiliary equipment and electric tools.

The airfield flight mechanics stood at the highest level in the worker hierarchy. The preparation of an airplane for release to the military representative and factory flight tests depended on them. Nikolay Nikolayevich Godovikov was the director of flight mechanics. There were legends about his ability to find defects in the propeller group by ear. From the nightshift duty electricians, he demanded failure-free operation of the powerful lights where the aircraft were parked outdoors and portable lights that could reach any place inside the aircraft.

During the first months on the job as the duty electrician, I gradually became steeped in the technology-worship prevalent at the factory. The shop walls were plastered with the slogans Technology Solves Everything!, Technology for the Masses!, Every Worker Should Be a Rationalizer and an Inventor! and What Have You Done to Master Technology? These slogans were not just empty appeals. They found a very lively response among the mass of young workers. Each proposal for improving production methods was quickly reviewed by the shop boards, and if it was accepted, it was encouraged with tangible material goods. Shock-worker and Stakhanovite cards were introduced, granting their holders the right to attend the best cafeteria.[13] A coupon was issued enabling those individuals to acquire clothing that was otherwise virtually impossible to obtain. Tangible monetary prizes were issued for especially valuable proposals.

I soon succeeded in this field. I made several suggestions about switching from transmissions to individual drives for the machines. Not a single machine had yet been converted along the lines of my proposal, but I had already gained possession of my very own shock-worker card. Soon I received an inventor's certificate for a photoelectric circuit breaker for electrical devices and was accepted into the USSR Society of Inventors. Some time later, I proposed that the building cradles be equipped with distribution sockets for plugging in drills and lights. This idea was accepted. Then I proposed that the voltage of the portable lighting be reduced to a safe level. The latter suggestion was far from novel, but it was not implemented at first because 12-volt lamps were not available. Having obtained the authority of the *Komsomol* committee, I assembled a delegation that set out for *Elektrozavod* in Moscow.[14] We were able to demonstrate to our *Komsomol* counterparts at *Elektrozavod* how vital their 12-volt lamp production was to the aircraft industry. They very rapidly arranged for the release of the lamps, which were in short supply, and step-down transformers. For this coup I was awarded a coupon for a leather jacket. During the civil war, black leather jackets had been in fashion among *Cheka* agents and Red Army commanders.[15] During the NEP

---

13. The Stakhanovite movement of the mid-1930s was a publicity campaign fostered by the Soviet government to increase productivity in factories. It was named after Aleksey Stakhanov, a Soviet coal miner who, it was claimed, used innovative work methods to greatly increase his personal productivity.

14. *Elektrozavod*—Electrical Factory.

15. The ChK or *Cheka* was the first Soviet security police organization and a predecessor to the more famous KGB.

years this fashion had spread to the proletarian youth. I was very proud of the jacket that I had honestly earned.

In the spring of 1931, there was a movement to transfer shops and even individual teams of workers to a self-financing operating basis. The *Komsomol* committee took the initiative and organized self-financing teams of young workers in several shops. Such a team was also formed in our electrical shop and they included me on the team. Our eight-man team was given the critical assignment of providing, within a very short time frame, all the electrical equipment for the automated canteen being built for the factory, which included four dining halls and a restaurant for Stakhanovites. We were supposed to put into operation a multitude of electrical processing equipment, lighting, ventilation, etc. We labored with an increased workload, and together with the construction crew, met the deadline to put Moscow's second canteen into operation.

Because of our team's labor victory, they sent me as part of a delegation from Factory No. 22 to a rally of shock workers from young Muscovite self-financing teams at the House of Trade Unions in the Hall of Columns. Moscow *Komsomol* leader Sasha Lukyanov asked our delegation to present a report about our accomplishments. They called on me to speak from the podium. My third time in the Hall of Columns I received the opportunity to pontificate at the top of my voice about the labor triumphs of the Factory No. 22 *Komsomol* members and of our self-financing team in particular.

I probably overdid it. My voice cracked. Later they told me that I squeaked, and suddenly instead of normal speech there was high-pitched childlike babbling. The hall fell silent and then thundered with laughter. Without losing his cool, Lukyanov started to applaud and the audience affably joined in. I regained my composure, and encouraged by a shout from the audience, "Come on, keep going, don't be shy!" I finished my fiery speech in a normal voice. Under repeated applause I proudly left the podium. Back at our company they had to make fun of me: "Great idea. They'll forget all the other speeches, but the fame of Factory No. 22 is going to spread all over Moscow."

Fame did not come easy. To run the power and lighting cables we manually chiseled grooves and holes in the concrete walls. My left hand swelled from being hit with the heavy hammer. Sometimes I missed and hit it rather than the chisel. The fingers of both hands were permanently coated with the black resin that we used to separate the high-voltage cables. "Now when you enter the institute, they won't even have to look at any questionnaire information—it's obvious that you are a real proletarian," Mama joked as she applied salve and bandaged my work injuries.

It would be two more years before I could realize my dream to enter the school of electrical engineering. However, if I had had the time back then to make the most fantastic predictions concerning my fate, I never would have dreamed that seventy years later, sitting in a luxurious restaurant on the third floor of that same canteen, Academician Chertok, Hero of Socialist Labor, holder of many orders and honored titles, that is to say, I, would be delivering a speech before the elite of the former Soviet military-industrial complex reminiscing about my work as an electrician.

It was April 2002 and I had been invited to an evening celebration in honor of the seventieth birthday of Oleg Dmitriyevich Baklanov.

Baklanov's career path had taken him from foreman at the Kharkov Instrument-Building Factory to minister in the rocket-space sector, and then CC CPSU Secretary of the Defense Industry. It was likely that after the collapse of the USSR he would have occupied high posts in the government of the new Russia. However, his participation in the GKChP putsch, which attempted to wrest power from Mikhail Gorbachev before Boris Yeltsin could seize it, precluded the possibility of a subsequent political career.[16] Baklanov has remained a symbolic figure for the former elite of the Soviet Union's military-industrial complex. When compared to the catastrophic degradation of the defense industry in the period of criminal market reforms, the Soviet past seems great and heroic. And there is nothing surprising about that.

In the first-class restaurant, the assembled crowd included former defense ministers, the former chairman of the military-industrial complex and deputy chairman of the USSR Council of Ministers, the former minister of the once powerful shipbuilding industry, pilots, cosmonauts, and many others who could rightfully consider themselves the true creators of the former superpower's might.

I began my toast by commenting that we were in a building that was opened the year our birthday boy was born. I provided this testimony as one who participated in the construction of the historic canteen and in the construction at Factory No. 22 of an air force that even then, seventy years ago, was the most powerful in the world. The thunderous applause that followed my toast, with its nostalgic remembrances of the former canteen, reminded me of the first applause I had earned after its construction at the Hall of Columns. Then I was twenty years old and now I was ninety! You couldn't dream this up if you tried!

THE COUNTRY WAS ENDURING a difficult period of transformation from being technically backward and primarily agrarian to being industrially developed. The government supported and encouraged all measures to shorten the time required to complete new aircraft in every possible way.

In June 1931, the newspaper *Pravda* published a USSR TsIK decree on the awarding of Orders of Lenin to a group of workers from our factory. Among them were Malakhov and Gorbunov. They received awards for fulfilling the five-year plan in two and a half years! The slogan Technology Solves Everything! was joined

16. The GKChP was the State Committee on the State of Emergency, which orchestrated the coup against Soviet leader Mikhail Gorbachev in August 1991. Its members included First Deputy Chairman of the Defense Council O. D. Baklanov, KGB Chairman V. A. Kryuchkov, Prime Minister V. S. Pavlov, Minister of Internal Affairs V. K. Pugo, Defense Minister D. T. Yazov, and acting President of the USSR G. I. Yanayev.

by the new slogan Personnel Who Mastered Technology During a Period of Socialist Reconstruction Solve Everything!

Halfway through 1931, KOSTR began to develop drawings for the series production of the TB-3 airplane. KOSTR handed the drawings over to TEKhNO, the process preparation department.[17] Here, the drawings were divided up according to the appropriate shop; process documentation and instructions were developed at PRIPO—the department that designed appliances, construction cradles, and all manner of engineering riggings.[18]

Political hoopla surrounded the program designed to produce hundreds of TB-3s. Voroshilov came to the factory twice. In one of the shops they even conferred on him the title Honorary Riveter, Class Five.

In the spring of 1931, VKP(b) Central Committee Secretary Pavel Petrovich Postyshev attended the factory *Komsomol* conference. He spoke with a characteristic *okaniye,* and called for the factory *Komsomol* to lead a movement to "Catch up with and pass up the leading capitalist nations in technology and science."[19] Postyshev said, "The aircraft industry is the one that is capable of pulling with it the other industries that lag so far behind the world standard. Only you young people are capable of this great historical feat." I am citing the gist of Postyshev's speech from memory. It was received with sincere, youthful enthusiasm. I was introduced to Postyshev as part of a group of especially distinguished youth. Twenty-five years later, we learned of Postyshev's tragic fate in Nikita Sergeyevich Khrushchev's speech.[20] He was liquidated as an enemy of the people in 1939 without any publicity.

Between 1929 and 1933, the extremely complicated problem of creating a domestic material and technical basis for aviation was being solved in practice. The decisions of the Politburo and Party conventions stemmed from the possibility of a military attack by capitalist nations on the world's first proletarian state. Competent, rational military intellectuals who had the power and the right to make decisions developed the doctrine for the Air Force's development and use during that time. They included, first and foremost, Mikhail Nikolayevich Tukhachevskiy, Petr Ionovich Baranov, Yakov Ivanovich Alksnis, and the elite of the Air Force Academy instructors.

The design workforces united around Tupolev at TsAGI did not just create airplanes. Tupolev used the specialists from his work force to organize the management of the aircraft industry, which set its sights on the massive production of heavy airplanes. "Personnel Who Mastered Technology" really would solve a prob-

---

17. TEKhNO—*Otdel Tekhnologicheskoy podgotovki* (Department of Technological Preparations).

18. PRIPO—*Otdel Proyektirovavshiy Prisposobleniya, Stapeli i Vsevozmoshnuyu Tekhnologicheskuyuostnastku.*

19. *Okaniye* refers to a Russian regional accent in which an unstressed (short) "o" is pronounced as a stressed (long) "o."

20. In his famous speech in 1956 at the Twentieth Party Congress, Khrushchev for the first time publicly exposed Stalin's crimes.

lem on a historical scale. In the aircraft industry's upper echelons of power, managerial staffs were rearranged. All aircraft factories were subordinated to Ordzhonikidze. He started off by convincing first Voroshilov and then Stalin to release Baranov from the Red Army Air Force command and make him head of the First Main Directorate of the All-Russian Council of the National Economy (VSNKh), which managed the aircraft industry, and at the same time appoint him First Deputy People's Commissar for Heavy Industry. Baranov's deputy, Alksnis, took his place as head of the Air Force in 1931.

The strengthening of the upper management structures did not eliminate the main problem: the rate of development in the aircraft industry was being held up by an acute shortage of personnel. There were not enough workers, skilled workers, designers, and engineers. An unusual situation developed at our factory: construction workers built new, turnkey shops earmarked for the production of TB-3s, but there was no one to work in them. The personnel department hired dozens of new workers everyday. These were primarily young people who had fled the hard life of the villages and had absolutely no clue what factory work or work discipline was, much less aviation technology. Two-month courses were set up to provide them with accelerated training. After this they were distributed among the job openings. The low skill level of the new workers caused a sharp reduction in quality and a drop in productivity, directly threatening a breakdown in the projected rates of TB-3 series production.

Having received a green, poorly trained bunch of workers, the shop chiefs demanded skilled workers, production engineers, designers, technicians, and engineers. It was not possible to train such leaders over the course of two or three months right in the shops. The personnel training problem needed to be solved at heretofore unprecedented rates. Baranov promised to assist Gorbunov in every way possible with the placement of specialists who were graduates of the Air Force Academy, the new aviation institute, and technical institutes. But this took time and more time! Having become head of the factory, Gorbunov turned to the Party and *Komsomol* committees, demanding that all means at their disposal be used to help educate the young workers. He also demanded that they set up their own system for training managerial personnel.

That same August 1931, the new secretary of the factory *Komsomol* committee, Petya Petukhov, approached me. He was a man who was quick to start up unexpected initiatives and ardent about the reorganization of personnel. Behind his back they called him Petushok.[21] Petushok was overflowing with new ideas. His predecessor, the intelligent and serene Sasha Vasilyev, was promoted to the post of secretary of the Frunze district committee of the *Komsomol;* he was destined for the Moscow committee. Vasilyev frequently came to the factory and did not always support Petushok's intrepid actions.

---

21. Russian word for young rooster.

Alluding to the authority of Postyshev, Vasilyev, and Gorbunov, Petushok announced that TB-3 production was a matter of honor for the factory *Komsomol*. But holding rallies would not help matters. He proposed that I become his deputy and at the same time hold the position of production-economic department manager that had been instituted in the committee. "Your task," Petushok persuaded me, "in the development of the TB-3 is essentially to become the *Komsomol* deputy to the technical director. It's guys our age who should develop this airplane. That means we should be production managers! You're an inventor. You're a quick study and you know all the shop leaders. You'll take those shop secretaries in hand. Come on, you can always go back to the electrical shop!"

I had only worked at the factory for a year and a half. I earned good pay in the self-financing team. In a year I would complete my three-year stint as a worker and I had dreamed of entering the Moscow Power Engineering Institute's school of electrophysics. Surely they would admit a worker, inventor, and *Komsomol* member who already had a start in a factory. I did not doubt this. I justified my refusal pointing out that a mere class-5 electrician would be ejected with a bang from the *Komsomol* for the inevitable failure of such a critical assignment. Petushok did not back down and promised me a serious talk with the Party committee.

Two days later the chief of the electrical shop, the terror of all social activists, approached me. He announced that after work I should report to the director.

That evening Gorbunov received me immediately. I did not know why he needed to see me, much less in his office one-on-one. He stared straight at me with his intent, dark eyes. The straight, handsome lines of his face were accentuated by the side parting in his thick, dark, carefully combed hair. The sky-blue gorget patches of his dark blue military tunic displayed the diamond-shaped insignias of his high military rank of brigade engineer. A brand-new Order of Lenin sparkled on his chest. People had told me, "We have a young, strict, well-built, and handsome director." Now I was convinced they were right.

Petushok had set up my meeting with Gorbunov. He had calculated correctly that the director's authority would overcome my stubbornness. Gorbunov said that I would receive an engineering education in three to four years. He intended to create a higher technical institute affiliated with the factory. Capable workers who had completed their comprehensive education would obtain a higher education without leaving the factory. My first task was to familiarize myself with the specific production plans and the key tasks of the main shops. Next I needed to organize *Komsomol* activities so that they would reach every young worker.

"I don't want to deal with a faceless mass of young people and a bawling bunch of Party cell shop secretaries," said Gorbunov. "There are talented people in every shop and we need to identify them. They should be role models in terms of productivity, innovation, and quality. Let the young convince the young. After work we need to keep as many people as possible in technical schools. Find and showcase people who knew nothing yesterday and today have become experts in their own job—shock workers. Fedor Shpak, the factory chief of production, will give you all the help you need. He will talk with the shop chiefs." Within thirty minutes, the young director had charmed and convinced me. He was a man who radiated energy rather than verbosity when he spoke with you.

The following day I was no longer an electrician, but a high-level *Komsomol* worker. Shpak acquainted me with production planning and provided me with graphic charts that modern science refers to as networks. Before my conversation with Shpak I arrogantly thought that I knew what was going on in every shop. Now I was convinced that I was a complete ignoramus. Shpak showed me the most critical places and patiently explained where the primary focus of *Komsomol* enthusiasm needed to be directed. I had to alter my whole way of life.

I was often at the factory for all three shifts. I had to meet with someone after a shift, conduct shop meetings with the young workers, discuss the content of the factory newspapers that were posted on the wall or the technical training programs, participate in executive production meetings, report my activities to the *Komsomol* committee, and travel to the *Komsomol* district committee and to other enterprises to exchange experience.

Gorbunov really tried to set up a factory institute that provided an education similar to that of a higher technical institute. For the sake of respectability he called it the special-purpose faculty or FON.[22] Almost all active Party and *Komsomol* members attended lectures on higher mathematics, physics, strength of materials, and philosophy four times a week in a semiconscious state until nine o'clock in the evening.

Training was under way at full speed in the shops, and parts and assemblies for the TB-3 were already being manufactured. In order to gain some idea of the new airplane, I was sent to KOSTR. Larisa Dobrovolskaya, the secretary of the KOSTR *Komsomol*, managed a group involved with interior special equipment. With her assistance, I became acquainted for the first time with the airplane's design and general layout. For me these were also my first lessons in reading the working drawings of aircraft designs and my first acquaintance with the general system of drawing documentation used in aircraft construction.

Often I had neither the strength nor the time to make my way home across the Moscow River, and I would spent the night in the *Komsomol* commune—in those days they had a sort of dormitory.

In spite of this stressful pace of factory life, we somehow found time to visit theaters and movies together. Such outings, distracting us from the bustle and burdens of life, rallied and unified the active *Komsomol* members. Vishnevskiy's *Optimistic Tragedy* at the Chamber Theater, Pogodin's *Poem about an Axe* at the Theater of the Revolution, Katayev's *Onward, Time!* at the experimental theater,

---

22. FON—*Fakultet Osobogo Naznacheniya.*

and the film *Vstrechnyy* all stuck in our memories.[23] Artists renowned throughout the country considered it an honor to come to our factory. For such occasions a temporary stage was set up in the new final assembly shop. Opera singers would give inspired performances of classical arias before an audience of several thousand, seated amongst half-assembled bombers.

Of the factory's 12,000 young laborers and white-collar workers, half had already become members of the *Komsomol* organization. The committee heading the *Komsomol* enjoyed great authority among the masses of young workers. Compared with the *Komsomol*, the factory's Party organization was relatively small, a fact explained by the age demographic and workers' brief period of service. The old Bolsheviks viewed the *Komsomol*'s emergence out from under Party influence as a dangerous phenomenon. They alluded to Trotsky, who had flirted with the youth under the slogan Youth Is the Barometer of the Revolution.

Party committee secretary Aralov was one of the factory veterans who had worked there since the time of the Junkers company. He was promoted to that responsible post by the Party organizations of the shops, which saw him as one of their own, a man who had emerged from the working masses, who knew about production, and who understood the needs of the workers. Aralov was respectful of the *Komsomol* initiatives. Most of the members of the *Komsomol* committee, including myself, were always invited to the Party committee meetings. We even prepared for them in advance so that we could participate in the discussion of issues that worried the young workers rather than just listen passively.

At one of these Party committee meetings, Aralov proposed to all the *Komsomol* committee members and secretaries of the leading shops that they join the Party. In his words, this was supposed to ensure Party influence and enable the actions of the active *Komsomol* members to be monitored. So it was that in 1931 I became a candidate and in 1932 a member of the VKP(b).

One day in early 1932, I was summoned to the Party committee during work—which was quite unusual. The active members of the Party and *Komsomol* were assembled. District committee secretary (at that time it was the Frunze district) Ruben had come to see us. We had seen him at the factory on a number of occasions, and moreover, I had also been in the district committee building on Zubovskiy Boulevard when I was received into the Party. A woman unknown to us accompanied Ruben. She was tall with short hair and wore a dark, English-style suit. She appeared to be over forty years old, and had the aura of a strict school matron—the terror of girls' preparatory schools.

Ruben said that we had been assembled at his request to familiarize ourselves with a secret Politburo resolution. He read a brief excerpt from the document

---

23. Vsevolod Vitalyevich Vishnevskiy (1900–51) was a famous Soviet playwright whose play "Optimistic Tragedy" opened in 1933. Pogodin was the pseudonym of Nikolay Fedorovich Stukalov (1900–62), another famous Russian playwright. Valentin Petrovich Katayev (1897–1986) was a Russian novelist and playwright who was famous for many satirical works on post-Revolutionary conditions in the Soviet Union.

signed by Stalin. The resolution stated that for the defense of the nation, Factory No. 22 had no less significance than an army corps. To strengthen the factory leadership and ensure control of its work by the Central Committee, the Politburo had deemed it necessary to introduce the position of Central Committee Party organizer. This position was also given the responsibilities of the factory Party committee secretary. By decision of the Politburo, Olga Aleksandrovna Mitkevich was appointed VKP(b) Central Committee Party Organizer for Factory No. 22.

"As regards Party committee secretary comrade Aralov, he is being transferred to work at our district committee," concluded Ruben.

This report was totally unexpected for Aralov and for everyone assembled there.

"Are there any questions?" asked Ruben.

Everyone was dumbfounded and silent. Suddenly our *Komsomol* secretary, Petukhov, got up and demanded:

"Tell us about yourself!"

Mitkevich (now we knew who Ruben had brought with him) stood up, and serenely moving her intent gaze from one person to another, began to speak.

She was born in 1889 to a broken family of the nobility. She had taken part in the revolutionary movement since 1903. In 1905, she joined the Russian Social-Democratic Workers' Party (RSDRP) (Bolsheviks) and began the illegal life of a member of an underground organization. She studied at the Moscow Commercial Institute and later graduated from the chemistry section of the Moscow Higher Courses for Women. In 1917, she participated in the October battles in Moscow.

During the conflict with Denikin she was named commissar of the Eighth Army, Thirteenth Division.[24] As a member of the Red Army, she marched from Voronezh to Groznyy. In 1920, she was transferred from military fronts to economic ones. She worked in the Donets Basin, Nikolayev, and Kharkov. She managed a mining office and was director of a textile factory in Yaroslavl. In April 1927, as a member of a Soviet delegation, she took part in the first Pacific Conference of Labor Unions in Canton. She spent a long period of time in China as a representative of the Comintern Executive Committee.[25] She participated in the Party committee congress in China. In 1930, she was called back to Moscow from China. Now she was deputy head of the Party organization department (*Orgotdel*) within the Central Committee.

"And by decision of the Politburo I have been sent to you at the factory. I am counting on your help." And with that, Mitkevich ended her speech.

---

24. Anton Ivanovich Denikin (1872–1947) was the Russian general who led the anti-Bolshevik "White" forces on the southern front during the Russian Civil War (1918–20).

25. The *Comintern* (Communist International) was an international organization dedicated to fostering Communist leadership among socialist movements across the world. Lenin founded the organization in 1919 as the "Third International." The Comintern gained strength through the 1920s but noticeably weakened in influence by World War II.

We dispersed from our meeting convinced there would be subsequent personnel reassignments. I told Dobrovolskaya my thought that a woman invested with Stalin's confidence must promote other women.

"There were only two women at the meeting today, and Mitkevich gazed especially intently at you," I remarked.

"Judging by her biography, Mitkevich will be a real commissar for us," replied Larisa. "It is difficult to predict how she will use her military experience. Too bad she doesn't possess the feminine charm of the heroine in *Optimistic Tragedy*."

Indeed, in outward appearance, Olga Mitkevich and Alisa Koonen, whom we had recently seen on stage at the Chamber Theater in the role of a female commissar, had nothing in common.

ZIKh museum archives.

Olga Aleksandrovna Mitkevich, Red Army veteran commander, VKP(b) Central Committee Party Organizer, and director of Factory No. 22 from 1933 to 1935.

Mitkevich began to make personnel changes in the Party leadership very quickly. New people arrived from the Central and Moscow Committee staffs and even from the Institute of the Red Professoriate. They replaced the secretaries of the Party organizations in the main shops. New faces appeared on the Party committee staff.

Next it was time for changes in the *Komsomol*. Mitkevich replaced factory *Komsomol* leader Petya Petukhov with Nikolay Bogdanov, whom she brought over from the Moscow Committee. Bogdanov decided that the main shops needed to have *Komsomol* cell secretaries who were freed of all other obligations. He reduced the committee staff, took over the managerial duties of the production department, and proposed that I be the secretary of the largest shop organization—the final assembly shop (OS).[26] I was well acquainted with the shop and the young staff gladly received me.

My transfer from general factory activity to work in the OS shop coincided with a rush to produce the first TB-3 airplanes. Even at the beginning of the year, work was already intense but steady, as Shpak had scheduled it. At the end of March, Gorbunov called a meeting at which he announced that Baranov had just called him. At Stalin's instructions, Baranov had requested that several TB-3s be made ready for an air review on 1 May 1932. Not a single one of the airplanes had yet made it to the assembly shop, and in just a little over a month they were expected to fly over Red Square! Nothing like this had ever happened! We agreed

26. OS—*Tsekha Okonchatelnoy Sborki.*

to assemble and make three flights ready for the review: nine aircraft total and a tenth as backup. KOSTR was given the assignment to urgently simplify everything possible. Get rid of the armaments, remove the side fuel tanks in the wings and any equipment for night flights, and whatever else.

The rush job throughout the entire factory was unusual even for veterans who had seen all kinds of production spurts. By the time it reached the OS, shop chief Morozov, who was always very calm and unflappable, had carefully thought through the assembly process, prepared a closed cycle team, and assigned the individuals responsible for each aircraft. When the assemblies began to arrive at the OS shop, the assemblers worked around the clock. A week before the May Day celebration, they started performing test flights over the factory airfield. All the airplanes took off and landed without incident although the pilots were not at all accustomed to these behemoths. On the eve of the celebration, all ten aircraft were transported to Khodynka, the Central Airfield.

On May Day, a column from our factory traditionally led a festive stream of demonstrators from the Frunze district. We were the first to enter Red Square after the troops had finished marching through. Numerous biplanes and sesquiplanes—reconnaissance planes and fighters—rumbled overhead. Next, several squadrons of TB-1s flew by. We tried not to hurry. Where were our TB-3s? We had already come alongside Lenin's Tomb when we let out a frenzied "Hurrah!" Stalin himself was greeting us and pointing to the sky. Overhead a rumble such as had never been heard before came rolling in from the direction of Tverskaya Street. It was very risky to fly nine four-engine giants in precise formation, in rows of three, at an altitude of no more than 500 meters. I saw neither Lenin's Tomb nor the reviewing stand nor the military attachés who were later mentioned in the newspapers.

I felt a lump rise in my throat out of sheer delight. I slapped someone on the shoulder and someone slapped me. We were all pushed and asked to move on quickly. I saw that fellow marchers throughout the shop column were wiping away tears. These were tears of joy that were impossible to conceal. I, too, needed a handkerchief.

OS *Komsomol* members initiated the establishment of a factory flight school and glider and parachute clubs. With Gorbunov's active support, a full-fledged flight school began operation in 1931. The chief of the school, Semyon Zalmanov, was a very energetic member of the *Komsomol* committee. Along with him and the new secretary, we assembled a noisy *Komsomol* committee delegation and burst in to see Hero of the Civil War, Army Commander Eydeman. At that time he was chairman of the powerful *Osoaviakhim*, the All-Union Society for the Promotion of Aviation and Chemical Defense. He oversaw hundreds of flight, glider, and parachute schools. Many graduates of these schools filled openings at aviation institutes, having received good flight training.

Eydeman granted our request for two U-2 training planes (subsequently the famed Po-2). In addition, he instructed the well-known pilot and parachutist Moshkovskiy to be the sponsor of our parachute club, and he sent sport parachute instructor Lyamin to work there permanently. There was no end to the number of people who wanted to fly. Petr Lozovskiy, the factory test pilot, was an instructor at the school on a volunteer basis.

The school trained pilots without taking them away from the production line. The *Komsomol* Committee devoted particular attention to the flight school, supporting the initiative of the school's chief in every possible way. Zalmanov enjoyed particular favor with Gorbunov, who found the means to acquire flight suits, parachutes, training planes, and everything that a flight school needed. By the end of 1933, the school had trained more than forty amateur pilots. Many of the school's graduates became professional pilots.

One of the factory school graduates was Aleksey Godovikov, the son of Nikolay Godovikov, to whom I said farewell forever before his flight in the N-209 across the North Pole to the United States on 12 August 1937. Aleksey Godovikov died in 1942 in a ram attack on an enemy Ju-88 bomber. Not far from Academician Korolev Street, where I live, is Godovikov Street, named in memory of pilot and Hero of the Soviet Union A. N. Godovikov. He died during a war when the death of pilots was seen as an unavoidable law of nature.

THE FACTORY FLIGHT SCHOOL was also touched by tragic events. On a hot July day in 1931, during a demonstration flight executing an aerobatics maneuver in an I-4 fighter plane, Petr Lozovskiy didn't pull out of a spin and crashed straight into the ground. The death of the pilot, the favorite of the *Komsomol* members, stunned us all. The very existence of the flight school was threatened. By then the school already had *Osoaviakhim* staff instructors. Gorbunov stood up for the school, and it continued to train young pilots.

More accessible and popular than aviation was sport parachuting. Parachute jumping from a special tower and then from airplanes became a real craze. Being a leader in the *Komsomol* organization of the OS shop, it was my duty to act as a role model to entice the shop's *Komsomol* masses into the parachuting school. The school's parachuting classes began under the leadership of Lyamin, who had made more than 500 jumps, including delayed jumps. After learning how to pack a parachute and executing several jumps from the tower, we went on a fly-around in a U-2, impatiently waiting for the real jumps to begin from an altitude of 800 meters.

We gathered for the first jumps on a Sunday at the factory airfield. Lyamin himself had lined up the first five to jump. I was fourth on his list. The airfield flight mechanic jumped first. Second was a female *Komsomol* activist from our shop. The third to jump was a shock worker—an aircraft assembler. The first two jumps went off without a hitch. On the third jump, the parachute failed to open. Completely shaken, we ran to the site where our comrade had fallen. He was lying in the tall grass on the bank of the Moscow River. His right hand tightly clutched the ring of the main parachute, which he simply had not pulled. What kept him from pulling the ring? Lyamin cautiously freed the ring from his firmly clinched, still warm fingers. We helped him put on our dead comrade's parachute. He persuaded the pilot to take off right then and there, and he jumped, demonstrating that the parachute was completely functional.

Classes at the school were called off. A month passed before we once again gathered at the factory airfield and jumps were permitted to continue. Lyamin decided to make a ranging jump to allow for a correction for the stiff wind. As the airplane took off and reached the site he had selected, the wind intensified. After the parachutist jumped, we saw that he was being carried away to the far corner of the airfield toward the Karamyshevskaya Dam. He came down in the middle of the river. The parachute canopy covered him completely, preventing him from breathing, and his flight suit pulled him to the bottom.

Several of the prison laborers at the dam rushed to the water. One even managed to enter the water and started to swim. The guards' shouts of "Back!" and the firing of warning shots followed. We ran to the bank at the same time a launch was leaving the factory pier. By the time they had dragged Lyamin out, brought him to shore, and freed him from his parachute gear, about a half hour had passed. He could not be saved. After this event, the parachute school virtually ceased to exist.

Central Committee Party Organizer Olga Mitkevich and Factory Director Gorbunov often made rounds through the shops and held production meetings. Production of the TB-3 was at a fever pitch. A large number of changes in the design and equipment makeup had been made. They changed the number of wheels on the landing gear from two to four; they changed the material of the landing gear half-axle and the engine mount frames. I have already mentioned the replacement of the M-17 engines with a modified version of the more powerful M-34. They altered the glazing of the Mosselprom—their name for the navigator's cockpit in the nose of the TB-3.[27] The TsAGI designers called the nose section "Arkhangelskiy's snout."[28] They also decided to remove the two retractable machine gun turrets on the underside that TsAGI workers called "Nadashkevich's trousers."[29] They replaced the low-power external electric generator that had a wind turbine with a generator that was driven from an aircraft engine. There were many other changes, including alterations made in response to the negative assessments that poured in from the Air Force in abundance.

During a routine quality conference, Gorbunov stated that, compared with the standard set by TsAGI, our factory's aircraft were overweight by 800–900 kilograms. Such a loss of bomb payload could not be tolerated. Therefore, among other measures, he announced a weight-reduction competition. A prize of 100 rubles per kilogram was set, if a suggestion was accepted. Soon KOSTR and other factory services were flooded with a stream of suggestions. Over a

---

27. The cockpit was named after the Mosselprom building in Moscow, which had a distinctive, angular design with many windows.

28. The nose section was named after Aleksandr Aleksandrovich Arkhangelskiy, a designer under Tupolev.

29. The turrets were named after Aleksandr Vasilyevich Nadashkevich, the leader of airborne weapons development working under Tupolev.

period of six months, the airplanes managed to "slim down" by a little more than 800 kilograms.

The transfer of our production—bombers ready for Air Force aviation units—took place at the factory airfield in a rather humdrum fashion. Crews from military units conducted the technical acceptance over a period of two to three days. On hot days, they found time to swim in the Moscow River. In the end the aircraft flew away to military airfields without any particular fanfare. The majority went west—to the Kiev and Byelorussian districts.

During the summer of 1932, when the handover of several dozen aircraft to the Special Far East Military District was coming up, Mitkevich proposed to the *Komsomol* leadership that the humdrum handover process be turned into celebrations to mark the transfer of our powerful airplanes to the heroic Soviet Air Force pilots.

The hero of the first legendary Moscow-to-New York flight, pilot Shestakov, was in charge of the aircraft acceptance. He commanded an Air Force formation of TB-1 bombers in the Far East. Now they were going to replace the TB-1s with TB-3s. By the time the Far East contingent arrived, the eighteen aircraft had logged extra test flight time and had undergone supplementary inspections to eliminate any defects that could possibly be found.

The handover festivities opened with a rally at the airfield. After the first aircraft inspection, Shestakov and his crews were invited to a fun-filled picnic that included boating on the Moscow River. The pilots from the Far East spent several days becoming familiar with their new airplanes and doing the paperwork for the acceptance. Mitkevich introduced Shestakov to a team of *Komsomol* members who had supervised the airplanes for the Far East. I took advantage of this opportunity to tell him that the factory was storing the "Land of the Soviets" airplane that had returned from America completely covered with American autographs. To Shestakov's great chagrin, no one could tell him exactly where the airplane was. Not just then, but also in subsequent years, we fecklessly lost material evidence that had ever-increasing historical value. Our museums of aviation history cannot show our descendants the "Land of the Soviets" and many other Soviet airplanes that have entered world aviation history.

The aircraft handover festivities ended with a magnificent banquet in the canteen. Early the next morning we accompanied Shestakov and his crews to the airfield. The airplanes took off one after another, fell into formations of three, and departed for the east. The TB-3s could manage the distance to Khabarovsk with four stopovers. Mitkevich reported to us that she had received a telegram from Shestakov thanking us for the warm reception and the powerful airplanes.

THE SUPER-INTENSE RATES OF INDUSTRIALIZATION during the period between 1931 and 1933 required that vast human resources be lured into the ranks of industry. Unskilled and often barely literate people were pumped out of the villages. In spite of socialist competitions, the introduction of ranks for shock workers and Stakhanovites, and moral and material incentives, labor discipline and production quality remained low.

One of the severe measures taken to improve discipline in industry was the law of 1932 on waste and absenteeism. According to this law, if one was more than 20 minutes late to work, one was fired. Persons guilty of absenteeism and waste were prosecuted. Even taking a smoke break during the workday was considered the same as tardiness if it exceeded 20 minutes. Such severe laws were far from necessary at our factory. Administrative measures, keen interest in technical training, effective incentives, and the active involvement of social organizations maintained a spirit of factory patriotism in our work force. The overwhelming majority of our people worked enthusiastically.

In the first weeks after the law was made public, many innocent people suffered when it was applied out of the zealous desire to set an example. There was a *Komsomol* member, a fitter in the propeller group named Igor; I forget his last name. He possessed the talent of a real inventor. Many of his suggestions concerning the assembly and design process for the TB-3 airplane were implemented. His last suggestion, which brought him fame throughout the factory and a substantial prize, was to simplify the complex system for the manual control of engine revolutions and power. They used to cite this suggestion at all kinds of meetings as an example of the inexhaustible potential of a worker's creativity. One not-so-fine day, I was taken aback when the shop chief told me, "Our renowned inventor was absent a half day. I am obliged to fire him, and if they go after him, he'll be prosecuted. This is your *Komsomol* member. That's why I'm filling you in." The stunned Igor explained everything to me very simply. "Yesterday I got married. We celebrated. I just overslept. My wedding night . . . you understand."

I decided to go to the director and tell him the whole situation. Gorbunov heard me out, the whole time staring off somewhere to the side. For a long time he was silent and then he snapped, "We didn't make this law and it's not up to us to abolish it." Without wasting any time I headed to the Party committee. Mitkevich said, "The director is obliged to set an example for the execution of laws, no matter how cruel they might be. I can promise that they will not prosecute Igor. Today you will convene an open *Komsomol* shop meeting. Igor must be dropped from the *Komsomol*. Show by this example that the law is the law for everyone. And explain to Igor that in around three or four months, after he has worked somewhere at a construction site without violations, if he feels like coming back to the factory, we will take him." Igor did not return. We lost a man who really had a divine spark.

In spite of all the difficulties, by spring of 1933 the TB-3 had been put onto the production line. During the military parade on 1 May 1933, squadrons of TB-3s flew over Moscow at an altitude of 800 meters, rousing an elated hurrah from the hundreds of thousands of spectators and a stressed reaction among the military attachés invited to Red Square. In the British Parliament's House of Commons, Harold Balfour came out with a special inquiry, "I have read the Russian newspaper *Izvestiya* dated 4 May, which contains a description of the May Day celebration in Moscow. There is a photo of the Russian parade in which the number of four-engine Tupolev aircraft in the field of view exceeds 158, and there were probably some that didn't fit in the shot. Unfortunately, this photograph cannot be considered a fake. The Russians have demonstrated not only the might of their fleet of bombers but also the capabilities of their industry."

In August 1933, by decree of the USSR TsIK Presidium, the factory was awarded the Order of Lenin for mastering the construction of special heavy aircraft and for the effective organization of factory operations. At that time this was a great honor. Orders of Lenin and the Red Star were awarded to eighteen of the most distinguished workers, including Gorbunov and Godovikov, who received Orders of the Red Star, and Mitkevich, who received an Order of Lenin. A large meeting was held in Fili Park for this event. RKKA Air Force Chief Ya. I. Alksnis spoke at the meeting.[30] In their speeches Alksnis and Voroshilov both appealed to the aircraft industry workers to produce aircraft and engines that were superior to foreign ones. In 1933, the TB-3 had no equals in terms of carrying capacity. But in terms of speed and altitude, it could have been much better were it not for the lagging of our engine industry.

That same August 1933, Aviation Day was celebrated for the first time. This celebration was used to the full extent to inspire young workers to new production feats. At all of our mass events we said that the Land of the Soviets must have airplanes that will fly "higher, faster and further than all others." Factory No. 22, the aircraft industry's best factory, would bear direct responsibility for realizing this task.

The factory experienced a boom period. Traveling out to regional and Moscow-area rallies and conferences, we always felt that a special respect was shown to the delegates of Factory No. 22. The continuous, single-minded public relations campaign was working. The new organization secretaries who joined the Party leadership in the main shops, Mitkevich's people, had created an atmosphere of unity and a continuous, festive labor upsurge. During this period Gorbunov and Mitkevich enjoyed deserved authority. For us in the *Komsomol*, they were teachers and idols—models whose lives we should copy.

Unexpectedly, these idols struck a blow to their own authority. In the beginning, whispered rumors had circulated about an intimate relationship between the director and the Central Committee Party organizer. In *Komsomol* circles, Gorbunov's authority and cult of personality were so great that the first reaction was, "Nonsense and gossip—it can't be. Gorbunov has a young, beautiful wife, his friend from *Komsomol* work in Zaraysk. Mitkevich is not so much a woman as a Party commissar with a long record that predates the revolution. She is fourteen years his senior. If he does something crazy, she should have enough self-control and common sense not to give cause for all kinds of gossip."

But the gossip and rumors turned out to be true. Gorbunov left his wife the apartment in the factory settlement in Fili and moved in with Mitkevich, who

30. RKKA—*Raboche-Krestyanskaya Krasnaya Armiya* (Workers' and Peasants' Red Army).

lived in a new government house on the riverside. This was quite incredible, especially since Gorbunov was exceedingly scrupulous in matters concerning nepotism at the factory. His younger brother Vladimir, who was later involved in the production of LaGG (Lavochin, Gorbunov, and Gudkov) fighters, complained that Sergey had not hired him just because he had feared accusations of nepotism.

This impasse was quite suddenly broken. Gorbunov died in the crash of an R-6 airplane. This model of aircraft was obsolete. KOSTR came up with the idea to convert it into a transport/passenger plane. The factory manufactured a single prototype. They removed all the aircraft's weaponry. Eight passenger seats were placed in the fuselage. The crew cabin was glassed in. The airplane made its first test flights in the area of our airfield. Unexpectedly, the order came to install additional fuel tanks to prepare the R-6 for long-range flight.

On 5 September 1933, this airplane set off for the Crimea carrying Gorbunov, *Glavaviaprom* Chief Baranov, Council of Ministers Civil Aviation Directorate Chief Goltsman, his deputy Petrov, Gosplan Presidium member Zarzar, and Baranov's wife, who had asked her husband to take her with him to visit their children vacationing in the Crimea.[31] *Glavaviaprom* senior pilot Dorfman and flight mechanic Plotnikov piloted the plane. The weather that day was not suitable for flying, but Baranov insisted on departing. At nine o'clock in the morning, the overloaded aircraft headed south. Twenty minutes after departure, the airplane crashed in the Moscow suburb of Podolsk. All of the passengers and crew were killed. According to the findings of the aviation board, the cloudy conditions forced the plane to fly very close the ground. The aircraft had no instruments or radio equipment for flying blind. Dorfman was forced to fly the plane so that he did not lose sight of the ground. While flying over Podolsk, the aircraft's landing gear ripped the wire of an amateur radio antenna off the tall poles to which it had been secured and dragged it behind. Then the aileron of the left wing grazed the top of a tall willow. The left wing panel fell off, the airplane's nose hit the ground, and the plane disintegrated.

The news of the death of Gorbunov, Baranov, and the civil aviation leaders stunned the work force. On the morning of 6 September, a sorrowful meeting was held at the factory. There were words shared by a man from Gorbunov's home district of Zaraysk, future CPSU Central Committee Secretary and future Academician B. N. Ponomarev. At the meeting, an appeal was made to the government to change the factory's name to the S. P. Gorbunov Factory. Now, instead of Factory Named After the Tenth Anniversary of the October Revolution, it would be called Gorbunov Factory No. 22. Gorbunov's name was also conferred on the Palace of Culture and a street in the Kuntsevo area of Moscow. The dead were laid in state in the Hall of Columns. Baranov and Goltsman were buried in Red Square. The others were buried at Donskoye Cemetery.

---

31. *Glavaviaprom* (Chief Directorate of the Aviation Industry) was formed in December 1931 to oversee the entire Soviet aviation industry.

It was up to Ordzhonikidze to propose a new director candidate for the country's best aircraft factory. Everyone expected Mitkevich to tell us about potential candidates. Someone who had mingled with officials from the People's Commissariat of Heavy Industry (*Narkomtyazhprom*) shared the rumor that Mikhail Moiseyevich Kaganovich, brother of Politburo member Lazar Moiseyevich Kaganovich, would be the director. The rumor proved false. Mikhail Kaganovich was appointed to replace the deceased Baranov as *Glavaviaprom* chief. At the end of 1933, Olga Mitkevich herself was appointed director of Gorbunov Factory No. 22. For experienced workers in the aircraft industry, this was a second shock. They respected Mitkevich as an intelligent, strong-willed, and strict Party leader. But experienced aircraft builders could not imagine her in the role of directing the largest aircraft factory in Europe. Moreover, general opinion had it that her immediate superior, Mikhail Kaganovich, was also an aircraft industry dilettante.

Those most aggressively disposed toward the director on account of her lack of credentials in aviation technology were the test pilots, airfield flight mechanics, factory flight-testing station (LIS) specialists, and old foremen of the final assembly shop.[32] They had not minced words with Gorbunov when he first came on the job, when the low quality of the shops' work or delays in providing components exasperated them. The tardiness and absenteeism law was a direct threat to the airfield's free spirits. Mechanics felt no compunction when they got a chance to bask a while in the sweet-scented grass or swim in the river on a hot day before climbing back into an airplane that had become scorching hot in the sun. For them Mitkevich personified a Party that had intensified their stress and invented a new, brutal law. Head flight mechanic Nikolay Godovikov tried to suppress the most zealous critics, but was not always successful in this.

When the weather was not suitable for flying or simply for other reasons, LIS pilots and flight mechanics thought up various practical jokes. Dobrovolskaya related to us how they tried to make her the object of their ingenuity in January 1932.

The chief of KOSTR got a call from the airfield requesting that he send out designer Dobrovolskaya to explain a drawing that she had signed. Her drawings had absolutely nothing to do with the activity of the flight station. "But you can expect anything from these degenerates," said her boss.

Possessing an attractive appearance and flushed from the freezing cold, the young woman entered the smoke-filled room where the airfield elite was shooting the breeze. Known for his weakness for pure alcohol, Khrisantov, the LIS chief and a former military pilot, turned to the flight mechanic. "Barabanov, look what a beauty they sent at your request! Explain what the deal is."

Flight mechanic Barabanov, a man who loved all kinds of practical jokes, explained, "There was a design defect in the aircraft, but the trouble is that I just

32. LIS—*Letno Ispytatelnoy Stantsii.*

forgot the number. A propeller got stuck on the skid. But we looked into it ourselves and fixed it. We can go out to the flight field and check!"

The assembled company could not contain themselves and burst into gleeful laughter. The joke was so old it had a beard, and Dobrovolskaya herself felt like laughing, but she stood firm. "At the next quality conference, I will without fail, ask for the floor to recount how successfully you eliminate design errors." She turned and left, slamming the door behind her.

One who loved practical jokes and awkward situations even more was the new LIS Party organization secretary, Klevanskiy. When he heard about the "propeller and skid" episode he was amused, but he persuaded Dobrovolskaya not to speak at the conference. If one were to tell such a story in front of Mitkevich, who knows, she might in all seriousness ask, "Can you show me how a propeller can get stuck on a skid?" Klevanskiy warned the flight mechanics that if Mitkevich were to make a routine visit to the airfield and somebody got it into his head to complain to her about quality with jokes like that, he would see to it that they were fired.

Test pilots, flight mechanics, and engine mechanics did not display a friendly attitude toward the Party and *Komsomol* leadership. Calls for mastering technology, for socialist competition, "using all 420 minutes in a shift"—none of this was for them. Mitkevich, who had a great deal of experience taming the obstinate masses, understood that for the post of LIS Party leader she needed to find an individual who would command trust and respect by the force of his or her special human qualities.

She had hunted down Klevanskiy at the Moscow *Komsomol* committee. He had a well-developed sense of humor and always had jokes, proverbial sayings, anecdotes, and funny stories in reserve. When the timing was right they could defuse the most strained situation. He rallied the LIS work force, eliminated drinking binges, increased discipline, and achieved a high evaluation for his activity from the Party committee. However, Klevanskiy stunned us all.

The Party leadership headed by Stalin did not permit any slackening of intra-Party tension. Under the slogan of a struggle against the vestiges of capitalism in the economy and in the peoples' consciousness, the leadership demanded heightened vigilance; they called for the exposure of harmful activity perpetrated by the bourgeois technical intelligentsia, Trotskyites, and those "deviating to the right."[33] In addition to learning the technology of their own job, Party members were also required to systematically expose the ideology of hostile classes. A campaign to purge the Party under slogans of freeing its ranks of alien and degenerate elements followed the general appeals. The atmosphere of the widespread labor boom was poisoned by the need "to expose, stigmatize, and uproot."

---

33. The "Right Deviation" was one of the last major Party factions to oppose Stalin before he assumed absolute power in the late 1920s.

The factory Party organization still did not know that Klevanskiy's father, a prominent economic planner, had been expelled from the Party for his ties with the Trotskyites. The exclusion of his son "for failure to provide information" was preordained. Klevanskiy, the soul of Party/*Komsomol* society, the merry optimist, shot himself. In his suicide note, Klevanskiy wrote that he could not live if he were expelled from the Party. Klevanskiy's suicide stirred up a campaign to expose "hidden class enemies" in all of the factory's organizations. This campaign also affected my subsequent fate.

Among various sorts of statements, the Party committee received a denunciation wherein the author asserted that I had concealed the truth about my parents when I joined the Party. It stated that my parents lived abroad, that my mother was an active member of the Menshevik party, and that during the NEP years, my father was a bookkeeper in a private enterprise.

A Party purge was looming ahead, and it was proposed that in my situation this personal matter be dealt with so that it would not reach the purging commission, in which case the factory Party committee would be considered guilty—Why hadn't they been aware of this? Why did they wait for the purging commission? Outflanking the Party committee, Bogdanov convened an enlarged session of the *Komsomol* committee and gave a speech exposing me. "Chertok is not a class enemy, but we cannot tolerate anyone in our ranks who is not completely candid and who hides his past." The majority of those who spoke were people with whom I had almost no contact on the job. My close comrades sat in dispirited silence.

In my own defense, I said that my mother had left the Menshevik party three years before I was born. My father had worked at a state-owned factory that the state had leased to a private company in 1922. All the workers and office workers had kept their jobs. My father had not been deprived of his right to vote, and they had even recommended that my mother join the VKP(b) during the Lenin enrollment. I had told my parents' entire history in detail to Wasserman in 1931 before I was received as a VKP(b) candidate. He was the standard of the Party conscience and a model technical leader with a prerevolutionary record dating from 1905 in what was then the OBO department. After attentively hearing me out and consulting with the shop Party cell secretary, Wasserman said that I should not expand on that subject in detail in the meeting. "Your parents are honest people. You do excellent work. You're a shock worker and an inventor. The fact that you were born in Poland has been written in all the questionnaires, and everyone who needs to, knows this."

Having succinctly recounted this history, I concluded by saying that I could not imagine life outside the *Komsomol* and the Party. The majority voted for my expulsion from the *Komsomol* ranks, but my close comrades voted for a severe reprimand and a warning. Formally, I was expelled from the *Komsomol*, but remained a member of the Party. This took place during very troubled times for the factory. Gorbunov had died, and Mitkevich had not yet been appointed. My situation was put on the back burner. At home, without holding back, I told my parents what had happened. Without warning me, my mother headed for Fili and had a meeting with Mitkevich. Neither one of them told me then about this encounter.

The Party committee finally found the time to discuss the *Komsomol* committee decision and unanimously voted for me to receive a "severe reprimand with a warning." This was also to be "noted on my registration form." In a private decision, it was recommended that I be transferred to a job in my specialty. Mitkevich gave instructions placing me in the same OS shop where just the day before I had been the *Komsomol* boss.

And so I once again became an electrician; no longer an industrial electrician, but one who worked on avionics. My fall from grace and my brush with expulsion from the Party was a very painful experience at the time. Many years later, I assessed everything that had happened not as a blow, but as a gift of fate.

They knew me well in the shop, and for "re-education" they sent me to Lidiya Petrovna Kozlovskaya's female shock-worker team. This team's job was to wire ignition systems. They were releasing them to flight mechanics on the airfield to be checked out on working engines.

Kozlovskaya was happy. "Finally I have at least one male subordinate, and one of the former leaders, to boot." She kept her young female subordinates in a state of fear. Incidentally, it was not just subordinates who were afraid of Kozlovskaya. She had been hired by the factory after several years in a "correctional" camp on the infamous Solovetskiye Islands.[34] The labor discipline and quality of work in Kozlovskaya's team was exemplary. It was hard to believe that she had a criminal past. She made very sure that every member of her team devoted 420 minutes per shift to production. Always "together," affable, and sociable with her shop comrades, she even knew how to win over carping bosses.

The most spirited troublemakers, the airfield flight mechanics, who carped about every piddling thing to the foremen and assembly shop team leaders, were careful not to clash with Kozlovskaya. She said exactly what she thought about unfair faultfinding, using language that would make a veteran engine mechanic's jaw drop. If somebody still needed to be taught a thing or two, she had developed a high-voltage shock technique using the manual magneto used to start up aircraft engines. When work was being done inside the spacious fuselage of a TB-3, the opportunity presented itself to unexpectedly touch an offender with an ignition wire. The sparking 20,000-volt discharge made a hole in his clothing, and while it didn't cause burns, it did give a brief jolt to the nerves.

A portrait of Kozlovskaya stood out vividly on the Honor Board, and in 1935 she was awarded the Red Banner Order of Labor. It is probable that the personnel department where the biographical data from questionnaires were stored and the people whom Kozlovskaya trusted knew about her criminal past. Her parents

---

34. The Solovetskiye islands, located in the White Sea, were home to one of the more notorious GULAG camps.

had been very wealthy and as a child she had grown accustomed to luxury. The revolution took away everything.

Having lost everything, including her parents, Kozlovskaya became an active member in a gang of robbers. Naturally gifted with organizational skills and an enterprising nature, she soon became the gang leader. Answering for serious crimes, she was threatened with the death penalty, but considering her age and the fact that she had no prior convictions, the court sentenced her instead to eight years in prison. During her stay at the correctional labor camp on the Solovetskiye Islands, Kozlovskaya's capabilities enabled her to obtain an early release with a reference attesting to her rehabilitation. She was even given the right to work at an aircraft factory.

On one of the hot days when the next batch of aircraft was being released, I was waiting to present the completed wiring of the electrical equipment to the quality control department (OTK) controller. Usually, a former Baltic Fleet sailor, whom we called Sasha-Bosun, conducted the acceptance process. He would carefully examine the aesthetics of how the cable bundles were run and check the security of the clamps holding them against the corrugated construction of the skin.

Instead of Sasha-Bosun, Kozlovskaya climbed into the roomy fuselage. With her was an unknown woman wearing a civil aviation uniform tunic. Kozlovskaya introduced the shapely, young, green-eyed woman saying, "This is our new quality control foreman, Katya Golubkina." Kozlovskaya immediately noticed that the new controller had conquered me at first glance. "Chertok will explain everything to you," she told Golubkina, and left the plane.

I began my interrogation and learned that Katya had just graduated from an aviation technical school with a specialty in aircraft special equipment. They sent her to work at Factory No. 22. Here she had ended up working for the chief of the final assembly shop quality control department, Nikolay Nikolayevich Godovikov, who had assigned her to do quality control on Kozlovskaya's team.

After a week, controller Golubkina announced to Kozlovskaya that she would not sign the form on the ignition system wiring until the faults she had noted had been eliminated. In order to meet deadlines, the team had to work overtime to correct the defects and again present its work for release. The disputed situation ended late in the evening and I asked the faultfinding controller for permission to walk her home.

It turned out that Katya lived with her brother and cousin in the studio of their deceased aunt, sculptor Anna Semenovna Golubkina. Anna Golubkina began her creative activity in her native city of Zaraysk. She studied in Moscow and St. Petersburg, and also in Paris under the renowned Rodin. The Soviet government presented her with a studio in the Arbat area on Great Levshinskiy Lane. Anna Semenovna died suddenly in 1927. Her nieces and nephews became the heirs and proprietors of her many sculptures and her studio. The oldest of them, Vera Golubkina, turned to Boris Ponomarev, a man from her home district of Zaraysk who held a senior Party post, asking for his assistance to create an Anna

From the author's archives.

The author's wife, Yekaterina Semenovna Golubkina (Katya), Zaraysk, 1929.

Semenovna Golubkina Memorial Sculpture Museum in her Moscow studio. All of her relatives signed a declaration for the gratuitous transfer of all of the artwork created by Anna Golubkina to the Soviet state. The USSR All-Union Central Executive Committee (VTsIK) established personal pensions for Golubkina's minor relatives, and Vera Golubkina was named the director of the new museum. Katya, who had moved from Zaraysk to Moscow to help set up the museum, entered Tushinskiy Technical School. After three years of study she ended up at Factory No. 22. Three years later, after our first meeting in the TB-3, senior control foreman Yekaterina Semenovna Golubkina became my wife. That is why the TB-3 bomber had a decisive impact on my personal life.

To a great extent, my work on ignition systems and aircraft electrical equipment determined my subsequent engineering fate. I had to release an aircraft ignition system at our factory airfield to the flight mechanics, military representatives, and crews who had accepted the airplanes. We had a lot of trouble with this system. Our industry had only just started to put a magneto, a complex instrument, into production. Until then, all aircraft engines had been equipped with magnetos produced by the Swiss firm Scintilla or the German firm Bosch. If an engine contained our magneto, the flight mechanics complained that it was unreliable and didn't generate a powerful enough spark.

Perhaps influenced by those conversations, I decided that it was time to stop using such a complex and capricious assembly as a magneto, and to put an end to the Scintilla and Bosch monopoly. Today, I myself am surprised at the daring and naiveté of this 21-year-old electrician. Through my ham radio activity, I had heard a lot about the remarkable properties of piezoelectric crystals. However, the quartz crystals that were already widely used back then provided such a paltry amount of electricity that there was no chance of using them to produce a spark in a spark plug. I had read in some radio journal about the miraculous properties of potassium sodium tartrate crystals.[35] These crystals had an electrical effect almost 1,000 times

35. Such crystals are also known as Seignette salt or Rochelle salt after Pierre Seignette from La Rochelle, France, the first to produce it in the late seventeenth century.

greater than quartz crystals. I began to look for literature. I stumbled onto a booklet that had just been published, *Seignette Electricity*.[36] I tried feverishly to take it all in—there was a lot of physics there that I did not fully understand (after all, I still had only a comprehensive education—in everything else I was self-taught), but I understood the main principle. Potassium sodium tartrate was going to cause a revolution in ignition technology. This was the beginning of the end for the age of the magneto and the European firms' monopolies. Within a week I had devised the entire system, and on my day off I had drawn it in India ink on Whatman paper and on tracing paper, and had also written up a description in the prescribed format of "the invention of the century." That was the factory inventors' council assessment.

A month later, I received a certificate of application and six months later an inventor's certificate. Next came a copy of an expert's positive findings, which contained the following:

> In contrast with all the existing electrical ignition systems of the fuel-air mixture in the cylinders of internal combustion engines, the proposed method is based on generating an electric spark in a cylinder's spark plug using a direct piezoelectric effect rather than a magneto, storage battery, or dynamo . . . . Numerous experiments have shown that potassium sodium tartrate crystals provide the greatest piezoelectric effect. This effect, which is enormous compared with all piezoelectric crystals, is used in the proposed method for the electrical ignition of the fuel-air mixture in an internal combustion engine.
>
> The proposed method has the following advantages over existing ones. Specifically, it is simple. Consequently, because of the absence of rotating magnets, windings, collectors, and other elements, there is a great degree of operational reliability, which in turn simplifies the electrical circuit. Its light weight and compactness due to the absence of the bulky magnet system or storage battery are exceedingly valuable for aircraft engines. The instruments' simplicity and low manufacturing cost make the proposed method more economically advantageous. (Expert V. M. Malyshka, Editor A. A. Denisov).

There were other positive reviews, but this was sufficient for me to begin my search for the miracle-working potassium sodium tartrate crystals, which until then I had never laid hands on. I was only familiar with their properties through literature.

A RUSH JOB TORE ME AWAY from my absorption with the new principles of ignition. This one entailed wiring the bomb release system and installing radio systems in the airplanes. They transferred me to a general special equipment and weaponry

---

36. The discipline of Seignette electricity is more commonly known as ferroelectricity, a term that became widespread after World War II.

team. We were faced with starting up production of navigational electrical equipment and a state-of-the-art (for those times) electrical bomb-release system.

Bombs weighing up to 100 kilograms were placed inside the TB-3 fuselage in special cassette bomb racks. Heavier bombs were suspended under the wing center section and the wing so as not to interfere with the opening of the bomb bay doors and the dropping of the interior bombs. The bomb rack's mechanical lock needed to be opened in order to drop the bombs. The navigator performed this operation using the bomb release, which was connected to the lock of each bomb rack through a system of several steel cables. Looking through a sight, the navigator would instruct the pilot how to fly in for the bombing run, and at the calculated moment, he would pull the handle of the mechanical bomb release with all his might. Dropping all of the bombs required the application of considerable physical force, alternately turning the handle back and forth. In so doing, the cable drive opened the locks of the bomb racks one after another. The navigator had to know how to drop the bombs one at a time and how to execute stick bombing. Stick bombing required that various time intervals be maintained between drops. Salvo bombing required that all the bombs be dropped simultaneously. Performing these operations using a mechanical cable drive was very difficult. When the cables got stretched, the dropping process became fouled up. Sometimes all of the navigator's calculations were in vain and the bombs did not hit the target.

In 1932, at the request of the Air Force, an electric bomb release was developed. Instead of cables, this model had bundles of wires running to the bomb holders. The mechanical locks in the bomb holders were equipped with a pyrotechnic device. An electrical impulse detonated a pyrotechnic cartridge, and the explosion generated gases that thrust a piston to open the bomb rack lock.

The bomb release invented by the Special Design Bureau and transferred for series production to the Aviapribor Factory was an extremely complex electromechanical device. The first electric bomb releases (ESBRs) barely passed testing at the Air Force NII, and the batches that we received were subject to malfunctions and failures that threw the LIS workers into a rage during a test drop of dummy bombs.[37]

The aircraft radio systems added new troubles for the special equipment shop section. Our powerful bomber aircraft did not begin to receive air-to-air transceiver radio stations (RSB) and intercom systems (SPU) providing communications between the seven crewmembers until 1933.[38]

The landing at our airfield of a TB-3 equipped with a top-secret autopilot caused another sensation. The pneumatic autopilot was frightening in its complexity and so it was never introduced on series produced airplanes. The first radio compasses and radio range beacon signal receivers were also installed on commanders' airplanes for night-time navigation and for flying under difficult conditions.

---

37. ESBR—*Elektrosbrasyvatel.*

38. RSB—*Radiostantsii Samoleta-Bombardirovshchika*; SPU—*Samoletnoye Peregovornoye Ustroystvo.*

All of this new equipment added so many failures that the airfield was covered with dozens of aircraft that had not passed the acceptance process. I was promoted to the position of radio engineer and was responsible for installing, tuning, and releasing radio equipment.

IN JANUARY AND FEBRUARY OF 1934, the Seventeenth Bolshevik Party Congress took place. It was declared the "Congress of Victors." According to all indicators, the Second Five-Year Plan had been fulfilled ahead of schedule. Stalin's policy for converting the nation from a backward agrarian country into a mighty industrial power received the Congress' unanimous support. Mitkevich was a congressional delegate from the Moscow municipal organization. She reported the results of the Congress and her impressions at a Party meeting.

I was in disfavor and did not attend the Party meetings. But soon at the airfield, while Mitkevich was making her routine rounds, having assembled the foremen and team leaders, she reamed us out about the delays in releasing the airplanes. Citing the congressional resolutions, she demanded the most heroic efforts from all of us to clear the flight field. After Mitkevich's general talk, she pulled me aside and asked why I had not appealed the Party committee's decision to reprimand me. She explained, "The thing is that the purging commission will not take the risk of removing such a recent and severe pronouncement. They would be more likely to simply expel you from the Party."

Now I had something new to worry about. Having studied the procedure for appealing the decisions of a primary organization, I proceeded step by step. First was the district *troika* for the review of personnel matters.[39] Its decision was approved by the district committee office located in the famous building on Zubovskiy Boulevard. This proved to be insufficient, and I appeared before the greater Moscow *troika* on Old Square. To the three old Bolsheviks, I appeared to be an inexperienced baby. Having moralized, they magnanimously replaced the line "severe reprimand with a warning and notation . . ." with simply "severe reprimand." After familiarizing itself with all the protocols and hearing my confession and the laudatory testimonials of the non-Party foremen, the purging commission finally cleared me, rescinding the Party's punishment.

Now, whenever I wasn't working, I was busy with the problem of inventing a simple and reliable electric bomb release. It is difficult to explain how the light bulb comes on in an inventor's head. First and foremost, he must desire at all costs to think up something that does not yet exist or replace an existing thing with something much better. The second condition is competence. He must have an absolute understanding of the end product and knowledge of the subject of his invention. I would call the third condition the erudition that saves the inventor from inventing the "perpetual motion machine." All the while, the inventor needs

39. Here, *troika* denotes a committee of three.

to be able to critically assess his proposed alternative before going into foolish raptures about his own genius. Finally, when everything is already clear to the inventor, when everything has been written and calculated, when the invention has even gone before a panel of experts and there are no doubts as to its innovation and advantages, that is when the most difficult stages begin: realization, experimental modification, and introduction. Therein lies the fundamental difference between the creative work of an inventor and engineer and that of a poet or artist.

I referred to my proposed electric bomb release as "the electronic." In place of the complex electromechanical clockwork used for the selection of time intervals during stick bombing, I devised an electronic time relay with a broad range for regulating the initiation time. I also needed to think up a circuit to convert the time relay to a pulse generator. In place of the highly complex mechanism used for selecting the required bomb racks, I used the simple and inexpensive switch used in an automatic phone switchboard. When everything, including a mercury-vapor thyratron, was assembled and soldered together on a sheet of plywood, it began to work. A factory military representative reported about my initiative to the armaments department of the Air Force Directorate.

After explanations at Varvarka, where the Air Force Directorate headed by Alksnis was located at the time, they sent me, with the appropriate reference, to the building on Red Square that housed the Directorate of Military Inventions. A few days later, Directorate Chief Glukhov sent for me and reported that Tukhachevskiy himself had approved funding for the development and manufacture of experimental prototypes of the bomb release system according to an agreement that the Air Force Directorate was to conclude.

IN AUGUST 1934, having endured the competitive examination, I finally became a student at the V. M. Molotov Moscow Power Engineering Institute (MEI). For material reasons, I did not want to quit work, and I began my studies in night school without leaving production. MEI had split off from MVTU about three years earlier and was scattered among various buildings in the area of Koroviy Brod and Radio Street. The subway was still just being built. After 420 minutes on my shift, I made the hour-and-a-half journey from Fili to MEI, satisfying my hunger on the streetcars on the way.

Classes in the night school began at six o'clock and finished at ten o'clock. Each student had already worked in industry for three to four years. From our personal life experience, we were convinced of how important it was to gain systematic knowledge in the fundamentals of science. It was precisely this desire to rise above everyday mundane cares that made us listen attentively to lectures, not fall asleep during seminars, and introduce improvements to our laboratory work. Our professors and instructors understood that they were not dealing with school kids, but rather with skilled workers, technicians, and designer-practitioners. A distinct community of interests was established in the educational process. We would let down our hair a little during the short breaks. By all indications we were adults, but we used to literally stand on our heads, in spite of appeals from the dean's office not to act like hooligans and to conduct ourselves with more restraint.

At the end of the year, having become acclimated to the institute, I discovered the scientific-research work section, which was fulfilling various orders for industry. After brief negotiations, the chief of the large electric machine laboratory agreed to set up a small special laboratory in his area to develop my electronic bomb release. This work had nothing whatsoever to do with powerful electrical machines, but it was attractive because of the contract with what was then the highly prestigious and influential Air Force Directorate.

At the request of the Directorate of Military Inventions, I was detached from the factory to the Air Force Directorate and headed the MEI special laboratory. Besides me, the laboratory staff included one mechanic. The mechanic had what they call "golden hands." He was over fifty years old and dealt with me in a patronizing manner. He started off by recommending to me that I not show up at the laboratory every day at all. He would take care of everything himself, and if I could just explain where those bombs hang that are supposed to be dropped, either one at a time or all at once. Ultimately, we found a common language. In the daytime we worked together in a small room that had a sign on the door reading Special Laboratory: Entry Strictly Forbidden. In the evening I switched to another building and buried myself in student cares.

During this period I had to visit many scientific institutions in Moscow and Leningrad for consultations and to order new parts. I had meetings with very interesting people. My dealings with them brought forth new ideas and sometimes made me reconsider previous endeavors.

In the electronic devices laboratory of the All-Union Electrical Engineering Institute (VEI), at the Svetlana and Krasnaya Zarya factories, in the offices of the Air Force Directorate, and at the Institute of Telemechanics, I met people who gave advice and helped me selflessly.[40] As it happened, three or four years later I once again needed to meet with some of these people. Svetlana director Yasvoin; VEI laboratory chief Olekhnovich; Air Force Directorate military engineer Vazinger; and Glukhov, chief of the Directorate of Military Inventions—they all disappeared in 1937. When I was told, "He doesn't work here anymore," I understood that I shouldn't ask any more questions.

I needed a thyratron that was particularly resistant to changes in temperature that was being developed at the Leningrad Telemechanics NII. This was a tube filled with argon. Having received the necessary letters from the Department of Military Inventions of the People's Commissariat of Military and Naval Affairs, I headed for Leningrad to get a thyratron, and at the same time, to have consultations concerning potassium sodium tartrate.

---

40. VEI—*Vsesoyuznoy elektronicheskogo institut.*

In spite of my lack of status in terms of age and service, the letters from high-level military officials were effective. The letters were printed on the letterhead of the Directorate of Military Inventions under the auspices of the Red Army Chief of Armaments. Anyone to whom such letters were addressed knew that the Chief of Armaments was the deputy to People's Commissar Marshal Tukhachevskiy. Thus, the directors of institutes and factories whose offices I began to visit in search of components for my bomb release mechanism never once refused my requests.

The thyratron developer, Yegorov-Kuzmin, a well-known specialist at that time, spent two days testing the suitability of his thyratron for my design. He provided me with a pair of ready-made thyratrons and promised to manufacture another five that were more reliable.

It was during this trip that I met with the young Kurchatov.[41] This meeting was accidental. If Kurchatov had not become a world-famous scientist twenty years later, I would not have mentioned our meeting, just as I am omitting hundreds of other meetings. But let me begin at the beginning.

The Leningrad Telemechanics NII was located in Lesnyy, next to the Physics Technical Institute. At my request, Yegorov-Kuzmin took it upon himself to accompany me to the Institute to find the right people for me to consult with regarding piezoelectric crystals for my proposed new ignition system. He introduced me to someone in some office. I presented Kurchatov's *Seignette Electricity* monograph and explained that I only needed a consultation. They began a search for the authors. It was soon determined that one of the authors, Nemenov, was not available at all, and Igor Kurchatov was away, but would be in the next day.

The next day a meeting actually took place. They did not invite me to see Kurchatov in the laboratory. He came out and talked to me in the lobby. The tall, well-proportioned, dark-haired, very modestly dressed man regarded me thoughtfully with his serene, dark eyes. He was curious as to who I was and why I needed a consultation.

I had already become quite an expert in all kinds of explanations before panels of experts and specialists at various levels. Pulling a tattered electrical diagram out of my folder, I began to explain the principle and advantages of the proposed ignition method. When my presentation had reached the crux of the matter—potassium sodium tartrate crystal—Kurchatov interrupted me. He had understood everything a lot sooner than I had anticipated when I was preparing my verbose report. He asked, "So this device is supposed to operate trouble-free on an aircraft or an automobile engine and withstand first heat, then cold? Is that it?" I confirmed that, yes, that was it.

I am not able to reproduce from memory the verbatim verdict that he pronounced on my already universally recognized invention. But this was the gist.

41. Igor Vasilyevich Kurchatov (1903–60) was, of course, the so-called "father" of the Soviet A-bomb.

The committee on inventions was quite correct to have given me an inventor's certificate in recognition of my priority for the practical use of the piezoelectric effect in aircraft and automobile technology. In principle, he did not have any objections to the fundamental possibilities for such usage. But the subsequent practical realization of a laboratory prototype was senseless. Potassium sodium tartrate crystals were very unstable and sensitive to changes in temperature and humidity. My proposal was premature. When new piezoelectric materials emerged, then it would make sense to work on realizing this idea. But at present, it would only be compromised.

Kurchatov praised the principle and buried my hopes to realize my invention. He did this without any malice, and calmly assured me that the time for developing such a system was still to come. I wasn't too distraught. During that period I had been so submerged in the development of my bomb release device that I had gradually cooled off toward the piezoelectric ignition system. It wasn't until I needed to write lists of my inventions on questionnaires that I dug through papers in search of the number of my inventor's certificate. And indeed, I should have pursued it. Kurchatov was right. Ten or twelve years later, new materials did emerge—piezoceramics. Now, at home, anybody can light a gas stove with the spark from a trouble-free piezoelectric igniter. I never had the sense to get an inventor's certificate for a kitchen range igniter. When I told my grandson this story while explaining the operating principle of the kitchen stove spark igniter, all he could say was, "Oh, Grampa, Grampa."

The names of the developers of our nuclear weaponry and of all the scientists that dealt with the problems of nuclear science were kept safely classified until the early 1950s. Even missile specialists such as myself, who had high-level clearance for secret projects, did not know for a while who was involved and what was going on in the nuclear kingdom. Only after the official announcement in August 1949 concerning the unveiling of our atomic bomb did I hear the name Kurchatov. Soon thereafter, Korolev and Mishin received an invitation to one of the regular atomic tests at the Semipalatinsk test site. Upon his return, Korolev, full of impressions, told us that "Black Beard"—Academician Igor Vasilyevich Kurchatov—was managing all the problems. At that time he had already been named a Hero of Socialist Labor twice.

Only after this did I finally recall that I had met with a certain Kurchatov who worked at the Leningrad Physics and Technical Institute. At home, in my heap of books, I tracked down the thin monograph *Seignette Electricity* by authors I. V. Kurchatov and L. M. Nemenov, which had miraculously remained intact after many moves. Everything added up. This was the very Kurchatov with whom I had obtained a consultation regarding my invention in 1934.

MEANWHILE AT THE FACTORY, a sensation of international scale was in the making. In November 1934, the International Air Show opened in Paris. Export versions of the TB-3 were prepared to fly to the show in Paris. All the weaponry was removed from them. New radio equipment was installed. The pilots' and flight mechanics' panels were replaced with ones that were more elegant. Every conceivable surface of the interior equipment was chromed or covered with frosted lacquer. In a word, our dismal, dark green bomb hauler was converted into a comfortable orange and sky-blue airplane with the most up-to-date navigational equipment.

Olga Mitkevich, the leader of the Soviet delegation, a former Comintern worker with a command of three European languages, and the director of the largest aircraft factory in Europe, roused tremendous interest in all levels of French society. For the French communists, this was a wonderful source of visual agitation and propaganda. The Soviet pavilion enjoyed the greatest attention. Mitkevich conducted numerous press conferences, visited workers' clubs, and met with the representatives of business circles. These were the finest hours of her life. But twentieth-century Russian history did not tolerate talented female leaders.

On 1 December, in Leningrad, Kirov was assassinated.[42] Mitkevich understood that this murder would have serious consequences for the Party and the nation. She cut short her stay in Paris and returned immediately to Moscow.

In the winter of 1935, at Mitkevich's direction, I was called to the factory. I had been included in a commission to determine the causes for the mass failures in the bomb-release system. Aleksandr Nadashkevich, director of aircraft weaponry development at the Tupolev design bureau, was appointed commission chairman.

Mitkevich herself assembled the entire commission at the factory airfield. She showed us the flight field crammed with dozens of aircraft and said, "We cannot release these planes because the bombs either don't drop when they are supposed to or they drop spontaneously. Do something! The factory's schedule has been disrupted. Never before has there been such a disgrace. In this commission I have assembled developers, theoreticians, and practitioners. Are you really not capable of understanding what needs to be done? Chertok, you're inventing a new device, but that will take a while. Figure out what to do with these planes today. Help the factory!" Now it wasn't the formidable Central Committee Party organizer addressing us, but a factory director who was in trouble. You could hear notes of despair in her appeal.

Always elegantly attired, with a professorial goatee, Nadashkevich had a reputation as a prominent specialist in aircraft weaponry. Having picked out three practical workers, among whom I found myself, he said, "Guys! Inspect all the wiring. From the distributor units on the electric release to each pyrocartridge. We *will* find the defects."

With two master electricians, Mayorov and Eyger, and a team of installers, we went from airplane to airplane and tested circuit continuity and probed each wire and each terminal box. We replaced several electric releases. We wrapped a bunch

---

42. Sergey Mironovich Kirov (1886–1934) was the popular Leningrad Communist Party leader whose assassination in December 1934 led to a series of events that culminated with the Great Purges in 1937–38.

of bare spots with insulation and rejected batches of pyrocartridges. The trouble was that pyrocartridge reliability and safety were theoretically difficult to achieve for the single-wire system of aircraft electrical equipment. Nevertheless, the scrupulous preventative repairs helped. About two weeks later the airplanes began to fly off to troop units.

I returned to my special laboratory, having earned the gratitude of Nikolay Godovikov, who at that time was OTK chief, and for whom the disruption of the production plan due to the low quality of the produced aircraft had been very hard.

That meeting at the airfield with Olga Mitkevich turned out to be the last. It was the beginning of dark days of repression, not only against monarchists who were innocent of Kirov's assassination, but also against many Party members who were suspected of sympathies toward Kirov. The problem was that at the Seventeenth Party Congress, many delegates had talked about promoting Kirov to General Secretary. Mitkevich was among them. The time had come to deal with all those suspected of being excessively sympathetic to Kirov.

The situation at the factory was grave. Party organization leaders, who through Mitkevich's intercessions had been transferred to the factory from the Central Committee offices and the Moscow Committee, disappeared from the shops one by one. Next, the arrests began even among the factory's leading specialists. In early 1935, Mitkevich fell ill. It was announced that she had been removed from her post as director and sent for studies at the N. Ye. Zhukovskiy Air Force Academy.

I made a slip and mentioned to my mother that Mitkevich was not our director any more. To my surprise, she took this news very hard. For the first time, Mother told me about her meeting with Mitkevich regarding my fate. During that single meeting they discovered that they had mutual acquaintances through underground activity. According to Mother, Mitkevich was quite an extraordinary, outstanding woman. "Such people ennoble the Party. But there are too few of them," she said.

During her studies at the Air Force Academy, Mitkevich tried to intercede on behalf of many so-called "enemies of the people." She considered them to be honest people who were devoted to the Party. In 1937, she once again became gravely ill. It is known that she sent a letter to Stalin and Beriya from the hospital. She was arrested immediately upon leaving the hospital. I am not aware of her subsequent fate. Mitkevich was rehabilitated posthumously after the Twentieth Party Congress. I was unable to determine the circumstances, place, or date of her death.

Chapter 6

In the Bolkhovitinov Design Bureau and KOSTR

At the end of 1933, an action committee of leading scientists from the N. Ye. Zhukovskiy Air Force Academy developed the design for a long-range, heavy four-engine bomber that in the near future was to replace the TB-3, which was becoming obsolete. Given the achievements in aircraft technology, it was proposed that a bomber be produced that would be a qualitatively new step in aircraft construction. It was to reach a speed of up to 330 kilometers per hour, fly at an altitude of 6,000–7,000 meters, and carry up to 5,000 kilograms of bombs with a maximum range of 5,000 kilometers.

Air Force Chief Alksnis approached Mitkevich, proposing that a special design bureau be set up at Factory No. 22 to develop the *Akademiya* long-range bomber, known also as the DB-A.[1] *Glavaviaprom*, Tukhachevskiy, and Ordzhonikidze approved the proposal, and the factory's own experimental design bureau was established. In contrast with the series production KOSTR, they called this design bureau KB-22, or the Bolkhovitinov KB. This was the last large enterprise that Mitkevich succeeded in setting up at Alksnis's initiative. For many years it played a major role in the history of rocket technology.

In late 1934, my electronic bomb release had begun to look like a finished product. I was in a great hurry, chasing after two hares. The first was my attempt to submit the device for state tests at the Air Force NII to obtain a "red book"—the official findings regarding clearance for flight tests. The second was my fervent desire to climb Mt. Elbrus during the summer of 1935 with Katya. Both hares got away from me.

While performing routine tests involving the detonation of dozens of pyrocartridges, the chairman of the factory inventors' society appeared in my special laboratory with a small group of active members, and behind them, two high-ranking military engineers, Viktor Fedorovich Bolkhovitinov and Mikhail Mikhaylovich Shishmarev. In their presence, I was presented an "Honorable Inventor of the USSR" lapel pin. The high-ranking guests had familiarized themselves with my invention and asked me questions, not so much about the technology as about bombing tactics, which they argued about between themselves. From their conversation, I understood

1. DB-A—*Dalnego bombardirovshchika 'Akademiya'* (Academy Long-Range Bomber).

that "the higher we aspire for our birds to fly," the more difficult it is to achieve precision bombing. It was quite hopeless for high-altitude, high-speed bombers to hit small targets such as bridges, buildings, or ships. New methods were needed for sighting and for combining the sight and bomb release mechanism into a single system. In short, they proposed that I transfer without delay to work at the new KB and head the special equipment team. The team had been almost completely staffed, but did not yet have a leader. They were designing a new bomber and needed to exercise the maximum initiative and inventiveness in order to increase to the maximum extent the aircraft's defense capabilities under fighter attack and its bombing precision. They suggested that we get to work immediately. The new bomber, the future pride and glory of our Air Force, was supposed to take off by March 1935! Bolkhovitinov took it upon himself to arrange everything required for the ESBR tests at the Air Force NII and also assigned me an assistant. As far as bonuses for the invention were concerned, he was also able to arrange everything through the Air Force Directorate.

I asked for twenty-four hours to think things over. I was faced with providing a very difficult explanation to Katya about our Elbrus plans, to MEI concerning the liquidation of the special laboratory, and to the mechanic about finding himself another job. Katya took it as a matter of course that the development of the new bomber was more important for me than climbing Mt. Elbrus, but she decided to go ahead with her plans. I suggested to the mechanic that he transfer with me to the factory and he agreed. At MEI they threw a fit and announced that they would demand payment for breach of contract in the event that the contract work was discontinued.

Twenty-four hours later, I met the team entrusted to me. There were already ten individuals on the team: four engineers, three technician/designers, two drafters, and a copyist. With the exception of the female drafters and copyist, this was the first time any of them had worked in aircraft design—right in a new KB. They put their full trust in me despite the fact that I, a second-year student, was supposed to teach them what was what. Bolkhovitinov had already called them together and told them that the team chief would be a skilled and experienced employee of Factory No. 22: Chertok.

My experience in industry and the general creative atmosphere of Bolkhovitinov KB helped me to acclimate myself to my new role of design team leader. Assembled here was a fellowship of diverse aviation enthusiasts, but they were all like-minded in their desire to break up Tupolev's monopoly on heavy aircraft. Bolkhovitinov and the scientists who had come with him—Air Force Academy professors—differed from most production workers in their unusual degree of culture and democracy. This trait created an atmosphere of good will, openness, and mutual assistance. There was no shouting or even conversations in an elevated tone; there was invariable civility, equal treatment regardless of rank, and respect for the opinions of others. Such was the psychological climate in our young work force.

Two engineer/electricians on my team, Anatoliy Buzukov and Yefim Sprinson, were developing the electrical system of the future aircraft. Both had experience with electrical equipment in industrial shops, and they considered the airplane a simple task. Semyon Gavrilovich Chizhikov, a former model maker at a foundry, graduated from Moscow Aviation Institute. He fearlessly designed the instrument panels and installation of the onboard equipment. I took on the task of developing the electrical system diagrams for the weapons and ignition, and issuing drawings for all the cable assemblies. Zhenya Ibershteyn proved to be indispensable for acquiring the documentation for all purchased instruments and equipment. We developed requirements for the new aircraft and made business contacts with the Moscow factories Lepse, Aviapribor, and Elektrosvet; with the Air Force Academy Department of Special Equipment; with the Teplopribor and Elektropribor factories in Leningrad; and with radio factories in Moscow and Gorky.

Right from the start, I became so immersed in this gripping, interesting work that I could barely tear myself away to Shchelkovo, where they were conferring about the testing of my electronic bomb-release device. The military testers dampened my enthusiasm with their findings. After listing the merits and shortcomings, they recommended that instead of a single central device for heavy bombers, several simpler, local devices should be developed for each caliber of bomb. In this vein, they hinted that it would soon be time to withdraw the TB-3 from service, and that there was still time to develop new heavy aircraft.

In Leningrad at the Elektropribor Factory, I first saw the American electrical sight produced by Sperry, as well as Elektropribor's own development: the vector sight. We agreed to develop the vector sight, combining it with an effector—an electronic bomb release. A young engineer from Elektropribor, Sergey Farmakovskiy, persuaded me to do this. Fifty years later, I would often meet with Doctor of Technical Science, Professor Sergey Fedorovich Farmakovskiy. We directed regular meetings with scientists at the Academy of Sciences to discuss problems of navigation and motion control, and we had even more scientific interests in common at that time than in the years of our youth.

AFTER RETURNING TO THE BOLKHOVITINOV KB at Fili after a business trip, I would fall into a regime of endless workdays with no days off. A struggle was going on, not only over the speed of the future aircraft, but also over the speed of its creation.

Three military engineers first class comprised the KB's nerve center—Bolkhovitinov, Shishmarev, and Kuritskes. It was risky for Bolkhovitinov, who was named chief designer, to take on the production of a heavy aircraft that would compete with Tupolev's ANT-6 without having the industrial experience of an aircraft builder under his belt. His striving for innovation was manifest particularly in the unrealized project of a superheavy aircraft that would be able to deploy tanks.[2] Bolkhovitinov was absolutely decent, technically erudite, and competent in design problems.

---

2. Author's note: They attempted to develop such an aircraft at the Air Force Academy without factory participation.

Kuritskes was a universally recognized authority in the field of aircraft aerodynamics. He had a negative attitude about the modernization of the TB-3, and believed that a qualitative leap needed to be taken immediately—having said A, we should go on to B. B was his dream—a bomber that, in terms of all its design parameters, surpassed the Boeing Flying Fortress. Kuritskes understood that to begin with they needed to win over a factory and rally a work force in the process of working on A. But this was dirty work and not something which he, a theorist, liked to do.

The most experienced man in this trinity was Shishmarev. He had already built airplanes that had been accepted as standard armaments. These were the R-3 and famous R-5 reconnaissance planes. I had many occasions to become convinced of his thorough engineering intuition. When people came to him for consultations he would—after reflecting briefly and drawing sketches—selflessly give unexpected and original recommendations that raised no objections.

All three men shared the psychological trait of originality of engineering thought. They cultivated in their work force this ability to think in nontraditional ways, sometimes to the great consternation of process engineers and production workers. The unorthodoxy and originality of thought in the production process disrupted schedules in the factory shops.

The Bolkhovitinov KB occupied the design hall, the series production KOSTR having moved to a new location. In the beginning, even Bolkhovitinov did not have a separate office. He, Kuritskes, and Shishmarev shared space behind a glass partition. There, team leaders and lead engineers regularly gathered to discuss common problems and orient each person toward a single, final goal. Sometimes we had fiery discussions. With the aid of a slide rule, Bolkhovitinov accepted or rejected proposals affecting weight characteristics and safety factors. Kuritskes guarded the aerodynamic contours and objected to any proposals that threatened to increase drag. With his inventiveness, Shishmarev demonstrated his ability to escape from dead-end situations.

There was no occasion for boredom at these meetings. I got to know other military engineers from the Academy there: Pesotskiy, Kan, and Frolov. But most of all, I was brought in touch with those who, like me, were workers who had transferred to the KB from factory shops and departments or who were sent by personnel departments.

Factory engineer-designers Saburov, Kirichenko, Alshvang, Arkhidyakonskiy, Gorelik, and Isayev bore the primary burden of issuing technical working documentation which, after a brief modification process, went straight to the factory shops. The one-to-one scale wooden mockup of the aircraft was a great help in our work. It was used to work out the layout of the pilots' and navigator's cockpits, the field of view, and the area of engagement, as well as to resolve debates on ergonomics. Before the final release of the drawings into production, a special Air Force Directorate mockup commission inspected and approved the mockup. After this, changes to the layout were forbidden.

Alksnis headed the first mockup commission, on which I was one of the developers' representatives. Until then, I had seen Alksnis only one time, at an all-factory meeting in 1932.

Our military chiefs had a high opinion of Alksnis. In Bolkhovitinov's opinion, Alksnis, the head of the Air Force since 1931, displayed rare persistence in the complex tasks of building an air fleet. He did not strictly limit himself to the military command sphere. Alksnis devoted great attention to drawing up proposals for the development of aircraft technology. He monitored its testing, organized long-range flights, and introduced military specialists into industry. He considered it necessary to personally head the mockup commission, enabling him to establish direct contact with the leading specialists of aircraft KBs.

Well-known Air Force NII test pilots Nyukhtikov, Stefanovskiy, and Air Force Chief Navigator Sterligov joined Alksnis on the DB-A mockup commission.

Before the mockup inspection, Bolkhovitinov made a general speech about the main features of the aircraft. He spoke quietly and calmly, as one accustomed to reading lectures at the Academy.

"The DB-A has a flight weight of 24 metric tons, 6 tons greater than the TB-3. As a result, the bomb load can be increased and the range increased up to 8,000 km. The DB-A has a smooth, rather than corrugated skin, and very importantly, after takeoff, the landing gear can be retracted into a special fairing or "trousers." We have designed for a speed of at least 330 km/h at an altitude of up to 8,000 m. In contrast with the TB-3, the aircraft's fuselage has a load-bearing monocoque construction—the entire skin contributes to its strength. The frames have no-load bearing bars partitioning the interior space. Therefore, the aircraft is very spacious and convenient for cargo transport."

Having attentively listened to Bolkhovitinov's account of the basic specifications of the new bomber, Alksnis began to size it up first-hand. He sat in the pilot's and copilot's seats, crawled through the hatch into the navigator's cockpit, which, given his heroic stature, was no easy feat, and meticulously interrogated us—the aircraft creators—sometimes asking the most unexpected questions. It seemed to us that he found the most fault with the degree of comfort afforded the pilots during prolonged flight, with the radio communications equipment, and with the defensive weaponry. Later, during a mockup commission meeting, he raised the last issue with Bolkhovitinov specifically: "The bomber should have powerful guns, to the extent possible, leaving no blind spots that would enable the safe approach of fighter aircraft. On your aircraft the rear and especially the lower hemispheres are poorly protected. This is a shortcoming of the TB-3. Although both the DB-A's altitude and speed are much greater, all the same, fighter aircraft will very soon be 100–150 kilometers/hour faster."

Over dinner, which was always held after the mockup commission's work was completed, the conversation turned to the role of heavy bombers. They did not forget to mention Douhet's doctrine. In this regard someone asked Alksnis if he would approve the production of Tupolev's six-engine, superheavy TB-4 bomber.

From the author's archives.

A DB-A bomber that incurred a broken landing gear strut during flight testing, photographed in 1936.

This airplane had four engines located in nacelles distributed over the wingspan and two engines installed on either side of the fuselage. It was a giant with no equal in the world until the construction of the ANT-20 (*Maxim Gorky*). Alksnis responded extremely negatively about getting carried away with these slow behemoths and asked Bolkhovitinov to prepare proposals for a higher-speed, higher-altitude, and longer-range aircraft than the DB-A. Nevertheless, over the course of this dinner we gave toasts to the successful flight of the DB-A. On 2 May 1935, the DB-A completed its first flight.

The summer holidays freed me from my evening journeys to the institute. My colleagues and I usually spent the evening hours in the still-bustling factory shops. An additional commitment for me was my involvement with the rowing school. I got hooked on this rigorous sport in 1934, when I saw the captivating races of shells on the Moscow River near the Borodinskiy Bridge. Katya rowed stroke in the women's quad at the Wings of the Soviets sports club. I couldn't be outdone by my beloved in this sport. In 1935, I was stroke in the men's quad. The instructor complained that the women's crew had fallen apart because stroke Golubkina had decided to conquer Mt. Elbrus. Now the club pinned its hope on the men's quad, filled out with the broad-shouldered, strapping men of Factory No. 22.

THE PRODUCTION OF THE FIRST TWO DB-A AIRCRAFT coincided with a rush job to introduce a new Tupolev aircraft into series production—the SB (ANT-40) high-speed bomber. The shops were freed from the large-scale TB-3 assemblies and learned to deal with the production process for the relatively small SB. The usual production of the TB-3, which had caught the fancy of the entire work force, was temporarily halted.

They assembled the first SBs in the OS shop, which had stood empty after the release of the last TB-3s. The assemblers, accustomed to working inside the TB-3s' spacious fuselages and thick wings, had a difficult time adapting to the SB's tight layout. One evening in the shop I met with Kozlovskaya, who together with Katya was studying diagrams of the new aircraft. Katya, singed by the mountain sun, had returned with a "Mountaineer of the USSR" pin for scaling Mt. Elbrus. She had been dazzled by the beauty and grandeur of the Caucasus. I gave my word that next year, without fail, we would go to the mountains together. With her trademark directness, Kozlovskaya commented on the difficulties of installing equipment in the SB and did not miss the chance to pass on what flight mechanic Barabanov, who had participated in the flight testing of the first series-produced SB, had said. "Flying in an SB is a lot like kissing a tigress—it's all fear and no fun."

In the summer of 1935, the OS shop started work on the second DB-A. By this time, the shop was full of the light blue SBs. The most nerve-wracking operation after assembling the SB was checking out its retracting landing gear. They put the aircraft on trestles and repeated the landing gear retraction and lowering operation over and over. During this operation the electrohydraulic system howled like sirens.

The lead designer of the SB aircraft was Tupolev's deputy, Aleksandr Arkhangelskiy. In order to introduce the SB into production as quickly as possible, Tupolev transferred Arkhangelskiy to the factory along with a large workforce of designers. Arkhangelskiy's KB was located a floor below Bolkhovitinov's bureau. We joked that this had been done because Tupolev had decided to undermine us from below, not yet having the capability to attack us from above.

The SB aircraft was classified as a tactical bomber. It had streamlined, even elegant contours which were unusual for Tupolev's airplanes. It had a fully retractable landing gear and a smooth skin with flush rivets. The SB reached a speed of 450 kilometers per hour and an altitude of 10,000 meters. Soon the SB created a sensation when it appeared in the sky over Spain during the Civil War. There for the first time planes produced by Factory No. 22 were widely used in combat.

My second meeting with Alksnis, also at the Bolkhovitinov KB, took place in early 1936. Alksnis arrived with the specifications for the American Boeing long-range bomber, the B-17 Flying Fortress. Its flight tests had already begun. Bolkhovitinov and Kuritskes arranged various design drawings of our next bomber, referred to as B, on the desk for comparison. The specifications were to our advantage. But the deadlines! "The B-17 is already flying," Alksnis said, "when will your 'B' fly?" Bolkhovitinov replied that if Factory No. 22 undertook this project with all due speed, then B could be produced in two years. The factory workforce of designers, process engineers, and production workers would go on to prove that we were capable of this feat.

Our strategic offensive air forces included a fleet of more than 800 four-engine TB-3s. In 1936, these airplanes were already obsolete and needed to be removed as standard armaments and used only for military transport, or at most, for airborne operations. However, during the first days of the war, in the general confusion, squadrons of TB-3s were fecklessly hurled against attacking German columns. They became easy and safe prey for Me-109s. By that time, Me-109s were equipped with 20-millimeter guns and had a maximum speed of 570 kilometers per hour, as compared to the 7.62-millimeter machine guns on the TB-3s, whose maximum speed was 250 kilometers per hour! Even the new Tupolev TB-7, which could reach speeds up to 430 kilometers per hour at an altitude of 8,600 meters, could not break away from Me-109s at middle altitudes. Our DB-As were also incapable of flying away from Messerschmitts at low altitudes.

The bombers could be saved at a high altitude, where bombing accuracy sharply deteriorated, or under cover provided by our own fighters. Our fighters, however, were not capable of accompanying the heavy bombers round-trip, or of fighting along the way. The flight range for bombers by 1935 had reached 2,800 kilometers (for the Boeing B-17), but the range for fighters was barely 600 kilometers. That is why the discussion was extremely heated as to whether there was a need for a large bomber fleet comprised of heavy and long-range bombers. Subsequent events showed that, in and of themselves, bombers were not capable of providing supremacy in the air given a well-organized anti-aircraft defense. Moreover, the strategic bombing of Germany did not have a decisive influence on the course of the war. Douhet's theory did not prove its value during World War II. Supremacy in the air was gained using fast, well-armed, high-speed fighters; high-speed bombers; dive-bombers; and ground attack aircraft.

The discussion then at Fili with Alksnis three and a half years before the beginning of World War II essentially had to do with the role of the strategic air force in a future war. For some reason, the fact that a war was inevitable did not disturb any of us. We all believed that this was the natural development of the historical process of the world's first proletarian dictatorship's struggle with the hostile, capitalist world.

Bolkhovitinov and those around him, and for that matter the entire Air Force leadership, understood that the 800-plus TB-3s had seen their day and that the future was for aircraft like the Flying Fortress, the American giant with four 1,200-horsepower engines and a takeoff weight of 26,760 kilograms. It was armed with thirteen large-caliber (12.7-millimeter) machine guns and carried over 2,700 kilograms of bombs at a range of 2,730 kilometers. The crew included five gunners who were supposed to provide 360-degree protection against fighters. The intention was that when a large number of these Fortresses were flying in formation, supporting one another, they would create a curtain of fire that was impenetrable to fighters. That was the reasoning of the Americans who designed the B-17, and that is what our theoreticians thought.

Agreeing that the obsolete TB-3s needed to be replaced and that there was clearly no need for such monsters as the TB-4, Alksnis then spoke in favor of combining the properties of a bomber and a fighter. This was the first time I had heard this. His wish was not met with enthusiasm in our company. And as far as aircraft B was concerned, he said that we needed to consult with *Glavaviaprom* since Tupolev was now building the TB-7 (ANT-42). We were hardly in a position to have several models of heavy bombers in series production.

The TB-7 was constantly plaguing our thoughts. It was a thorn in our side. Besides Tupolev being Tupolev, he was also very close to Mikhail Kaganovich and Sergo Ordzhonikidze, who at that time, together with Alksnis, primarily determined which aircraft would be put into series production. Nobody opposed the construction of one or two experimental aircraft. On the contrary, KBs were multiplying in the mid-1930s, and each had its own conception of air supremacy on which it based this or that type of aircraft. But the road to a series production factory, much less to acceptance as a standard armament, was much more complicated and more difficult than developing an aircraft in one or two years that fit the latest trends in the capricious world of aircraft design.

At the factory, our new DB-A aircraft was nicknamed *Annushka*. Its assembly ran behind schedule, reflecting the young design collective's lack of production process experience. In the OS shop we had to fit many parts on site. The fitters' wealth of experience smoothed over the young designers' mistakes.

The semiretractable landing gear mechanism, Isayev's first independent work in the world of aviation, required various modifications. The enormous wheels were supposed to be synchronously pulled under the fairings of the special "trousers" to their supports, and in reverse, they were to be reliably locked in place and a signal sent indicating readiness for landing. A hydraulic system controlled the landing gear's retraction and lowering. Compressed air from tanks, which could be filled using a special compressor, produced pressure in the power cylinders. In case of a hydraulics failure, a manual winch provided an emergency lowering system. Nikolay Godovikov had attempted to lower the landing gear manually and said that you need to take a strongman with you into the air. Isayev himself was not equal to the task, either.

From the author's archives.

Model of the TB-4 heavy bomber designed by A. N. Tupolev, photographed in the N. Ye. Zhukovskiy Museum.

Before coming to Bolkovitinov KB, Isayev had had nothing to do with aviation. He had tried being a mining engineer in the coal mines and a builder in Magnitogorsk and Zaporozhstal, before deciding that his future was in aviation technology. He submitted his factory acceptance application to Director O. A. Mitkevich. She sent the future renowned rocket engine chief designer for interviews at the Bolkhovitinov KB. Bolkhovitinov believed in him, and under Shishmarev's supervision, entrusted him with the development of the landing gear for the DB-A. In the OS shop, the testers of the series-produced SB, demonstrating the speed and ease with which the landing gear of this aircraft retracted, bad-mouthed *Annushka* quite a bit because she could not learn to pull the gear up into the fairings quickly. Isayev devised various ways of installing homemade terminal contacts. He repeatedly altered the design of the manual winch. Chizhikov designed a special signal panel showing the position of the landing gear. Process engineers struggled with the system's pressure integrity. Bolkhovitinov himself checked the strength analyses of the landing gear struts. Highly experienced factory mechanics brought their own insights and modifications to the table.

At the airfield, the grueling process had begun of running the engines, adjusting the variable-pitch propellers, calibrating the gas gauge, and making endless modifications to the exhaust manifolds. The exhaust pipes burned on the new Mikulin M-34FRN augmented, geared, and supercharged engines. They were constantly being modified.

My workday started at the airfield. Together with engineers from other factories, I took advantage of the engine testing to check and adjust the voltage stabilizers and radio equipment.

Factory test pilot Nikolay Grigorievich Kastanayev, who just barely squeezed into the SB's tight cockpit, patiently waited for our plane to be ready for its first flight. Finally it was time for flight tests and landing runs. During one of the approaches the landing gear broke. This was a critical situation and the first time we heard Isayev's expression, "I should blow my brains out!" Depending on the situation, Isayev used the exclamation "blow one's brains out" as an expression of extreme disappointment, delight, indignation, or anger. It all depended on his tone.

In 1940, we went on vacation with Isayev to Koktebel. While playing croquet, he got into a bitter dispute with his opponent and threatened to "blow his brains out." In November 1941, while we were freezing on the platform of the special train that was taking us from Moscow to the Urals, Isayev fired a burst from an aircraft machine gun into the night sky, hoping to "blow the brains out" of some Messerschmitts. On 15 May 1942, after Bakhchivandzhi's first flight on the BI aircraft, when the landing gear strut broke during landing, he exclaimed, "Blow my brains out! Just like *Annushka* five years ago." Even when he had become the second ranking chief designer of rocket engines after Glushko, he continued to "blow brains out" on the occasion of any serious error in the dozens of systems in the liquid-propellant rocket engines produced at his famous KB.

Winter set in, and for *Annushka*'s flight tests we had to switch the landing gear to skis. The first flights went well—if you don't count the usual leaks in the engines' lubrication systems.

I participated in the flights, checking the new, high-powered aircraft radio station along with the radio engineer Traskin from the Gorkovskiy Radio Factory. The primary radio equipment was located in the tail section of the spacious fuselage, in the sound-proof radio operator's cockpit. The transmitters and receivers of that time were crammed with dozens of electronic tubes that were very sensitive to the jolts and vibrations generated by the aircraft's engines during takeoffs and landings. To protect the equipment from overloads, we suspended all the instruments containing tubes in spring shock absorbers. Nevertheless, during a rough landing, the heavy transmitter would hit against the frame. In these cases Traskin would replace the powerful transmitting tube.

After ten flights or so, the aircraft was broken in. During the frosty days of February 1936 they scheduled a show of new aircraft technology at the airfield in Monino. Having brilliantly demonstrated *Annushka*'s rate of climb and maneuverability, Kastanayev decided to end the show by making an impression on the high-ranking military leaders in attendance. He came up with the idea of executing a low-altitude fly-by of the reviewing stand where Tukhachevskiy, Alksnis, and many other high-ranking chiefs were sitting. Once he had gained altitude, he began to dive at the airfield to gain high speed. Suddenly one of his landing-gear skis, which were drawn up to the fairings, turned 90 degrees under the effect of the approach flow. The aircraft now had an uncontrollable elevator that threatened to drive it into the ground.

I will not presume to describe the feelings that seized the spectators—the aircraft creators—at such moments. I wanted to close my eyes tightly so that I would not see the horrible end. Seconds remained before the inevitable crash, but suddenly the engines revved down and the aircraft began to cock its nose. Its speed decreased sharply. At the very ground the ski grudgingly went back into place. Now Kastanayev was headed for a wall of pine trees. But he managed to give it the gas and accelerated into a climb. He flew right over the forest, circled, safely landed, and taxied to a stop.

During the investigation of this critical incident, they discovered that the shock-absorbing cord that held the tip of the ski against the aircraft had not been designed for the torque generated by dynamic pressure. Isayev had personally selected and calculated the diameter of the cord. Reporting to Bolkhovitinov, he explained, "Blow my brains out! I calculated on a slide rule and slipped up by one mark."

The aircraft was saved thanks to Kastanayev's composure and exceptional physical strength. He pulled the control wheel toward his body with tremendous effort, using the elevator to overcome the nose-down moment generated by the skis. He succeeded. By spring, the factory tests were completed and flights for the state test program and record-breaking flights had begun. At that time, setting world records was more than just a matter of prestige for aircraft designers and pilots. Each new aircraft had to set some kind of record. *Annushka* had set many Soviet and four world records. Nikolay Grigoriyevich Kastanayev, Georgiy Filippovich Baydukov, and Mikhail Aleksandrovich Nyukhtikov completed record-breaking flights. In the 1936 May Day parade, our *Annushka* flew full speed over Red Square. Kastanayev executed a steep ascending banking turn. This gripping spectacle provoked a storm of delight. Once again, as in 1932 when the first TB-3s flew over the square, I got a lump in my throat from sheer joy. Now I was walking not at the head of a factory column with workers from the vanguard OS shop, but somewhere in the middle with a crowd of design intelligentsia.

AT FACTORY NO. 22, after Mitkevich's departure, acting directors came and went without leaving a mark. In the summer of 1936, Boris Nikolayevich Tarasevich took over as chief engineer. Since the factory was lacking a strong director, he took all the power himself. Tarasevich had been director of the Kolomensk Locomotive Building Factory until 1930. In 1930, he was arrested and convicted for his involvement in the "Industrial Party Affair."

In the history of Russia throughout the pre-revolutionary, Soviet, and post-Soviet times, the Industrial Party was the only organization of the technical intelligentsia that professed truly technocratic principles. The leader of the Industrial Party, a scientist and thermal engineer with a worldwide reputation, Professor Leonid Ramzin, sincerely believed that the nation's power should be in the hands of scientists, engineers, and industrial leaders. The Industrial Party carried out no direct sabotage. Its members dreamed of replacing the dictatorship of the proletariat with a dictatorship of scientists, engineers, and industrialists.

They dealt with the Industrial Party members humanely (for those times), but cunningly. They gathered together all those who were sentenced to long prison terms and proposed that they work under the leadership of their chief ideologue, Leonid Ramzin, in a special closed institute for thermal engineering focused on the creation of new types of steam boilers. Here Ramzin developed the design for a straight-through boiler, which was called the Ramzin boiler. The intellect of the Industrial Party members was used with a high degree of efficiency.

In view of the fact that the convicts worked with complete devotion on behalf of the government of the proletarian dictatorship, they were freed and assigned to managerial work. That is how one of the members of the Industrial Party, a former "saboteur and servant of world imperialism," was entrusted with the technical leadership of Factory No. 22.

Our short, gray-haired, slightly balding, agile, and very lively chief engineer began to introduce order and very strict discipline—not into the ranks of the workers, but among the technical intelligentsia. He sometimes flaunted his allegedly anti-Soviet past. "I have already recovered from that case of influenza. That's why you can complain about me and accuse me as much as you want. Just remember that I will not tolerate idlers, illiterates, or demagogues."

Tarasevich quickly showed what a real factory chief engineer was, not to mention one who had obtained the authority of a director. He politely disregarded Party and social organizations and lavishly dispensed reprimands for blatant errors in the work of designers, process engineers, and shop chiefs. He placed people in managerial positions only after assessing their professional qualities and technical competency, showing no regard for Party membership.

Differences of opinion immediately arose between Bolkhovitinov and Tarasevich concerning DB-A series production. The factory had been assigned a very heavy schedule for SB production. In addition, they were demanding that TB-3 production start up again. Tupolev had already started his public relations campaign to prepare the TB-7 (ANT-42) for series production. Tarasevich convinced *Glavaviaprom*, where Chief Designer Tupolev was also chief engineer, that DB-A series production should be set up at the Kazan Aircraft Factory. The decision was made and sanctioned by the People's Commissariat for Heavy Industry. They proposed that Bolkhovitinov and his entire workforce transfer to Kazan.

This was a heavy blow to our young KB. Bolkhovitinov could not refuse—that would have meant the refusal to introduce an aircraft into series production. All of the KB workers received orders to leave for Kazan, initially for a prolonged temporary duty, and from then on, we would see.

Our nerve center immediately fell apart. Kuritskes, Shishmarev, and the other Academy instructors would not agree to leave the Academy. After a long conversation, Bolkhovitinov agreed not to take me to Kazan—I needed to finish my studies at the institute. Another twenty employees refused to go to Kazan due to family circumstances. In autumn 1936, the Bolkhovitinov KB at Factory No. 22 ceased to exist. Tarasevich transferred the KB's remaining staff to KOSTR, the series production KB. I was appointed chief of a special equipment and weaponry team and once again was the leader of a new workforce. But this time it had already been formed and had worked for an extended time in series production. My activity began with a factory-wide reprimand ordered by Tarasevich for the unauthorized alteration of the design of an SB relay regulator fixture. I had introduced the alteration at the request of a military representative.

"I am in charge here, not the military rep," Tarasevich explained to me. "The military rep liked it that way and you stopped work in the shop. For each change please think seven times and then get my approval. This is series production, not Bolkhovitinov's band of freewheelers." My subordinates smiled cunningly. They liked the fact that their innovating chief had immediately been put in his place. The fewer changes, the better!

But there was one more entity in charge of the SB series—Arkhangelskiy's KB. They had the right to issue changes on behalf of the chief designer. All the changes came to our KOSTR. After receiving the chief designer's changes, each team chief reported personally to Tarasevich. He was the only one who, after speaking with Arkhangelskiy, could decide the series into which the change would be introduced.

After the spirit of romance and creative enthusiasm that had permeated the atmosphere in the Bolkhovitinov KB, work in the KOSTR seemed boring and commonplace. However, here I was schooled in the discipline of designing for large-scale series production and also in the system for drafting management. This experience, which no institute in the world provides, was very beneficial to me when we were putting our first missiles into series production. And of course there were flight incidents, which injected a little excitement into our monotonous work with series production drawings. LIS had not run out of practical jokers hatching risky schemes. I will tell you about one such incident.

The Spanish Civil War began in the summer of 1936. During this conflict the Soviet Union rendered assistance to the army of Spain's antifascist Republican government with military advisors, "volunteers," and weapons. The Republican Army did not actually have its own air force. German and Italian pilots flying Messerschmitts and Junkers fought on the side of General Franco. Soviet volunteers in *Chaikas* (I-153s) and *Ishaks* (I-16s) fought on the Republican side.[3] These were fighters produced by Polikarpov in the early 1930s. They were inferior to the Messerschmitt-109 in speed and weaponry. Factory No. 22 received a combat assignment to strengthen the Republican Air Force by increasing SB production and manufacturing dozens of SBs for Spain. This required the aircraft to be modified so that it could be packed in dismantled form into a large wooden container, and upon arrival on site, be quickly assembled under field conditions. This was a super rush job. During the flight acceptance tests, problems arose that threatened to hold up the airplanes' shipment to Odessa, where they would go by sea to Spain.

One of the military representatives particularly annoyed LIS when filling out the acceptance documents. His formal attitude and constant nitpicking even infuriated Tarasevich, who appealed to the chief of military acceptance. "Which one of us is the saboteur?" Tarasevich asked, "I, who was convicted in 1930, or your military engineer second class, whose whims keep our planes from leaving for Spain?"

The LIS degenerates and hooligans, as they used to call them in KOSTR, thought up a risky scheme. They proposed that the capricious military rep himself take a seat in the navigator's cockpit and satisfy himself that firing from the forward firing position machine gun in flight was completely trouble-free. The military rep really did fire in flight, having no qualms about wasting cartridges, and he reported to the pilot over the intercom that he was accepting the aircraft, and they could now go in for a landing. Then the pilot suddenly announced that the landing gear lowering device had failed. He would be forced to land the plane on its belly with the landing gear retracted. In this case, the rep could not stay in the forward cockpit. During landing it was certain that "in the best case scenario your legs will be injured, and in the worst . . . ." To make a long story short, the pilot made a deci-

3. *Chaika* (seagull) and *Ishak* (donkey).

sion and ordered the obstinate military rep to open the lower emergency hatch and, at his command, jump and parachute to safety. Up till then, no one had any experience parachuting from an SB. The unfortunate military rep was badly injured when he jumped out of the airplane and broke his leg upon landing. Relieved of his passenger, the pilot maneuvered over the airfield as though to help lower the landing gear. When he had made sure that the ambulance had reached the parachute of the military rep, he lowered the landing gear and safely landed. The military rep was laid up in the hospital for about three months. The handover of the airplanes proceeded without delay.

At first, only a few individuals knew that the landing gear had actually been fully operational. A third crew member—the flight mechanic, who sat in the center wing section in the gunner/radio operator's seat—confirmed that during attempts to lower the landing gear the electric motor of the hydraulic pump was humming loudly. Several more witnesses appeared at LIS saying that this was not the first time the electric motor had hummed, even without a command to lower the landing gear. The pilot who had ejected the military rep suggested, "Let an electrician fly and figure out what's going on there."

To establish the truth, Tarasevich suggested that I fly. I prepared cables and instruments, and with the help of factory electricians we installed everything in the rear gunner/radio operator cockpit. Before the flight, the pilot was obviously disappointed that I could not be confined in the forward navigator's cockpit. The flight mechanic refused to fly in the forward cockpit. He justified this by saying that the designer would mess up somehow and the landing gear wouldn't extend, and he had no desire to parachute out of a high-speed aircraft, all the more so since he had never even jumped from a tower.

We flew without a flight mechanic. During the flight, I did not find any malfunctions or humming. However, the pilot was in no hurry to land and proved to me that flying on an SB was "all fear and no fun." After landing, Godovikov, with whom I was on good terms, disclosed to me the cause of the landing gear "failure." "Thank God, the military rep survived, but his nitpicking was making our life impossible. You think up something to tell Tarasevich." Godovikov was the kindest person. He was really upset when he found out about the escapades of the LIS flight mechanics after the military rep had already been dispatched to the hospital. Katya, who was working as a foreman in quality control, was subordinate to Godovikov. "I can't imagine a better boss than Nikolay Nikolayevich," she used to say. In my report to Tarasevich, I said that everything was fully operational and that we had been unable to reproduce the incident that occurred previously. Evidently it was a very rare instance of a foreign particle getting under the electric motor's switching contacts.

AFTER OUR INTENSE WORK, Katya and I spent our days off and two evenings a week at the Wings of the Soviets boathouse near the Borodinskiy Bridge. Katya once again took her place as stroke sweeping in a four. Rowing proved to be a very rigorous and hotly contested sport. In August during competitions, Katya's crew won first place in Moscow in their class, while my crew only managed to hold out for third place.

In order to move up from a training quad to a racing shell, we had to train every other day and, in October, again enter the All-Moscow competitions. But I had already given Katya my word a year before that we would make a trip to the mountains. In spite of our coaches' protests, we both announced that we were dropping off our teams until the following year and set off for the mountain camp. From the standpoint of tourism and mountain climbing adventures, there was nothing particularly outstanding about the trip that we made to the Caucasus in autumn 1936. The trip left me with truly beautiful memories of nature, "the only thing better than a mountain is more mountains," as well as wounded masculine pride. Katya, the owner of a "Mountaineer of the USSR" pin, felt considerably more confident in the mountains than I. We traversed from Kabardino-Balkariya to Svanetiya together with a group of Austrian mountain climbers. They had hired a guide who loaded all of their backpacks onto his donkey. When we had climbed to an altitude of around 3,000 meters, it felt like our backpacks had doubled in weight. A gallant Austrian approached Katya and suggested that she place her backpack on the donkey.

"In our mountains only fops use donkeys," she announced, but no one could translate the word "fop" into German. In the pass, a Svan carrying an antediluvian Berdan rifle met us, and under his protection, we safely made our way to Mestiya, the capital of mountainous Svanetiya. In 1996, if anyone had tried to retrace that route, passing through Kabardino-Balkariya, Svanetiya, Georgia, and Abkhaziya, the chances of returning safe and sound would have been slim due to the international armed conflicts in that region.

For the Bolkhovitinov KB, 1936 ended with the tortuous work on the series production of the DB-A in Kazan. The mood in the collective was pessimistic. The series was limited to six aircraft, justified by the need for preliminary troop trials. *Glavaviaprom* was trying to buy time in order to begin the ANT-42 flight tests (this was the TB-7, later renamed the Pe-8). This aircraft surpassed the DB-A on all parameters. At an altitude of 8,000 meters, it could reach speeds in excess of 400 kilometers per hour. The new four-engine bomber actually had five engines—an engine with a compressor that supercharged the four engines for altitude performance and could supply air to the pressurized crew cabins was installed on the center wing section. Subsequently they did away with this fifth engine, having installed Mikulin AM-34FRNV self-supercharging engines. The TB-7 had powerful defensive weaponry. Based on all tactical flight parameters, the ANT-42 surpassed the Boeing Flying Fortress which appeared a year later. The creation of such an aircraft was a very great achievement for Tupolev's collective and the Soviet aircraft industry.

Bolkhovitinov understood that the DB-A could not compete with the TB-7. He began work on two new unusual designs: the "I" fighter and the "S" high-speed fighter-bomber. Isayev designed the "I" as a twin-fuselage, twin-fin aircraft with remote controlled machine guns to protect the rear hemisphere. There were two ShKAS machine guns and new 20-millimeter ShVAK aircraft guns for frontal attacks.[4]

Isayev was very absorbed by this project. He subsequently pulled me into the development of the remote control for the rear hemisphere moveable machine gun pair. I had long dreamed of developing a remote-controlled machine gun mounting using slaving systems. This problem had been solved on naval ships. It was time for aviation to adapt a similar technology.

The matter never reached the point of practical realization. Several years were needed for development and the war intervened. But in the process of working on the system, which interested Bolkhovitinov very much, I became closely acquainted with the developers of slaving systems, Andronik Gevondovich Iosifyan and David Veniaminovich Svecharnik. Andronik Iosifyan was already well known at that time for his unusual design of an "electric helicopter." My first acquaintance with Iosifyan in 1936 turned into many years of collaboration and friendship that continued until the end of his life.[5]

The KB devoted primary attention to the design of "S"—the paired-engine aircraft. The innovation of the "S" was the mounting of two engines one behind the other. The long shaft of the rear engine passed between a V-shaped split in the cylinders of the front engine. The two propellers, each one operated by its own engine, rotated in different directions. This configuration reduced head resistance by the resistance value of one engine, enabling a twenty-five percent increase in speed over the conventional two-engine configuration. The "S" design reached speeds of 700 kilometers per hour at a range of 700 kilometers.

The idea of pairing the engines, which was realized on the "S," was used in the design of the "B" bomber. Instead of the conventional four-engine aircraft configuration, with the engines in wing nacelles, the design called for two sets of paired engines, with a speed in excess of 550 kilometers per hour and a maximum altitude of up to 11 kilometers.

In spite of these new designs, *Annushka*'s modernization continued. They installed boosted M-34FRN engines with turbo compressors and variable-pitch propellers. Instead of pulling the landing gear up into "trousers," it was fully retracted. The turret in the center portion of the fuselage was equipped with a ShVAK machine gun and a drive. Two machine guns providing a 360-degree field of fire were installed in the cockpits under the center section of the wings. The crewmembers increased from six to eleven. These substantial improvements did not help, however. Tupolev's TB-7 began to fly in the spring of 1937, and based on all parameters, immediately moved far, far ahead of our *Annushka*.

---

4. The ShKAS was the Shpitalniy-Komarnitskiy rapid-fire aircraft machine gun, and the ShVAK was the Shpitalniy-Vladimirov rapid-fire aircraft machine gun.

5. Andronik Gevondovich Iosifyan (1905–93) served as chief designer of VNII EM, the organization that designed Soviet weather and remote sensing satellites such as Meteor.

Chapter 7

Arctic Triumphs and Tragedies

In the late 1920s and early 1930s, exciting events took place in connection with the exploration of the Arctic. The Arctic was still an area where heroism could be fully displayed. The press and radio widely publicized the work of Arctic stations and expeditions, particularly emphasizing the romance associated with conquering the Arctic. The enormous significance of the Arctic regions to the economy of the Soviet Union was so obvious that no one questioned the expenditures necessary to open them up.[1] Events associated with Arctic explorations aroused passionate feelings in the most diverse social strata. The excitement generated by every Arctic adventure was also of great political significance. Public attention was somewhat distracted from the difficulties of daily life, the repressions, and the food crisis that had developed as a result of collectivization in the countryside.

The successes in opening up the Arctic raised the Soviet Union's international prestige. The intelligentsia, isolated from cultural and scientific interaction with the outside world, saw in Arctic research a hope for international collaboration. Society was united in the fact that the Arctic must be Soviet.

This outward appearance of solidarity was very advantageous for the Stalinist leadership. The heroic feats of icebreaker crews, polar pilots, and men who wintered at Arctic stations; the record-setting flights of Soviet aircraft; and the rescue of expeditions in distress were a graphic demonstration of the unity of our whole society for the common goals of mankind.

The headlines of all the newspapers—and the radio broadcasts, which were only just becoming popular—reported on the rescue of the Italian Nobile expedition by the icebreakers *Krasin* and *Malygin* in 1928, as well as the international Arctic expedition of a German Zeppelin in 1931, and the voyage of the icebreaker *Aleksandr Sibiryakov*, which in 1932 completed the first nonstop voyage from the White Sea to the Bering Sea along the North Sea Route.

---

1. The Soviet Union had both economic and strategic interests in the Arctic. The Northern Sea route that links Murmansk with Vladivostok was the shortest sea lane between the European and Far Eastern regions of the Soviet Union and also provided easy access to northern Siberia. Additionally, the Arctic contains vast untapped hydrocarbon reserves.

The Main Directorate of the North Sea Route, *Glavsevmorput*, was established in 1932. It was directly subordinate to the Council of Peoples' Commissars. Otto Yulyevich Shmidt, a well-known scientist and leader of polar expeditions, was appointed director of *Glavsevmorput*. His deputy and chief of polar aviation was Mark Ivanovich Shevelev. In 1933, Shmidt attempted a nonstop voyage on the North Sea Route from Murmansk to Vladivostok in the steamship *Chelyuskin*. In February 1934, the ship was trapped and crushed by ice in the Chukchi Sea, and its many passengers, including women and children, were taking refuge on an ice floe. People of all social strata feared the fate of the expedition, and were thrilled by the heroism of the pilots who saved every last inhabitant of the Shmidt ice camp.

The Soviet pilots who participated in the rescue of the *Chelyuskin* disaster victims were the first in the USSR to receive the title Hero of the Soviet Union. Arctic pilot Sigizmund Aleksandrovich Levanevskiy was also among the first heroes. He came to fame in 1933 by rescuing American pilot Jimmy Mattern, who was trying to complete a solo round-the-world flight and crashed his plane near the Anadyr Range.[2] Levanevskiy transported Mattern back to the United States.

In the spring of 1935, Levanevskiy approached the government with a proposal to organize a transpolar flight from Moscow over the North Pole to the United States. He proposed using the Tupolev ANT-25 aircraft in which pilot Mikhail Gromov had set a world record for the longest nonstop flight. The Council of Labor and Defense adopted a resolution to arrange for a flight from Moscow to San Francisco. On 3 August, Levanevskiy, copilot Baydukov, and navigator Levchenko took off in the ANT-25 aircraft from the Air Force NII's airfield in Shchelkovo, intending to land 63 hours later in San Francisco. The entire world followed the flight preparation and the flight itself with enormous interest. A successful flight would help establish closer political and economic relations with America. According to foreign press assessments, this flight promised to be the most dangerous and the most remarkable in the history of aviation.

Ten hours after takeoff, having flown only as far as the Barents Sea, the crew requested permission to terminate the flight on the designated route and return to the nearest airfield due to a defect in an oil line. Oil had spilled over the wing, flowed down onto the cockpit canopy, and leaked into the cockpit.

The failure of the flight was a heavy blow to the prestige of the Soviet Union. Stalin decided to personally hear a report from the crew about the causes of the failure and their suggestions for the rehabilitation of our aviation. During the meeting with Stalin, Levanevskiy suddenly announced, "I don't trust Tupolev! In my opinion, Tupolev is a saboteur. I will never fly one of his airplanes again!" The details of the discussion concerning the causes of the emergency, of course, never made it into print, but aviation circles learned about Levanevskiy's declaration

---

2. The Anadyr Range is located in the extreme northeast of Siberian Russia. Its easternmost part ends in the Chukchi Peninsula near the Bering Sea.

Archives of Y. P. Salnikov.

The pilots awarded the title Hero of the Soviet Union for the Chelyuskin rescue in 1934. Standing, left to right: N. P. Kamanin, I. Doronin, V. S. Molokov, S. A. Levanevskiy, A. V. Lyapidevskiy; sitting, M. T. Slepnev, M. M. Gromov, M. V. Vodopyanov.

from meeting participant Baydukov, who was soon appointed chief pilot at Factory No. 22.

At the instruction of Air Force Commander Alksnis, Baydukov combined his work as a factory pilot with the testing of the ANT-25 aircraft that the Tupolev designers and mechanics had modified at the Central Airfield in Khodynka. Tupolev and his supporter Alksnis believed that the defect involving the oil leak was accidental, and they were convinced that in the near future the ANT-25 would be the only aircraft capable of executing a flight from Moscow to the United States over the Pole.

The work to finish the Tupolev aircraft was very intensive, in spite of the fact that at their previous meeting Stalin had instructed Levanevskiy to travel the United States and select and purchase an airplane that would be substantially more reliable than the single-engine RD (ANT-25).[3] Stalin told Levanevskiy, "No matter how expensive your chosen airplane is, we will pay any amount of money." I cite this quote from Stalin from Baydukov's memoirs.[4]

---

3. RD—*Rekord dalnosti* (Long-Range Record).

4. Author's note: Georgiy Baydukov, *Komandarm krylatykh* (Winged Army Commander) (Moscow: Zvonnitsa, 2002), p. 233.

Levanevskiy and Levchenko departed for America. They did not succeed in finding an aircraft that was suitable for such long-range flights, but they bought three hydroplanes.

Now it wasn't Levanevskiy, but Baydukov who could make accusations of sabotage, if not against Tupolev himself, then against his coworkers. They had been responsible for optimizing and preparing the aircraft.

This he did not do. By then, Baydukov's father had already been repressed.[5] They had not touched the younger Baydukov, acting on the principle that "the son is not responsible for the father." They observed this principle until 1937.

From the author's archives.

Soviet aviator Sigizmund Aleksandrovich Levanevskiy, photographed in 1937.

The black mark placed on Soviet aviation by Levanevskiy's failed flight had to be erased as soon as possible.

On 20 June 1936, in a modified ANT-25, a crew comprised of Chkalov, Baydukov, and Belyakov executed a 9,734-kilometer flight over the north of the Soviet Union in a time of 56 hours and 20 minutes. This was an outstanding flight in the history of Soviet aviation. The crew landed on the island of Udd (now named Chkalov) then returned from the Far East with stopovers for festive receptions along the way back to Moscow. On this notable occasion, a command came "from the top" to find Baydukov's father, who was in one of the NKVD's camps, and bring him to his son's native Omsk. They found him among prisoners building the railroad branch line from Khabarovsk to Komsomolsk. Baydukov barely recognized his haggard father, who had been strictly warned never to tell anyone anything about the camp. Georgiy Baydukov, through his participation in the heroic flight, saved his own father from death in the camp.

RESEARCH WAS BEING CONDUCTED in the Soviet sector of the Arctic on a scale never before seen. After setting up continuously operating bases—weather stations on the mainland and northernmost islands—the problem of establishing permanent stations in the central portion of the Arctic became more and more pressing. In the opinion of scientists, the "weather kitchen" of the entire

5. In Russian/Soviet language and culture, it has been common to use the word "repressed" to denote the arrest, incarceration, and often execution of citizens during the Stalinist era.

Arctic Ocean basin was located there. But how were they to deliver the personnel to the Pole who could man the Arctic stations, along with their scientific equipment, food stores, radio stations, tents, and fuel for many months of work? This was one of the main problems.

One of the first to use the TB-3 aircraft in the Arctic and attempt the delivery of an expedition to the North Pole region was polar pilot Mikhail Vasilyevich Vodopyanov. He was one of the seven Heroes of the Soviet Union who received this title for rescuing the victims of the *Chelyuskin* disaster.

In those years, proposals from famous pilots made directly to Stalin or high-ranking government officials were accepted sooner than the initiatives of Peoples' Commissars or scientists. Shmidt and Shevelev decisively supported Vodopyanov's proposal. Factory No. 22 received the assignment from *Glavaviaprom* to manufacture at least four TB-3 aircraft with special Arctic modifications. In mid-1936, KOSTR began to issue the drawings. We made modifications after consulting with the polar crews. Thereby we became acquainted with Vodopyanov, Babushkin, Mazuruk, Alekseyev, Golovin, and Moshkovskiy.

The radio equipment was completely updated. We installed a new radio station in the tail section, in the insulated radio operator's cockpit. It could operate not only with the aircraft's electric system but also with an emergency autonomous electric generator, which served as a source of energy when the plane was on the ground. True, the generator's startup reliability in the cold left a lot to be desired. Due to fire hazard, the crew was directed to use it only after deplaning.

A fixed-loop radio compass and a receiver for beacon navigation were installed in the nose section. Radio communications could be conducted using two types of antennas: a rigid antenna extended over the fuselage for use when the aircraft were standing on the ground; and an extendable, weighted antenna reeled out from a special winch in flight. The pilots' cockpits and the machine gun turret hatches, which were open to the wind, were glazed to prevent drafts in the fuselage. Engineers devoted particular attention to making the transition from wheels to skis a simple process. The operation would have to be conducted not in a heated factory, but in winter conditions at a northern airfield. Absolute reliability had to be guaranteed. The aircraft were painted a bright orange color for the best visibility against the snow cover.

The workers labored over the arctic version of the TB-3 with the same enthusiasm as their shop comrades who labored over the SBs for the Spanish antifascists.

During January and February 1937, the four orange giants were to undergo test flights and acceptance by Chief of Polar Aviation Shevelev. He arrived with a *Glavsevmorput* radio engineer—a tall, gray-eyed blond—who demonstrated enviable knowledge of the specifics of polar conditions for radio communications. He criticized parts of the radio equipment layout and demanded alterations. My designers and the factory assemblers were offended by this young radio operator for the lecturing that he did with the aplomb of an experienced polar explorer.

And this was how in the spring of 1937, I met for the first time with Boris Mikhaylovich Konoplev, the future chief designer of R-16 intercontinental guidance systems. Our subsequent close collaboration continued until his tragic death on 24 October 1960.[6]

Four aircraft were ferried to the Central Airfield for further preparation before their flight to the North. I can boast that I was part of a flight to the North Pole, that being the initial 6-kilometer segment from Fili to Khodynka.

On 22 March 1937, one by one, the airplanes took off from the Central Airfield: N-170, with pilots Vodopyanov and Babushkin; N-171, with pilot Molokov; N-172, with pilot Alekseyev; and N-169, with pilot Mazuruk. N-166 had taken off three hours before them. This was one of our factory's R-6s that had been transferred to polar aviation.[7] Its pilot was Pavel Georgiyevich Golovin. He was to act as an ice scout in the Arctic.

With that, Arctic activity at the factory was temporarily finished. We waited for news of the landing of our airplanes on the North Pole.

The first Soviet pilot to fly over the North Pole was Golovin in N-166. We celebrated; Factory No. 22's reconnaissance plane was the first to fly over the Pole. This took place on 5 May. On 21 May 1937, the TB-3 N-170 that we manufactured was the first to land on the ice floe that became known throughout the world as the *Papaninskaya*.[8]

The world did not learn of this historical event until twelve hours after the landing! The reason was our low degree of equipment reliability. While N-170, Vodopyanov's airplane carrying expedition chief Shmidt, was on its way to the Pole, the entire world followed radio reports on the flight's progress. When N-170 reached the Pole, Shmidt radioed, "We're descending, we'll look for a place to land." After this, communication with the expedition broke off. The aircraft did not respond to calls from ground stations. Could there really have been another tragedy in the Arctic and did we have to organize a rescue expedition once again? The whole world and every amateur short-wave radio operator knew the airplane's call sign, but nobody had received anything during the twelve hours since the last radiogram.

The reason was all very simple. The thermal mode of the transformer (motor-generator) converting the 12-volt voltage of the onboard network into high voltage to power the transmitter was not designed for prolonged operation. It is possible that it had some sort of manufacturing defect. To make a long story short, the transformer burned out.

---

6. The worst disaster in the history of rocketry occurrred on 24 October 1960, when over a hundred engineers, administrators, and soldiers were killed during prelaunch preparations of the R-16 intercontinental ballistic missile. The disaster is more commonly known in the West as the "Nedelin disaster" after the military commander, Mitrofan Ivanovich Nedelin, who was killed during the explosion.

7. The R-6 (or ANT-7) was a reconnaissance/escort fighter version of the ANT-4 bomber.

8. *Papaninskaya* was named after Arctic expedition leader Ivan Dmitriyevich Papanin (1897–1974).

The bomber radio station (RSB), which was manufactured especially for this Arctic expedition, and which had undergone a triple acceptance and flight test process, was dead. The aircraft had no backup. After celebrating their safe landing and shouting "hoorah," the Papanin team quickly unloaded the radio operator Krenkel's radio unit. This radio station was supposed to maintain contact with the world the entire time Papanin's station, dubbed North Pole, was in operation. But even Ernest Teodorovich Krenkel, the legendary radio operator of the *Chelyuskin* epic, was unable to quickly set up communications. During the flight, the radio station's storage batteries had frozen. They had to start up the small gasoline-powered generator and use the charging current to warm up the batteries. The generator didn't start up immediately and then coughed and sputtered for a long time. It was half a day later before Shmidt reported to Moscow that they had landed safely. Now Moscow could make the decision whether or not to send the remaining three airplanes to the ice floe.

IN THIS LARGE SYSTEM for coordinating the mission that was supposed to be reliable, radio communications proved to be one of the weakest links. Unfortunately, the celebratory commotion in the wake of the polar conquests prevented us from analyzing this incident and drawing conclusions from it.

In a ceremony on 6 June 1937, Shmidt gave the command to raise the flag of the USSR over the first station, North Pole-1. One by one, the four TB-3s safely took off from the ice floe. A festive reception awaited them in Moscow.

Official reports of the uncovering of "an anti-Soviet Trotskyite military organization" threw a pall over the nationwide rejoicing. High-ranking military officers Tukhachevskiy, Kork, Yakir, Uborevich, Putna, Eydeman, Primakov, and Feldman were accused of organizing a fascist military plot, preparing to overthrow Soviet power through armed uprising and defeat of the USSR in an ensuing war, espionage and sabotage, and forming terrorist groups to annihilate Party leaders and the government. The affair, which was fabricated by Yezhov's department, was reviewed on 11 June 1937 at a special session of the Supreme Court comprising Budennyy, Blyukher, and Alksnis, and chaired by Ulrikh.[9] The Court sentenced all of the accused to be shot to death. Subsequently, Blyukher and Alksnis were also shot to death. The "officers' affair" was the beginning of a campaign of massive repressions in 1938.

IN NEWSPAPER ARTICLES AND RADIO BROADCASTS, dispiriting demands for vigilance and the unmasking of Trotskyite espionage activity were intermingled with reports about the achievements of our aviation. The reports about new and outstanding flights were breaths of fresh air in the stifling atmosphere.

---

9. Nikolay Ivanovich Yezhov (1895–1939) was chief of the Soviet security police (NKVD) from 1936 to 1938, during the most severe period of the Great Purges, also known as the *Yezhovshchina.* Marshal Semyon Mikhaylovich Budenny (1883–1938), Marshal Vasiliy Konstantinovich Blyukher (1890–1938), and General Yakov Ivanovich Alksnis (1897–1938) were all famous Soviet armed forces officers.

On 18 June 1937, the transpolar flight of Chkalov, Baydukov, and Belyakov began. Sixty-three hours and 25 minutes later, the ANT-25 landed in the United States at Pearson Field near the city of Vancouver, Washington. This flight opened the shortest route over the Arctic ice from the USSR to the United States. The celebration of this historic event had hardly subsided before the world heard about the beginning of the next transpolar flight.

On 12 July 1937, in the same type of Tupolev single-engine aircraft, Gromov, Danilin, and Yumashev departed for the United States from the Air Force NII's airfield in Shchelkovo. After flying for 62 hours and 17 minutes they landed safely in the vicinity of Los Angeles, having broken the world records for nonstop flight along both a straight line and a broken line.

I witnessed the public jubilation when the heroic pilots paraded through the streets of Moscow after their return from the United States, and I would say that the festive receptions for the Chkalov and Gromov crews was comparable to the public rejoicing that took place on 12 April 1961.[10]

Officially, our factory had nothing to do with the transpolar flights by the Chkalov and Gromov crews. But Georgiy Baydukov, whom Chkalov persistently called Yegor, was the copilot in Chkalov's crew. As I have already mentioned, Baydukov was a test pilot at our factory in 1937. He not only tested the series-produced SBs, but also participated in DB-A flights along with Kastanayev and Nyukhtikov. In May 1937, Baydukov and Kastanayev set two speed records on the DB-A with a 5-ton load at ranges of 1,000 kilometers and 2,000 kilometers.

IT HAD NEVER OCCURRED to Viktor Bolkhovitinov, our KB chief, or any of the other specialists who worked with him in Moscow or Kazan, to conduct a flight over the Pole in an aircraft that had not yet been made sufficiently reliable.

I cannot in all certainty answer the question as to who was the first to come up with such an idea. According to the memoirs of Baydukov and Vodopyanov, Stalin was very favorably disposed toward Levanevskiy, in spite of the fact that he had relatives in Poland and his brother was a Polish military pilot. It is possible that Stalin was giving him his due for his past service during the Civil War, for his popularity in the United States after rescuing Mattern, or for his courage in his first attempt to complete a transpolar flight from Moscow to San Francisco.

Before Chkalov took off for the United States over the Pole, Levanevskiy was among those summoned to the Politburo. Stalin well remembered Levanevskiy's failed attempt to complete a transpolar flight in an ANT-25 in August 1935, but he was also well aware of the cause of the failure.

Baydukov had not forgotten that Levanevskiy originated the idea of flying over the Pole. He suggested that Levanevskiy introduce himself to Bolkhovitinov and have a look at the DB-A. Bolkhovitinov was immediately

10. On 12 April 1961, Yuriy Alekseyevich Gagarin became the first human in space.

called from Kazan and instructed to show Levanevskiy the airplane. After being introduced to Levanevskiy, Bolkhovitinov gathered his few remaining compatriots at Factory No. 22, including me. He was very against the idea of using the only DB-A that had undergone flight tests for a transpolar flight. When I told him about the modifications we had made on the TB-3 for the arctic version and estimated that it would take at least two months, his mood darkened. He said, "In any event, we will not get the airplane back. And how long will it be before we have another one?"

Bolkhovitinov also considered Tupolev's attitude toward the use of the DB-A for a transpolar flight. Tupolev did not support the idea of the DB-A, period.

From the author's archives.

Factory No. 22 Chief V. F. Bolkhovitinov (far left), consulting with S. A. Levanevskiy (far right), August 1937.

Moreover, after the incident at the meeting with Stalin, Tupolev had a very negative attitude toward Levanevskiy. At that time Tupolev was not only a chief designer, but was also the acting deputy chief of the Main Directorate of Aviation Industry within the People's Commissariat of Heavy Industry. Preparation for the proposed flight was impossible without the aviation industry's assistance. Bolkhovitinov had every reason not to agree, but word had it that secret pressure from the top came from Stalin himself.

In early June, the DB-A developers met for the first time with Levanevskiy at the factory airfield. At that time I was still unaware of Levanevskiy's difficult military biography. Sporting stylish clothes, with a thoughtful and intent gaze, he gave the impression of a well-bred aristocrat. During preparation for takeoff, he was very reserved and taciturn. Obviously Bolkhovitinov's stance pained him.

As the aircraft was being fueled and prepared, Bolkhovitinov, tightly buttoned into the military uniform of a brigade commander (in modern terms, that corresponds to major general), was somber, and slowly meandered through the sweet-scented grass of the airfield, lost in thoughts. The demonstration flight was assigned to Kastanayev. He emerged from the flight mechanics' hut along with Godovikov. Serious and unsmiling, both men approached Bolkhovitinov, briefly discussed something, and then climbed into the aircraft.

After a short takeoff run, Kastanayev lifted off, gained altitude, and then dove to gain speed toward the village of Mnevniki and made a very steep, banking turn over the airfield. After deafening us with the roar of the four boosted engines, he once again climbed steeply upward. The aircraft was empty and fueled only for a demonstration. It was easy for Kastanayev to execute showy maneuvers that were not typical for a heavy bomber.

Observing the flight, Levanevskiy was transformed. We never expected such a wild reaction from our taciturn guest. The airplane had not yet landed, and Levanevskiy was beaming, radiating delight, and literally throwing himself at Bolkhovitinov. "You've got to, got to give me this plane! We've got to show this to the Americans! They've never even dreamed of such a plane!" I was unable to hear what else was said between Levanevskiy and Bolkhovitinov.

We know that the following day Levanevskiy was at the Kremlin. Next, they also summoned Bolkhovitinov. A day later in Tarasevich's office, Bolkhovitinov gathered a team of designers and informed them that the government had approved Levanevskiy's request and was permitting him to fly the Moscow–North Pole–Alaska route.

We were subsequently called up and mobilized to adapt the DB-A for a trans-polar flight. We were given one-and-a-half months for the entire work.

The assignment was quite an honor. But if Bolkhovitinov was unhappy because he, the chief designer, was being deprived of his only DB-A aircraft, I was upset for a completely different reason. Preparation for the Arctic expedition, and now yet another flight, were depriving me of the opportunity to finish my third year of studies without taking incompletes. The semester had just begun, I had four difficult exams ahead of me, and I was at risk of being confined to barracks. Nevertheless, I hurried to take my exams in philosophy and machine parts before my total confinement. I still needed to take the electrical engineering and vector calculus exams. Professors Krug and Shpilreyn, the patriarchs of electrical engineering and vector analysis respectively, were not lenient with exams. A profound knowledge of the subject matter was required. Tarasevich promised me a note of excuse.

Bolkhovitinov and the design unit deputy, Saburov, relocated to the factory. A rush job was begun to issue the drawings transforming the armed bomber into a peaceful transport aircraft that would deliver gifts via the shortest route over the Pole: caviar for President Franklin Roosevelt and expensive furs from the Russian North for his wife Eleanor.

The moniker "patron" had already stuck to Bolkhovitinov. Word had it that the first person to call him that behind his back was Isayev. "Our patron has started to change his tune," said military engineer Frolov, who was flight test lead. "He has already sat down with me to consider how it would look if they convert *Annushka* into a cargo/passenger plane."

But it would be a long time before a cargo/passenger version for transpolar routes appeared on the scene. Based on preliminary calculations, if all of the weaponry was removed, then the weight of the empty aircraft would be 16 tons. In order to ensure a range of at least 8,000 kilometers, the aircraft would have to hold 16.5 tons of gasoline and 900 kilograms of oil. The crew, along with equipment and food supplies, would add up to 1.5 tons plus a minimum of miscellaneous baggage—in total, we had already exceeded 35 tons for the takeoff weight. Given the 840 horsepower for each M-34FRN engine at an altitude of 4,000 meters, this was the takeoff weight limit. But the first step is the hardest.

My group in the so-called "ground crew" was in the most difficult situation. In order to issue the electrical diagrams and installation drawings for the new equipment, we needed the initial data from the other factories involved. They still did not know anything about the decision that had been made. During the very first days, authorized representatives from *Glavaviaprom* and the Air Force were detailed to us on temporary duty. I only had to mention my difficulties in passing before everything I needed started to appear. The chief developers of all the factories received instructions to consider our jobs top priority.

Together with Chizhikov and the engineers of the Gorky and Moscow radio factories, we designed instruments and radio equipment. The new powerful *Omega* radio station was installed in the tail section in the specially insulated radio operator's cockpit. The *Omega* could operate on short and long waves in telegraph and telephone mode. In the navigator's cockpit, the sights, bomb release devices, machine gun, and forward gunner were removed. They decided to place the flight radio operator there along with the navigator. In the navigator's cockpit we installed a second, lighter radio station without the long wave range, and we equipped the radio operator's seat.

The crew consisted of Levanevskiy and a specially created flight staff. It was clear who the pilots would be: Levanevskiy and Kastanayev. Two flight mechanics were needed for the four engines, which could be accessed through the thick wings. The first flight mechanic was Grigoriy Pobezhimov, who had traveled with Levanevskiy to the United States to procure airplanes. Pobezhimov was an experienced polar flight mechanic, but he was not familiar with the DB-A and had no experience with M-34FRN engines.

Kastanayev recommended Godovikov as a candidate. He told Bolkhovitinov that he was the only man among the factory workers who knew that airplane inside and out. He had a wonderful feel for the engines, could instantly figure out the quirky lubrication systems and fuel lines, and if necessary would crawl into the most inaccessible place.

Nikolay Nikolayevich Godovikov did not want to fly. He aspired neither to fame nor to new awards. In his forty-four years, he had already received the Order of Lenin and the Order of the Red Star. What is more, he had already been to the United States with Vladimir Gorbunov in 1934. But the main thing was that he loved his family, which included seven children. But Godovikov could not refuse. He understood that he would be the only one in the crew who knew all of the aircraft's mechanisms and was capable of performing the duties not only of flight mechanic, but also of flight engineer. He was actively engaged in the modification process and helped us redesign the flight mechanics' instrument panels and install fuel gauges. He also spent a great deal of time mastering a new gadget: electric gas analyzers. These instruments made it possible to monitor the makeup of the exhaust gases in order to select the most efficient operating mode.

Crew navigator Viktor Levchenko rendered a great deal of assistance in the installation of the navigation equipment. The design of the astrodome for the solar heading indicator required the most attention. Next, we mastered the American Fairchild radio compass. The radio compass indicators had leads to the navigator, pilot, and copilot workstations.

The last person to join the crew was Leonid Lvovich Kerber. Due to his particularly nonproletarian origins—he was the son of a Russian Navy vice-admiral—Kerber had been unable to enter an institute in order to obtain a higher education in radio or electrical engineering. By the time he had reached thirty-four years of age, this talented self-taught individual had gone through on-the-job training as a military telephone operator, a Central Airfield communications radio operator, and an aircraft special equipment team leader at the Tupolev KB. Kerber's best reference came from Amik Avetovich Yengibaryan, one of Tupolev's specialists, whom he had promoted to a high-level post in *Glavaviaprom*.

Amik Yengibaryan himself was a colorful figure in the new field of avionics. A strong-willed, energetic leader in the rapidly developing field, he tracked the aircraft's preparation for the flight and helped us out in every way possible. Introducing Kerber to Bolkhovitinov, Yengibaryan said, "This man can do everything. If necessary, he can even replace the navigator. Besides being a brilliant radio operator, he is also a specialist in all aircraft electrical equipment."

I was quickly convinced of the truth of this testimonial. Buzukov and I familiarized Kerber with a diagram of the aircraft's electrical equipment. He grasped everything instantly, at the same time demonstrating a good sense of humor when he was dissatisfied with something or other.

Kerber, Buzukov, and I devoted a great deal of attention to the reliability of the electrical power system. The system, developed at the Lepse Factory, was comprised of two generators installed on internal motors, buffer batteries, and generator regulators, which stabilized the voltage and switched the batteries to charging mode when the motors reached high RPMs. We joked that Lepse Factory representative Goldobenkov, who fussed over the regulators the whole time, would "regulate" them to death.

After twenty days and nights of work at the factory, the aircraft, painted unusually with a dark-blue fuselage and red wings, was ferried to the Shchelkovo airfield of the Air Force NII and assigned the polar aircraft number N-209. This number was to enter the history of Arctic exploration forever.

THE ENTIRE "GROUND CREW" headed by Bolkhovitinov—flight planners, engine mechanics, designers, and even drafters—were housed in the NII service buildings, which had been converted into a design bureau and dormitory with full room and board.

Levanevskiy was urgently summoned to Sevastopol for the acceptance tests of three hydroplanes arriving from the United States. Kastanayev performed the first flights of N-209 in Shchelkovo without him.

The propeller-engine units gave Frolov, the lead engineer for the aircraft's general flight tests, the most trouble—the exhaust manifold pipes caught on fire, the fuel consumption exceeded the design value, and the gas analyzers indicated unintelligible readings. During all the ground and flight tests, the engines had to be used sparingly because of their limited service life, which amounted to 100 total hours. Given a planned flight duration of 35 hours, there was very little time left over for control flights and ground adjustments.

Maks Arkadyevich Tayts, the leader of the flight plan group, was a TsAGI employee. He grumbled that his slide rules were smoking from the continuous corrections and recalculations that he had to make, including computation of the distance margin based upon the dry weight of the structure and changes in the baggage makeup, emergency food supplies, and fuel and lubricant data, as well as changes resulting from meteorologists' recommendations regarding route and altitude selection. The flight plan group also had to provide for reserves based upon meteorologists' predictions of the most unfavorable conditions that could occur. But who could say what the most unfavorable conditions would be in August beyond the Pole? As of yet there were no statistics. The meteorologists' recommendations boiled down to one thing: the closer you are to autumn, the worse the weather—fly as early as possible.

After two weeks of work on breaking in the aircraft, and control flights from the Shchelkovo airfield, our first incident occurred. Kerber did not show up at the usual time. Twenty-four hours later they introduced us to the new crewmember who would be taking his place, radio operator Nikolay Galkovskiy. Time had already taught us not to ask questions concerning such incidents.[11] Galkovskiy worked in the Air Force NII. He was the flagship radio operator in the holiday air parades in Moscow and had participated in flights throughout Europe. In September he was to begin his studies at the N. E. Zhukovskiy Air Force Academy.

---

11. Leonid Lvovich Kerber (1903–93) was arrested and jailed by the NKVD in 1938.

Godovikov, Frolov, electrical system foreman Mayorov, and I were very upset. In the time remaining, it was going to be difficult for a new crew member to master the aircraft. Kerber had managed to make some alterations based on his own experience.

The last nonstop control flight over the route Moscow-Melitopol-Moscow, a distance of more than 2,000 kilometers, was scheduled for 28 July. Galkovskiy, having a total of three days experience with the N-209, asked me to participate in the flight to perform joint checks on all the radio equipment. I was therefore included in the crew of this control flight.

Kastanayev occupied the commanding pilot's seat on the left side for the entire route. Levanevskiy, sporting a pressed suit, snow-white shirt and bright necktie, smiling and joyful, strolled around the aircraft observing the crew members in action. From time to time he would sit down in the co-pilot's seat and see what it was like to steer the plane. During the entire flight, Godovikov and Pobezhimov went from engine to engine scrutinizing their operation. I spent most of the flight in the tail section in the radio operator's cockpit checking all the station operating modes while Galkovskiy worked in the navigator's cockpit with Levchenko the entire trip.

We had good weather at an altitude of 3,000 meters as far as Melitopol. When we turned around to head home, Godovikov laid out dinner, which consisted primarily of caviar and chocolate. I was a little greedy and was punished. An oncoming front of thunderstorms caused strong turbulence and forced us to climb to an altitude of over 5,000 meters. Levanevskiy, having noticed that I was clearly not feeling well, made me put on my oxygen mask. Meanwhile, in spite of the cold, he continued to saunter around in his elegant suit.

Having checked the emission of the *Omega* transmitter in all wavelengths from 25 meters to 1,200 meters, I suggested to Galkovskiy that he move back to the tail, but he did not have time. Levchenko was making him learn how to determine location based on the intersection of radio direction-finder bearings. Something with the orientation procedure wasn't clicking.

After this flight, Tupolev came to check out the state of preparations. He listened to Tayts, Yengibaryan and his engineer/engine mechanic Rodzevich, who had helped our team adjust the engine-propeller unit. He posed several questions to Bolkhovitinov and Kastanayev. Levanevskiy did not participate in this meeting.

The N-209 was scheduled to take off on 12 August. On the eve of departure, in order to facilitate takeoff, the aircraft was towed onto the concrete slope from which all aircraft heavily loaded with fuel for long-range trips started. On the morning of 12 August, the fueling process and the bustle of the last hours of preparation began. Godovikov, Pobezhimov, and Galkovskiy were almost constantly in the airplane. Together with the testing leader, they were checking the last packs of gear, the emergency supply of food, warm clothes, firearms, and the lifeboat.

Frolov let it slip that the day before, Levanevskiy had demanded that all manner of baggage and supplies be thrown out so that the plane could be filled with additional fuel. They rolled the aircraft onto the dynamometer scales. The takeoff weight exceeded the permissible limit of 35 metric tons.

Photo by I. Shagin.

The crew of N-209, photographed on 12 August 1937 before their attempted transpolar flight to the United States. From second left to right: N. Kostaneyev, S. Levanevskiy, G. Pobezhimov, N. Godovikov, V. Levchenko.

By the middle of the day, reporters, numerous cameramen, and well-wishers had arrived. A crowd of reporters surrounded Levanevskiy. Bolkhovitinov, with dark circles under his eyes from sleepless nights, was discussing something with Kastanayev and brushed aside the reporters. The anxious Godovikov argued with a group of factory engine mechanics and designers. I headed over to the airplane on the slope, intending to hand over to Godovikov or Galkovskiy the flashlights that had been tossed out along with other, supposedly unnecessary, baggage.

A tall serviceman was strolling near the airplane. He held a little boy about nine years old by the hand and, pointing to the airplane, explained something to him. As I drew closer and saw the four diamond-shaped insignias on his gorget patches, I finally realized that it was Alksnis. The meticulous reporters sensed something, but none of them approached and annoyed the RKKA Air Force Commander-in-Chief with questions.

After saying good-bye to the crowd, the first to approach the airplane was Godovikov. He appeared perplexed. When he saw me, he took the package, but he didn't seem to hear what I said about the flashlights.

"Now you see it, now you don't," Godovikov uttered his favorite saying, and after shaking hands, started climbing into the airplane.[12]

---

12. Author's note: "Now you see it, now you don't" is a saying originating from Chinese magicians who used to wander throughout pre-Revolutionary Russia doing sleight of hand tricks with a ball. In aviation slang, it refers to a defect that appears and then suddenly disappears. This never became part of rocketeer slang. It was a different time.

"Good bye and good luck, Nikolay Nikolayevich!" I shouted. Godovikov waved, turned back around, and disappeared into the fuselage.

Then he suddenly appeared in the dark entry hatch doorway; and after shouting, "Farewell, Boris. Now you see it, now you don't!" he finally vanished.

That is how I remember my unhappy farewell with Godovikov. I believe that he was convinced of the unfortunate outcome of the flight. How the remaining crew members approached and said their goodbyes, I do not recall. The excited and happy Levanevskiy was the last one to climb into the aircraft.

Bolkhovitinov came to an agreement with the pilots that Kastanayev would take off and fly the aircraft for the first several hours. The concrete strip was cleared. Everyone dispersed from the aircraft. One after another the propellers began reluctantly to spin. Finally all four were working. Red Army soldiers ran up to the wheels and pulled out the chocks. The engines began to roar and the airplane rolled down the slope. It ran down the runway for an unbearably long time. It seemed like it had reached the forest before it tore away from the concrete. Kastanayev managed to take off at the very end of the runway. It was 6:15 p.m. Then someone who had timed it announced that the takeoff run lasted 37 seconds. The N-209 slowly rose up over the forest leaving behind a smoke trail from the far right engine.

After such constant stress we didn't know where to go or what to do with ourselves now. No one left the field. After a while Alksnis received the first radiogram. He read it aloud:

> Ya—RL.[13] 19:40. Crossed Mother Volga, cruising speed 205 kilometers. Altitude 820 meters. I hear Moscow well on wavelength 32.8. All OK. Crew feels fine.

"Good radiogram," said Alksnis. He took his son, who had been clinging to him, by the hand, and giving no further instructions, left the airfield.

From Shchelkovo, the primary N-209 "ground crew" staff went to the Air Force communications center located at the Central Airfield. Here, Nikolay Shelimov, Air Force Deputy Chief of Communications, was responsible for radio communications with the N-209. Thirty sleepless hours still lay ahead of us before Levanevskiy's airplane was to land in Alaska. A festive reception awaited him in Fairbanks.

I will not describe in detail everything that happened during those hours. In that regard there have been many publications concerning the subsequent fate of the N-209.[14]

---

13. RL—*Radioliniya* (Radio communications link).

14. For the best Russian-language work on the mission, see Yu. P. Salnikov, *Zhizn otdannaya Arktike: o Geroye Sovetskogo Soyuza S. A. Levanevskom* (A Life Devoted to the Arctic: On Hero of the Soviet Union S. A. Levanevskiy) (Moscow: Politizdat, 1984). For English language works, see John McCannon, *Red Arctic: Polar Exploration and the Myth of the North in the Soviet Union, 1932–1939* (New York: Oxford University Press, 1998); Pier Horensma, *The Soviet Arctic* (London: Routledge, 1991).

We tried not to bother the communications operators. In the middle of the following day, they showed Bolkhovitinov a radiogram signed by Levchenko and Galkovskiy reporting the following:

> Latitude 87° 55′. Longitude 58°. Flying over clouds, crossing fronts. Altitude 6,000. We have head winds. All OK. Hardware operating excellently. Crew feels fine. 12:32.

Bolkhovitinov woke up the dozing Tayts, and together they got out their slide rules and began to calculate how much fuel would be used if the entire route were flown at an altitude of 6,000 meters with a headwind.

At 1:40 p.m. on 13 August, the entire crew signed the last radiogram, the complete text of which was received in Moscow:

> We are flying over the Pole. Going has been difficult. Constant heavy cloud cover since the middle of the Barents Sea. Altitude 6,000 meters, temperature -35°C. Cockpit windows covered with frost. Strong head wind. Report weather on other side of Pole. All OK.

When I heard that the temperature was -35°C, I shivered and began to consult with my comrades about the possibility of instruments failing and storage batteries cooling down. Chizhikov and Alshvang confirmed my apprehensions. In their opinion, ice plugs could form in the tubes of the pressure gauges, altimeters, speedometers, and fuel gauges.

Our conversations were interrupted by a new radiogram, which the duty officer placed in front of Bolkhovitinov.

> RL. 14:32. Far right engine failed due to lubrication system malfunction. Flying on three engines. Altitude 4,600 with solid cloud cover.

This was radiogram number nineteen. "Who signed?" asked Bolkhovitinov. "Galkovskiy," answered the communicator.

Someone requested that this be clarified, to inquire—but that no longer made sense. There was nothing we could do to help, except recommend that they go to a lower altitude. An altitude of 4,600 meters for three engines was the limit, provided the aircraft did not ice over. But icing was inevitable when the aircraft's body, cooled to -35°C, hit the moisture-saturated clouds. They needed to go as low as possible and thaw out. Bolkhovitinov agreed to a descent to 2,000 meters. At that altitude, according to calculations, the lightened aircraft could hold out even on two engines. That advice was sent to Galkovskiy.

Whether he received our radiogram still remains a mystery. Communications with the aircraft were broken off. Yakutsk, Cape Shmidt, and Alaska reported the reception on RL wave of broken, indecipherable messages. It was difficult to assess their authenticity.

Several hours later, a certain emptiness fell upon Bolkhovitinov and all of us. Even flight headquarters no longer needed us. It was up to the government commission and headquarters to organize the search and rescue of the crew, if they were still alive. We went our separate ways on the morning of 14 August. By that time, even all of the unrealistic time limits for flying to Alaska had run out.

I slept around the clock and appeared at the factory intending to request leave to settle my academic debts. Instead of this, the deputy chief engineer, smiled and proposed, "I can give you another twenty-four hours to dry out some crusts of bread.[15] But tomorrow we are receiving three polar expedition airplanes for repairs. They are going to fly search missions. Until we release them, the factory isn't going to release you."

We really did have to switch over to a barracks-like situation. The chaotic days of organizing the search for the N-209 crew had begun.

On 15 August, all the central Soviet newspapers published the decision of the government commission on the deployment of search missions:

> As reported yesterday, the flight of Hero of the Soviet Union comrade S. A. Levanevskiy on the aircraft "USSR N-209" took place under very difficult atmospheric conditions. Due to high-altitude unbroken cloud cover, the aircraft had to fly at a very high altitude—as high as 6,000 meters. At 2:32 PM one of its engines failed and the aircraft had to descend to an altitude of 4,600 meters. Since that time no complete radiograms have been received from the aircraft. From the telegram excerpts we have received, it appears that the aircraft continued its course for some time. One can surmise that, forced to fly in the clouds, the aircraft might have been subjected to icing, which might have led to a forced landing on the ice. The ice conditions in the polar region and beyond it are comparatively favorable for such a landing. All polar radio stations are listening continuously on the aircraft's wavelength. Several times radio stations have heard activity on the wavelength of comrade Levanevskiy's aircraft, but due to the weak signal they have been unable to receive anything authentic.
>
> The N-209 crew has been provided with food stores for one-and-a-half months, as well as tents, sleeping bags, warm clothing, and firearms.
>
> Having discussed the situation, the government commission has undertaken a series of measures to render immediate aid. This aid has been organized in two areas: in the Eastern and Western sectors of the Arctic. The following measures are being taken in the Eastern sector from the Chukotka Peninsula:

---

15. The phrase was said jokingly as a warning of impending "prison" conditions.

1. The icebreaker Krasin, located at the shore of the Chukchi Sea, has been ordered to head immediately for Cape Shmidt, where the Glavsevmorput base is located, to take on board three aircraft with crews and fuel, and head to the area of Point Barrow in Alaska. From there, it is to go as far north as the ice will allow, where it will serve as a base.
2. The steamship Mikoyan, located in the Bering Sea, has been ordered to head for the Krasin with a full cargo of coal.
3. The two-engine hydroplane USSR N-2 piloted by Zudkov, located in Nogayevo Bay, has been ordered to head immediately for Uelen and from there to the site of the Krasin.

In the Western sector, using the air base on Rudolph Island and Papanin's "North Pole" station, the following instructions have been given:

1. Prepare for the departure of the three ANT-6 aircraft that had returned to Moscow from the Pole. These aircraft, under the command of Heroes of the Soviet Union comrades Vodopyanov, Molokov, and Alekseyev, are heading for Rudolph Island, and from there to the North Pole region.
2. Papanin's polar station, located on the zero meridian at a latitude of 87° 20′, will be converted into an air base—the point of origination for flights. Fuel will be transported from Rudolph Island on the ANT-6 aircraft. Comrade Papanin responded to an inquiry by the commission, stating that his field is completely intact and aircraft landings are possible.
3. Hero of the Soviet Union comrade Golovin and pilot Gratsianskiy were given orders to fly the two-engine USSR N-206 and USSR N-207 to Dickson Island and to be on call to fly to the North, to either Western sector or the Eastern sector of the Arctic, depending on the need.

The entire network of radio and weather stations is continuing operation.[16]

We prepared the three TB-3s that had already been to the Pole. I participated in their test flight. The last airplane was ferried to the Central Airfield on the day of the celebration of International Youth Day. Alekseyev piloted the plane. We flew low over Petrovskiy Park and Tverskaya Street filled with those commemorating the event, then turned back and landed at Khodynka. Shevelev commanded these three heavy airplanes. It was a month before they reached Rudolph Island. From there, after another three weeks Vodopyanov made several fruitless flights to the center of the Arctic.

The disappearance of the N-209 was a tragedy that was widely reported in the world press. Dozens of proposals continued to pour in to the authorities about how best to conduct the search. Enthusiasts and Arctic researchers from the

---

16. Text of the government commission's decision cited from Yu. N. Salnikov's book *Zhizn otdannaya Arktike* (A Life Devoted to the Arctic) (Moscow: Politizdat, 1984).

United States and Canada participated actively in the searches for Levanevskiy and his crew. Through its embassy in the United States, the Soviet government financed the procurement of airplanes and expenditures for a search conducted from Alaska. American newspapers wrote that the scope of the measures taken to rescue the N-209 crew in terms of their dramatic nature and grand scale surpassed all historic precedent.

In early 1938, searches for the N-209 were renewed from Alaska. In the spring, Moshkovskiy, flying our TB-3, examined the icy expanses west of Franz Joseph Land and between that archipelago and the North Pole.

WHAT MORE COULD HAVE BEEN DONE in those times? Streets, ships, schools, and technical institutes were named in honor of Levanevskiy, Kastanayev, and Godovikov. To this day, the Arctic has not given up its secret. Journalists, historians, and mere enthusiasts have conducted enterprising research into the possible causes and sites of the loss of N-209.

In 1987, the management of the Moscow House of Scientists proposed that I chair a conference devoted to the fiftieth anniversary of the flight. The meeting proved to be very impressive. Presenting their own scenarios were Pilot Nyukhtikov, who tested the DB-A; polar pilot Mazuruk; journalist Yuriy Salnikov, who had gathered comprehensive materials concerning the flight; and aviation engineer Yakubovich, who recalculated the aircraft parameters under conditions of icing. Yakubovich calculated the N-209's flight range limit after the failure of the far right engine. The detailed calculations showed that given an emergency-free flight, it was possible on three engines to reach the nearest shore of Alaska. New hypotheses based on stories formerly told by Alaskan Eskimos, who allegedly heard the noise of an aircraft, concurred with this. Salnikov, while in the United States, even took a trip to Alaska. No traces of the aircraft have ever been found on the shore and neighboring islands. If one assumes that the flight continued with slight deviations from the shortest route to dry land before the fuel ran out, then the aircraft sank in the coastal waters.

I am still convinced that the scenario we discussed a week after takeoff is the most plausible. Having lost altitude, the aircraft quickly iced over. The icy coating could have weighed several tons. The aircraft's aerodynamics changed. The ice could have jammed the rudder and the aircraft could have become uncontrollable. Instead of a smooth descent, the aircraft began to fall rapidly. It is possible that incredible efforts at the very surface managed to correct the aircraft. During an attempt to land on wheels on the pack ice the aircraft was damaged, and Galkovskiy was injured or killed. Without a radio operator, they were unable to restore communications using the radio station in the tail section, even if someone in the crew had survived. If the aircraft had heavily iced over, it might have broken up while still in the air.

I support the scenario that the catastrophe took place one to two hours after the last radiogram. Judging by the time, this happened at a distance of 500 to 1,000 kilometers south of the Pole in the American sector of the Arctic. In the spring of 1938, the sea currents and the direction of the ice drift were already well known. One could assert with great probability that if the aircraft did not sink when it crashed, then it was carried along with the ice toward Greenland and from there to the Atlantic Ocean. The unexpectedly rapid drift of the ice floe on which the polar station North Pole-1 was located confirmed this hypothesis. In February 1938, the four members of the Papanin expedition might have perished off the shores of Greenland if rescue ships had not managed to reach them in time. A detailed description of the searches for Levanevskiy during the period from 13 August 1937 until the end of 1938 is contained in Yu. P. Salnikov's book *A Life Devoted to the Arctic.*[17]

In the Soviet press and radio broadcasts of 1937–38, the enthusiastic accounts of heroic feats by our pilots and polar explorers shared the spotlight with reports of uncovering hidden Trotskyites and other "enemies of the People." During those years the majority of my acquaintances and I had no idea of the actual scale of the arrests. After the trial and execution of the officers headed by Tukhachevskiy, the names of other "enemies of the People" were not officially reported. In October 1937, Tupolev, along with many of his closest collaborators, was arrested. In November 1937, Air Force Commander Alksnis, who had just been decorated and treated with much favor by Stalin, suddenly disappeared.

AFTER THE RUSH JOBS to prepare the rescue aircraft were over, I resumed my studies at the night school of the Moscow Power Engineering Institute. My night student classmates had not lost their sense of humor, and they recommended that I immediately "go get lost in the ice" on one of the search aircraft. Only dozens of years later did I become aware that it was only by pure chance that I did not share the fate of many repressed individuals that I knew. The operation of the NKVD apparatus at that time also had misfires.

August 1937 has remained in my memory as the month of the tragic loss of the N-209. Among the crew, the man I was most fond of was Godovikov. But over the month of the continuous joint preparation at the hot airfield, others had also become my good friends.

The feat of Levanevskiy and his crew has remained forever in the history of Arctic flights.

For the 1930s, the North Pole scientific-research station on the Arctic ice floe, the first transpolar flights from Moscow to the United States by the crews of Chkalov and Gromov, and Levanevskiy's flight ending in tragedy were great and heroic feats in the schema of history. Sixty-five years after these flights, a group of experienced Russian test pilots decided to repeat the trans-Arctic routes of 1937. Honored USSR test pilot and Hero of Russia Anatoliy Nikolayevich Kvochur,

---

17. Politicheskaya Literatura publishing house, 1984.

with the support of the Air Force Command, obtained the permission of the Russian government for a flight by two Su-30 fighters from Moscow to the U.S. The flight range of the fighters was to have been ensured by in-flight refueling using an Il-78 tanker airplane. This would not have been a sensation for the twenty-first century, but proof of the fact that Russia still had aircraft technology and pilots who were prepared to take risks not for the sake of money and fame, but for the prestige of their country.

But it turned out that such a flight could be spoiled in our time for ridiculous reasons. The U.S. government did not object to three Russian military aircraft flying into its territory, but it could not guarantee the return of these aircraft to Russia. It turns out that the aircraft could be seized and handed over to a certain Swiss company called Noga. Since 1992, the Russian government had owed this company either $50 million or $70 million for the delivery of baby food. The governments in Russia had changed, and each one had verbally recognized the laws of the free market. But the debt, which exceeded $100 million when interest was factored in, had not been repaid because the budget did not contain a debit item to indemnify losses to companies that had suffered from the looting of budgetary money—plunder which had gone unpunished. According to the decision of an international court of arbitration, Noga was granted the right to seize Russian state property to recoup its losses if this property was on the territory of a country that recognized the international arbitration decision. Thus, according to the laws of 2002, if two Su-30 fighters and an Il-78 tanker flew safely over the Arctic to the United States, they would be confiscated and handed over to the ownership of a baby food company. From the great and heroic to the ridiculous required not a single step, but sixty-five years!

Chapter 8

"Everything Real Is Rational..."

In accord with the 1937 curriculum at MEI, we took a compulsory course in philosophy. At the conclusion of the course they held departmental theoretical conferences. At one such conference I presented a report on the philosophical significance of Einstein's theory of relativity. My opponent was Germogen Pospelov, known as "Sonny Boy."

We never even got to the substance of the matter, having grappled at the very beginning with Hegel's renowned thesis: "Everything real is rational, and everything rational is real." After lambasting me from strictly materialistic positions, Sonny Boy cited another quotation from Hegel: "In its development, the real is revealed as necessity." Hence, a conclusion was drawn which at that time was uncontestable—that being that everything our wise leader and teacher did was rational and necessary.

I recall that we even argued about how rational and necessary it was to perform heroic feats in the Arctic and then secondarily demonstrate heroism by rescuing the primary heroes.

After the reports that the search missions for the N-209 had been terminated, the rumor circulated that Levanevskiy had not died at all, but had landed in either Norway or Sweden and asked for political asylum. In the atmosphere of terror and with the absence of trustworthy information, such rumors were natural. Many wanted very badly for the crew to be alive. The students were more open in their conversation in the evenings than at work. They teased, "You'd better start drying out bread crusts again. This time for your ties with Levanevskiy."

To this day I cannot understand the logic of the NKVD. Almost the entire primary staff of the Tupolev KB workforce that had supported the triumphant flights of Chkalov and Gromov to the United States was repressed. Meanwhile, in spite of the glaring loss of the N-209, not a single person from our Bolkhovitinov KB collective was touched.

During this same time, two engineers who had nothing to do with the Arctic project disappeared from teams subordinate to me in the KOSTR. Both engineers, the Ovchinnikov brothers, had enjoyed deserved prestige in the design work force.

The elder, Ivan Ovchinnikov, was an electrical engineer. In addition to his qualification as a general industrial electrical engineer, he had also obtained qualification in avionics. He was considered the bane of the TsAGI electricians because he found many errors in their work and never passed up an opportunity to criticize them about this. When I came to KOSTR as his boss, Ivan, being older and having more work experience, was extremely helpful to me.

The younger, Anatoliy Ovchinnikov, came to Factory No. 22 after graduating from the new Moscow Aviation Institute (MAI). The tall, dark-haired Anatoliy, always elegantly attired and smiling benevolently, was an adornment to the male contingent of our team. In the summer of 1938, Katya and I rented a room in the dachas in Bakovka.[1] Here we met Anatoliy Ovchinnikov, who was staying in the neighborhood with his wife. This beautiful couple was simply adorable.

Both Ovchinnikov brothers were arrested. The agent who searched their desks at work warned us, his coworkers, not to inquire about the fate of the arrested enemies of the people.

Almost all of the old Party members at the factory were arrested. To play it safe, the personnel department started to purge the entire workforce, thoroughly researching their questionnaires to look for any kind of blemish. Before being fired, each purged individual was called to the special "Room No. 16." There he was informed of the decision to dismiss him due to some suspicion of political unreliability. There were several hundred such suspects. Even the former "saboteur" Tarasevich revolted and announced that there was a saboteur in room No. 16 who was taking away the factory's best personnel. And to everyone's surprise, somewhere at the top they listened to Tarasevich. The rumor passed through the factory that the fearsome inspector from that room was arrested.

But before that happened, my wife Katya was summoned to Room No. 16. There they explained to her that during the years 1922–25, her father, Semyon Semenovich Golubkin, had owned a garden plot in Zaraysk and had used hired labor. Because of this, he was deprived of his voting rights. But Semyon Semenovich had two sisters. Anna Semenovna Golubkina became a famous sculptor, and upon the order of the Bolshevik party, she had created a beautiful bust of Karl Marx. Aleksandra Semenovna Golubkina was well known in Zaraysk as an experienced physician's assistant. The fact was that Katya was raised not by her own mother, who was only sixteen years older than her first daughter, but by her aunt, Aleksandra Semenovna. According to the data of the Zaraysk office of the NKVD, right after the revolution Aleksandra Semenovna recruited her brother into the anti-Lenin party of the Social Revolutionary (SR) socialists, who defended the interests of the peasants. According to the logic of Room No. 16, that meant that Yekaterina Golubkina had grown up in a family of SRs—a party that had been banned back in 1920. With that sort of genealogy, there was no place for Yekaterina Golubkina at the largest aircraft factory in the nation.

---

1. Dachas are summer homes in the country. They can range from rustic to well-appointed, depending on the status of the owner.

True, there were mitigating circumstances. Anna Semenovna Golubkina, the natural sister of Katya's father, was a well-known Soviet Russian sculptor. By decree of the Presidium of the USSR Central Executive Committee, all of her closest relatives had been granted personal pensions. This same decree had opened Anna Golubkina's studio as a memorial museum in Moscow. The second mitigating circumstance was that Katya's husband, Boris Chertok, was a Party member and esteemed inventor.

By virtue of these good reasons, she would not be fired, but to get out of harm's way they asked that she submit a statement of resignation of her own volition. Not wanting to get me, or anyone else who might stand up for her, involved in this incident, Katya wrote the statement without leaving Room No. 16.

AMONG THE THOUSANDS OF INCIDENTS associated with the repressions, there was one with a happy ending, where the arrest saved a life. In July, Kerber had been removed from the N-209 crew so that he could be arrested in August. After taking an abridged course in the infernal science of the "GULAG Archipelago," he returned to his favorite work in Tupolev's collective.[2] By that time, the entire primary staff of the KB's work force had been arrested along with their head, the world-famous "ANT."[3] At the Tupolev Special Design Bureau 29 (TsKB-29) on Gorokhovskaya Street, now called Radio Street, they set up a relatively comfortable special prison, with design halls and a factory for experimental designs, where the imprisoned specialists worked.

Kerber wrote his memoirs about the life, work, and customs in the prison that they called the *sharashka*.[4] In spite of all the tragic elements of the events described, Kerber's memoirs are overflowing with his characteristic optimism and sense of humor.[5]

During the time of the Khrushchev thaw, when all the prisoners of the *sharashka* were rehabilitated, I met with Kerber, who had come to see Korolev. After talking shop for a bit, we turned to our recollections of the N-209. I asked Kerber what he thought about the fact that, in contrast to thousands of other repressed individuals, he should be thankful to the NKVD agents for arresting him in August 1937. If they had delayed this action, he, Kerber, would not have escaped an icy grave in the Arctic. He categorically disagreed with me. "If I had flown, that would not have happened," stated Kerber so flatly that I stopped opening old wounds. In 1987, Kerber came to the Moscow House of Scientists for an

---

2. GULAG—*Glavnoye Upravleniye ispravitelno-trudovykh LAGerey* (Main Directorate of Corrective Labor Camps)—was the system of forced labor camps in the Soviet Union. The term 'GULAG Archipelago' was coined by the Nobel Prize-winning Soviet dissident Aleksandr Isayevich Solzhenytsin (1918–)

3. Tupolev's initials were "ANT", which formed part of the designation of his aircraft

4. *Sharashka* is the diminutive form of *sharaga*.

5. L. L. Kerber, *Stalin's Aviation Gulag: A Memoir of Andrei Tupolev and the Purge Era* (Washington, D.C.: Smithsonian Institution Press, 1996).

evening devoted to the fiftieth anniversary of the flight and loss of the N-209. Even fifty years later he remained of the same opinion—if he had flown, the flight would have ended successfully.

In 1993, as I attended the funeral service of former radio operator, Doctor and Professor Leonid Lvovich Kerber, I thought about how self-assured he was, yet I still maintained the same opinion—he would not have been able to ensure the failure-free operation of one of the engines. The NKVD agents saved Kerber from death in the Arctic and thereby granted him an additional fifty-six years of life.

That same August of 1937, Air Force Commander-in-Chief Alksnis was arrested. My meeting with him at the Shchelkovo airfield on 12 August was the last. The punitive agencies dealt swiftly and mercilessly with the Red Army's best command cadres. According to the official information of unclassified reference publications, Alksnis' life was cut short in 1938. In spite of his subsequent rehabilitation, no information has been reported as to the date or site of his shooting.

BETWEEN 1929 AND 1937, Soviet aviation made qualitative and quantitative leaps. Over a period of only eight years, the following aircraft that attained a world-wide reputation were developed, put into series production, and accepted as standard armaments of the Air Force and civilian aviation: the TB-3 (ANT-6), SB (ANT-40), DB-3 (TsKB-30), TB-7 (ANT-42), and DB-A (limited production) bombers; the R-5 and R-6 reconnaissance aircraft; the I-16, I-15, and I-153 fighters; and a U-2 (Po-2) trainer. In addition, many aircraft were produced in only one or two prototypes without going into series production because the aviation industry lacked the capacity. Among these were the ANT-25, the *Maxim Gorky*, the TB-4, and many other aircraft that were epochal in terms of their outstanding parameters.

All in all, the 1930s, which started with our aviation lagging far behind the level of achievement abroad, brought much fame to Soviet designers, and even more to Soviet pilots.

The names Tukhachevskiy and Alksnis influenced the "adolescence" of the chief designers' teams that ensured victory in World War II seven years later. These were talented and visionary leaders who cared about the integrated and comprehensive development of an air fleet, who maintained close contact with the aircraft KBs and industry, and who did much to improve the combat training of pilots and the mastery of new fields of technology.

Long before the appearance of missile launchers among the infantry troops, airplanes were armed with missiles. This was an unqualified contribution of the Air Force military leaders of those years. They were the first to appreciate the developments of the Leningrad-based Gas Dynamics Laboratory and RNII. Airplanes received missile armaments four years before the infantry forces.

By late 1937, our Air Force had 8,000 aircraft of all classes, including 2,400 heavy and high-speed bombers—more than half of which were produced by Factory No. 22. In Moscow, Kazan, Kuybyshev, Voronezh, Komsomolsk-on-Amur,

Gorkiy, Rybinsk, Kiev, and other cities, large aviation industry enterprises were built. Their status reflected the level of technical progress at that time. However, during the period when the Soviet Union really could have caught up with and passed by the industrially developed countries, mass repressions dealt a very heavy blow against scientific and technical progress.

In this regard, I will speak one more time about the role of the individual in our history. Alksnis was the first one, five years before the attack of Nazi Germany, who prophetically pointed out the danger of an aggressor's surprise attack on airfields at the very beginning of a military action, in order to gain complete supremacy in the air. In 1936, Alksnis wrote to People's Commissar Voroshilov: "Air Force airfields are the primary target of an enemy air attack in the first hours of a war."

The Germans proved Alksnis' case, destroying our airplanes on our airfields in the very first hours of the war. Several thousand (to this day there are no precise numbers) of our airplanes were put out of commission in the first days of the war, while German losses were negligible. After this, cities, strategic centers, and infantry troops were left defenseless against German aviation, which gained absolute air supremacy without a fight.

After Alksnis' arrest, Aleksandr Dmitriyevich Loktionov was appointed Air Force Chief in December 1937. Until then he had commanded the Central Asian Military District. Two years later, in November 1939, Ya. V. Smushkevich—two-time Hero of the Soviet Union, former aviation brigade commander, hero of the war in Spain, and veteran of battles on the Khalkhin-Gol River and the Karelian Isthmus—became head of the Air Force.

Within his own narrow circle, Bolkhovitinov, who regularly associated with high-ranking Air Force commanders, used to say: "There is an atmosphere of complete incomprehension, but a certain enlightenment is setting in. Smushkevich is a combat pilot with great prospects and a grasp of the future. It is interesting and useful to meet with him."

The testimonials about Smushkevich that I and some of my coworkers heard from Bolkhovitinov have been confirmed by the memoirs of Aleksey Shakhurin, who was appointed People's Commissar of the Aviation Industry in 1940. He wrote: "Among the many high-ranking aviation commanders that fate threw me together with, I never met a man who possessed such courage, such boldness of judgment, such charm, as Smushkevich."[6] Smushkevich bravely and persistently defended his opinion even in meetings with Stalin. In Shakhurin's opinion, these meetings determined the program for the expansion of aircraft production that played a major role in the war. Smushkevich headed the Air Force for less than a year and was then executed.

In the autumn of 1940, Pavel Vasilyevich Rychagov was appointed chief of the Air Force. He was a fighter pilot and Hero of the Soviet Union who also distin-

6. A. I. Shakhurin, *Krylya pobedy* (*Wings of Victory*), 2nd ed. (Moscow: Politizdat, 1985), pp. 14–15.

guished himself in battle in Spain, Lake Khasan, and in the war with Finland.[7] He had not finished flight school until the early 1930s, and could not comprehend the diversity and complexity of problems facing "the best aviation in the world" on the eve of the war. That approximates what the respectable Air Force Academy professoriate thought of their high-ranking chief. Rychagov met the same fate as the preceding Air Force chiefs.

Finally, in April 1941, Pavel Fedorovich Zhigarev was appointed Air Force Chief. He was the first Commander-in-Chief who had graduated from the N. E. Zhukovskiy Air Force Academy. Before his appointment, he had been Chief of the Air Force Directorate of Combat Training.

The Stalinist leadership understood very well how critically important science was in maintaining the nation's economic and political sovereignty and ensuring the historical interests of its peoples. And yet it acted as if it were striving to prove that "everything real is irrational." The intellectual advantage that the USSR had had up until 1937–38 gradually disappeared as a result of the extermination of the military intelligentsia's most progressive thinkers. This was one cause leading to the heavy military defeats that we suffered in 1941 and 1942.

During the three and a half years before the war, the Air Force had replaced five chiefs! Only Alksnis lasted longer than five years in that post. He alone had the opportunity to implement a specific strategic doctrine, remain in close contact with airplane designers, and know what to require from them. In the immediate period before World War II, there was virtually no one who had the sense of state responsibility required to seriously and competently consider and implement state policy regarding the most important types of weaponry. One after another, the military leaders of the Air Force were replaced, along with thousands of lower-ranking, experienced commanders, down to the commanders of aviation regiments and squadrons.

After the death of Baranov, the arrests of Tupolev and Deputy People's Commissar Mikhail Kaganovich, and the mysterious suicide of Sergo Ordzhonikidze, the rhythmic work of the factories became disorganized, and the development of new experimental and promising designs proceeded largely of its own accord, without strict monitoring by military contractors. Tupolev's TsKB-29 was now subordinate to Lavrentiy Beriya, and he, supposedly on Stalin's instructions, was dictating to the designers what aircraft they should develop. This process for managing the Tupolev collective is very vividly described in Kerber's memoirs concerning his work in the *sharashka,* which I mentioned previously.

THE SAD EXPERIENCE of the N-209 flight demonstrated the necessity of reliable radio communication for aviation. The repressions significantly weakened all of

7. The Soviet engagement with Japan in 1938 ended with the Battle of Khasan near Russia's border with China and North Korea.

our communications technology and radio engineering. During the war with Finland, scandalous inadequacies had appeared in our communications equipment, despite our scientists' achievements in the field of radio engineering.

At the beginning of World War II, we recognized how far behind the Germans we were in terms of our radio communications—both in general and with particular regard to aviation. Aircraft transceiver stations were complex and of poor quality. They primarily used short- and medium-wave ranges. Radio stations were installed only in squadron commanders' airplanes. For example, by 1 January 1940, in the Moscow Military District, radio stations were installed in only 43 out of 583 fighter aircraft. The primary forms of communication in the air were signal flares and rocking the wings. Ground radio support systems for flying in difficult meteorological conditions and at night were just being developed. The lack of radio facilities on the ground and in the air during the first year of the war contributed to additional losses. In many cases, it was impossible to direct flights during air battles or guide aircraft back to their home airfield at night or in bad weather. Aircraft control via radio within a group, guidance from the ground, and elementary radio navigation did not appear until after the war had started. I am dwelling on this problem because I dealt directly with aviation radio communications during the war.

In general assemblies of the Academy of Sciences, I often sit next to my comrade from my student years—Academician Pospelov. While digging through old papers, I found his notes on the philosophy of Hegel. After returning them to their author fifty years later, I asked what he now thought on that subject. "It's all rubbish!" replied Germogen Sergeyevich Pospelov—academician, general, and specialist in the design of artificial intelligence systems.

Using an example from my biography, Germogen defined the problem of artificial intelligence in his own way.

"Our living intellect sprang from a chaos, into which nature was trying to bring systemic order. We have not succeeded in creating artificial intelligence because we desire to exclude chance as the primary property of chaos and establish a strict, causal sequence in decision-making. At the same time, some chance incidents precipitating from a systemic conformity with law are fortunate. Remember 1937 when, after the loss of Levanevskiy, all of our classmates and I were convinced that you would be repressed. After all, you were the lead engineer for aircraft electrical equipment. According to NKVD logic back then, you should have had to do time. This would not be chance, but conformity with law. The fact that they did not touch you and we are now peacefully conversing is indeed a chance incident that is characteristic of chaos."

Chapter 9

Return to Bolkhovitinov

In the autumn of 1938, I became a full-time student and was completely freed from my industrial cares.

The MEI's professors and instructors for the different career specialties had close ties with industry. The activity of many instructors was combined with scientific-research work in industrial institutes and design bureaus. Our seminars and even our exams were sometimes like heated discussions.

Professor Andrey Nikolayevich Larionov taught a course on special electric machines. At his direction, I completed a course project in which I designed an alternating current generator stimulated by permanent magnets. Larionov was the chief developer of the electric power plant for the *Maxim Gorky*, the largest aircraft in the world, which tragically crashed. Larionov's scientific interests far exceeded the limited framework of his academic course on special electric machines. His contractors from the Tupolev KB lived and worked on the same street where he lectured us. But since they were now prisoners, he no longer had access to them and there was no opportunity for collaboration with them. He dreamed of the further practical development of his ideas for using high-frequency, high-voltage, alternating-current electric generators on airplanes. In me he discovered an ardent admirer of these ideas.

I had not lost contact with Viktor Fedorovich Bolkhovitinov, who spent the greater part of his time in Kazan after the loss of N-209. He reassured me that I would always be assured of work in his experimental design bureau if I did not return to Factory No. 22. Bolkhovitinov felt certain that his collective would be returned to Moscow or its immediate vicinity sometime in 1939.

Our patron's friends and compatriots at the academy had not stopped working on the design for the new high-speed, long-range bomber, referred to as B. In spite of the change of authority, there was still a glimmer of hope in the People's Commissariat of Aviation Industry and in high-level military circles that the funds would be found for the development of a new heavy aircraft, especially since the Americans and the Brits were not just maintaining, but developing this area.

Our patron, of course, was not an ordinary chief designer. Working in Kazan, he directed the diploma projects of MAI students and N. Ye. Zhukovskiy Air Force Academy cadets. In so doing, he selected and proposed the most insane ideas.

When I mentioned the alternating current (AC) system to him in passing, he lit up. A new idea! He guaranteed his support. Bolkhovitinov always found time to listen to and discuss an innovative idea. I never heard him say, "Rubbish, nothing will come of that." No one ever approached him with "rubbish."

I assured Larionov that it would be possible to find funding to design an AC system for a heavy aircraft. As I set about this work, I asked him to be my diploma project advisor. He agreed, and we began to develop the generator.

After that, it was like a fairy tale.

In early 1939, Bolkhovitinov really did return from Kazan. The entire KB settled down in Khimki at Factory No. 84. This factory had mastered the series production of two-engine transport aircraft under license from the American firm Douglas. Although the aircraft was a very precise copy of the Douglas DC-3, it was nevertheless called the Li-2, from the first two letters of the surname of B. P. Lisunov, who was the chief engineer at Factory No. 84.

As soon as Bolkhovitinov settled down at Khimki, I was included on his staff, working an abbreviated workweek. I was tasked with developing the new bomber's entire system of AC electrical equipment. At that point the aircraft was in the most rudimentary form, but according to the design, it was supposed to become the farthest- and highest-flying, fastest, and best-armed bomber in the world.

The Bolkhovitinov KB did not take root at Factory No. 84. For a series-production factory reproducing an American aircraft, the Bolkhovitinov KB was an alien organization.[1] Our patron did not like to wander the corridors of power and solicit elementary favors for his work force. There, where people elbowed their way around, it would have been easy for him to be forced out and shoved aside. When there were unexpected attacks from subordinates on the need to bring some matter or other to the attention of the People's Commissariat, he loved to cool off particularly enthusiastic individuals with sayings such as, "He who is in a hurry, let him hurry. But we will be working." Nevertheless, he got a decision through to build a new factory, No. 293, next to Factory No. 84.

And so, in Khimki we gained a new construction site, and Bolkhovitinov became Director and Chief Designer of Factory No. 293 of the People's Commissariat of Aviation Industry. The new site had previously been home to *Fotolet*, an organization that did aerial photography. The design teams managed to barely fit into *Fotolet*'s modest wooden buildings, so they began construction of the experimental factory right next to them. The new factory had already begun to operate in early 1940.

It is not in our power to see the future. Neither Bolkhovitinov, nor we, his co-workers who had begun to work at the new Factory No. 293 construction site, could imagine that twenty years later, missile systems capable of reliably shooting down aircraft flying considerably higher and faster than the airplanes of our dreams

---

1. Bolkhovitinov's KB was focused on design while the factory itself was geared towards production.

would be produced on this same site.[2] All of the most recognized aviation technology designers, including enthusiasts of Bolkhovitinov's most cutting-edge ideas, believed that missile flights were for those who dreamed of interplanetary journeys. Our task for the next ten years was to use the propeller and internal combustion engine as effectively as possible!

Getting down to our new work, I commandeered a small room in an old *Fotolet* building and obtained the right to establish contracts for the development and supply of elements for an alternating current system for the future bomber.

I selected the electric machine laboratory of the V. I. Lenin All-Union Electrical Engineering Institute (VEI) as my primary supplier. The laboratory's scientific director was Academician Klavdiy Ippolitovich Shenfer, a world-class specialist in the field of electric machines. During our first meeting, he immediately appreciated both the problematic nature and the great potential of the subject. Shenfer refused to personally participate in the work, explaining that he was loaded down with research on asynchronous electric machines, but he proposed that I give a report for the "youthful" contingent of the laboratory. The laboratory engineers, especially machine designer Boris Dmitriyevich Sadovskiy and machine design specialist Teodor Soroker, were very interested in my proposals. A year after our joint collaboration had begun, I was convinced that their creative input would give my fuzzy dream real definition. The laboratory collective developed generators, electric motors, dynamos, drive mechanisms in the form of reduction gears with built-in motors, remote switches, and many other devices.

Looking from today's perspective at the system we conceived in 1938, one must admit that on the whole it was an interesting and absorbing technical adventure. We were faced with solving hundreds of very difficult technical problems, some of which to this day have yet to be satisfactorily realized.

Having selected a triple-phase alternating current system with a frequency of 500 Hz and a voltage of 48 V, we spent a long time inventing stabilization methods. We devised and tested a voltage stabilizer relatively quickly. Frequency stabilization gave rise, so it seemed, to insurmountable difficulties. The generators were driven by aircraft engines, the RPMs of which varied over a very broad range. Given a nominal frequency of 500 Hz, the frequency of the current at the generator output had to be maintained with an error no greater than ±10 Hz.

How were we to achieve this? We studied a multitude of ideas. We settled on a hydraulic converter, which, having a variable RPM rate to the input shaft, ensured a strictly constant rotation rate at the output.

The most surprising aspect of this story was that primary prototypes of the main assemblies were not only designed, but also fabricated and tested in the laboratory.

---

2. Plant No. 293 eventually housed OKB-2 (or later MKB Fakel), the organization that, under General Designer Petr Dmitriyevich Grushin, designed and built several generations of the Soviet Union's best anti-aircraft and anti-ballistic missiles, including the missile system that shot down Francis Gary Powers' U-2 aircraft in 1960.

Soon my small workroom at *Fotolet* was filled with the company's green cases. Their contents were articles made almost completely out of iron and copper. My neighbors on the other side of the wall were Aleksey Mikhaylovich Isayev and his team. Once after the latest batch of green cases was loaded into my room, Isayev dropped in on me and said, "Now I understand why the floor in our building is sagging. No two ways about it, we'll need a superheavy bomber to lift your new system into the air, but there won't be any room for bombs!"

By this time, Isayev had staffed his team with outstanding young engineers, which he called his "*Wunderkinder.*" Each had been given a nickname in place of name and surname, but none of us were offended. The spirit of creative enthusiasm enhanced with good helpings of optimism and humor spread from his design hall throughout all of *Fotolet*. This jolly band was loaded down with continuous alterations of the "I" fighter design. The mechanisms of the movable wing flaps of the "S" aircraft kept them even busier. This equipment was designed to reduce the wing surface after takeoff and increase acceleration. During landing, the flaps reengaged to reduce the landing speed.

The first "S" aircraft was manufactured in Kazan. During the flight tests, which were conducted at the Flight-Research Institute (LII), this short-range bomber piloted by test pilot Boris Kudrin reached a speed of 570 kilometers/hour. At that time, that was the speed limit for fighters. However, the flight tests showed that the paired engines and mechanized wing required extensive modifications. None of the series-production factories wanted to get involved with the complexities entailed in the production of this aircraft.

In late 1939, I finally sat down to draw up my diploma project. In the project I criticized the shortcomings of existing electric power supply systems and proposed a new promising AC system for heavy aircraft. My project's strong suit was its references to electrical machines developed "per the author's specifications" for the future system and photographs of the finished all-purpose electric drive mechanisms. My advisor, Professor Larionov, announced that he had read the project as though it were an engrossing novel. He said it was too bad that it was stamped Secret. Such work deserved broad publication.

Having defended my project and received my diploma "with distinction," I nevertheless did not leave MEI. In the autumn of 1940, I was accepted in the post-graduate program "without leaving production." At the recommendation of department head Academician Viktor Sergeyevich Kulebakin, I was to present a course on "aircraft special equipment" at the night school during the 1940–41 academic year. I spent a great deal of time traveling the legs of a triangle: from my home on the Moscow River to Factory No. 293 in Khimki to MEI and VEI in Lefortovo.

In April 1939, I became a father. Katya had not yet returned to work and had submerged herself in maternal cares, trying as much as possible to free me from domestic concerns. It is amazing, but at that time, for some reason, in spite of all the very difficult everyday conditions, there was enough time for everything.

Having submerged myself in the problems of Factory No. 293, I discovered that our patron had created a special group of weapons experts who were working on purely mechanical remote-control methods, despite his enthusiasm for the electric remote-control of machine guns and guns using the slaving systems I had proposed. Young engineer Vasiliy Pavlovich Mishin stood out in this group. He developed something similar to a modern-day manipulator with a manual control. The gunner, located under an airtight transparent dome, was supposed to guide the gun or machine gun located several meters away using a purely mechanical manipulator. The kinematics of the remote manipulator were very clever. Mishin began his creative path with this work. He was the future compatriot of Chief Designer Sergey Pavlovich Korolev.[3]

They didn't manage to bring the manipulator developed in 1940 up to full standard. To compensate, the factory acquired several ShKAS aircraft machine guns and a pair of brand new ShVAK large-caliber aircraft machine guns for future use. In the autumn of 1941, Mishin mounted them on swiveling gun carriages, which we hoisted onto an open rail car to ward off possible attacks by German aircraft during our relocation from Khimki to the Urals.

Factory No. 293 was located three kilometers from the Khimki railroad station. During the autumn and winter, we made our way to the factory on public transportation. The same familiar faces were always waiting in line for the bus. One time, a young man dressed in unseasonably light and natty attire appeared among the passengers. A stylish felt hat that did not cover his high forehead was jauntily perched on his large head. His open face shone with a barely perceptible smile. It seemed that the line and conversations on the bus did not distract him from his own private thoughts, which brought a smile to his face.

I encountered this young man several times in our patron's reception area, and then he appeared in our *Fotolet* buildings. We found out that this was the new director of the mechanisms group, Aleksandr Yakovlevich Bereznyak.[4] Isayev's wise guys had jealously protected their monopoly on the development of new mechanisms. One of them commented, "This Bereznyak is a real enigma. He smiles like the Mona Lisa."

The mystery surrounding the appearance of Bereznyak in our circle was soon uncovered. Bolkhovitinov had been the advisor for MAI student Bereznyak's diploma project, which he had brilliantly defended in 1938. The project involved paired engines located in tandem on the fuselage, similar to the paired engines on our "S" aircraft.

Bolkhovitinov decided to test the paired-engine idea with the assistance of this talented student. Bereznyak made many original suggestions for the design of the

3. Vasiliy Pavlovich Mishin (1917–2001) served as Korolev's principal deputy for twenty years before succeeding him in 1966.

4. Aleksandr Yakovlevich Bereznyak (1912–1974) served as Chief Designer of OKB-2-155 (later MKB Raduga) from 1951–1974, overseeing the development of many generations of tactical missiles.

one-seat aircraft. One suggestion was evaporative cooling of the engines, whereby water was not circulated through radiators, but evaporated and vented into the atmosphere. The water supply was sufficient for one hour of flight. To reduce resistance, the fuselage had a modified windshield. The pilot was located under the glazing of the cockpit canopy, which was raised over the fuselage before landing. By design, the aircraft was capable of reaching a speed of 940 kilometers/hour at an altitude of around 7,000 meters. Such an aircraft could break the world speed record of 709 kilometers/hour. Every world record would then belong to Soviet aviation.

The project was reported to Yakov Vladimirovich Smushkevich, who had just been appointed Deputy Chief Air Force Corps Commander. He had received personal instructions from People's Commissar of Defense Voroshilov to submit proposals on the possibility of a Soviet pilot breaking the world speed record in a Soviet aircraft. The Air Force command consulted with Professor Vladimir Sergeyevich Pyshnov, who kept no secrets from Bolkhovitinov (Pyshnov and Bolkhovitinov were married to sisters). As a result, Bereznyak was sent to Bolkhovitinov for a detailed study of the project.

At that time, Bolkhovitinov's experimental-design bureau had moved from Kazan to Khimki, on the outskirts of Moscow. A fight for survival was going on and they were not in the mood for world speed records. A series of five DB-As in Kazan had not been completed; the flight tests for the paired-engine "S" aircraft had begun with difficulty; the "I" fighter design kept having to be redone; and the heavy bomber B was only in the beginning stages. For the sake of the B, I had interrupted my furious activity on the alternating current system.

Bereznyak took the plunge and took over leadership of the team. In the evenings he continued to select diagrams and do calculations for the cutting-edge, high-speed aircraft. We were having problems, as before, with the pairing of two piston engines. The engine-propeller unit with co-axial propellers required a prolonged modification process. The greatest doubts, however, related to the aerodynamics. They believed that speeds approaching the speed of sound were virtually unattainable using piston engine-propeller designs. The young engineer Berznyak had of course already heard and read about rockets and rocket engines, but no one had yet dared to develop proposals that completely rejected the use of the piston engine on an aircraft.

In 1956, chance brought Bereznyak and me to Kislovodsk. At that time, Bereznyak was the chief designer at the cruise missile OKB on the Moscow Sea.[5] We ended up at the Caucasus resort at the same time, but in different sanitariums. Having determined the schedule for our therapeutic procedures, we set out for the traditional Kislovodsk activities of climbing the Maloye and Bolshoye Sedlo and hiking along

5. The Moscow Sea is commonly known as the Ivankovo Reservoir, formed by the damming of the Volga River at the village of Ivankovo in Kalinin (now Tver) Oblast. At the time of the visit in 1956, Bereznyak was Chief Designer of OKB-2-155, an aviation industry design bureau that was a branch of the Mikoyan-Gurevich design bureau (OKB-155) that designed the famous MiG jet fighters.

the long tourist path.[6] While discussing our past, present, and future plans, I asked Bereznyak when he first got the idea to give up on piston engine-propeller units in his design and install only liquid-propellant rocket engines on the aircraft.

"You know," he said, "I can't answer that question precisely. Where and when did it first happen? There wasn't an exact time and place. This idea kind of came over me spontaneously. In thoughts and on paper, I drew and added all sorts of layouts, sometimes stupid ones. The liquid-propellant rocket engine had already been invented some time ago and it seemed very simple. There wasn't any intelligible information on turbojet engines, just ideas. I finally managed to break through to Likhobory, where I first saw a liquid-propellant rocket engine on a bench.[7] Its shape reminded me of a bottle. When I understood what a voracious bottle it was, and how much fuel it burned, I decided to just give it up and forget about it. But there wasn't anything else. Nothing at all!"

A stop at a snack bar during our tiring hike interrupted further elaboration of the idea for the famous BI.[8] To restore our strength we each had a glass of thick sour cream, washed it down with dry wine, and dreamed of one day sitting down and writing our memoirs. At that time we felt it would be a shame to waste time on history.

IN THE SPRING OF 1939, after the difficult sessions of the past academic year, Germogen Pospelov talked me into vacationing in Koktebel to restore our strength before our diploma projects. Koktebel had two advantages. First, there was an MEI boarding house there where students could stay for next to nothing. Second, Koktebel was the best place on the entire Black Sea coast.

I left Katya and our four-month-old son in Udelnaya, the home of her mother Kseyniya Timofeyevna, who doted on her first grandson. Pospelov swore to Katya that he would keep a strict eye on me—and if she would not allow me to go to Koktebel, then as a sign of masculine solidarity, he would not go either.

In the summer of 1939, I spent a wonderful month with Pospelov in Koktebel swimming and rock climbing along the Karadag. Our group also included chemistry student Mikhail Gavrilovich Slinko. Many years later, the three of us would become members of the USSR Academy of Sciences. At general Academy meetings we never missed an opportunity to tease one another about various things—our senile condition; the unattainable, beautiful "shores of Tavrida"; the radiant cliffs of Karadag in our student days; and our contact with the literary Bohemia residing at the legendary dacha of the poet Voloshin.[9]

---

6. The Russian word *sedlo* literally means 'saddle.' The Maloye (or Small) and Bolshoye (or Big) Sedlo refer to two mountains in the vicinity of Kislovodsk.

7. The Reactive Scientific-Research Institute (RNII) was located at the site of a tractor factory in Likhobory.

8. The BI (for Bereznyak-Isayev) was the first Soviet fighter with a liquid-fuel rocket engine.

9. Maksimilian Aleksandrovich Voloshin (1877–1932) was a famous Russian poet influenced by the French Symbolists and Impressionists.

At work in the spring of 1940, I was going on about fabulous Koktebel in the presence of Isayev. Right then and there Isayev asked me to get him passes to the MEI boarding house for his entire team.

NOWADAYS, I AM ASTONISHED by the confidence that we had then in our complete safety and our calm perception of the events that had shaken the world. On 1 September 1939, World War II began. In March 1940, our ignominious war with Finland ended. Over the course of two months in the summer of 1940, the Netherlands, Belgium, and France capitulated to Nazi Germany. The Germans occupied Norway, and Soviet troops marched into Lithuania, Latvia, and Estonia. In all of Europe, only England was opposing Hitler's Germany. And under these circumstances, Isayev and I were enthusiastically discussing the problems of taking a group vacation in the Crimea.

To clear our conscience we went to consult with our patron. He calmly responded to our request, having reasoned that it had been a trying year, that we needed to unwind, and he reassured us, "We will be going to war. But it won't be for another two years or so. And what we'll use—that's another matter. But, in any event you won't finish either the B or the "I" in a month." Our patron was an avid yachtsman and was himself getting ready for a sailing vacation. He couldn't say no. I talked Katya into joining us, leaving one-and-a-half-year-old Valentin in the care of his grandmother. Isayev did not risk inviting his wife Tatyana to the Crimea. Their first child, Vanya, was only four months old.

And so, during our country's last peaceful summer, Katya and I vacationed in Koktebel with Isayev, his design team, and Semyon Gavrilovich Chizhikov, who was tagging along with us. He had left Factory No. 22 in due course along with Bolkhovitinov's work force.

In Koktebel, Isayev introduced us to his friend Yuriy Beklemishev, known also as the writer Yuriy Krymov. His story *Tanker Derbent* had enjoyed great success. Krymov was vacationing there at the dacha of poet Maksimilian Voloshin, which had been converted into a home for Soviet writers. Krymov confided that he had taken the pseudonym in memory of the wonderful time he had spent in Koktebel and the Crimean adventures he had enjoyed together with Isayev.[10] Interrupting each other, the two of them regaled us with stories about their escapades during the time of their not-so-long-ago youth. Both of them were brilliant storytellers.

With magnificent humor they told us about their attempt in 1925 to sail from Koktebel to the islands of Tahiti. Their sea voyage in a sailboat ended when a border patrol cutter intercepted them several kilometers from the Crimean coast. They were both arrested and spent a week in an uncomfortable prison cell until their parents managed to free them. Krymov promised to write a semi-humorous,

10. *Krym* is the Russian word for Crimea.

semi-satirical story about these and other interesting adventures from the 1920s. But alas, Yuriy Krymov died on the front in 1941.

It wasn't until our time in Koktebel that I first heard Isayev's revelation of how difficult it had been for him to get the job with Bolkhovitinov. The Factory No. 22 personnel department rejected him because of his specialty—he had a degree in mining engineering. Then he wrote a moving application addressed to "respected comrade director," in which he assured the director that the risk was small, and that he would become an aircraft engineer within a year. In late 1934, the "respected director" was Olga Mitkevich. Isayev's application touched her with its directness and its expression of a passionate desire to work in aviation. Mitkevich gave the order to hire Isayev and send him to the Bolkhovitinov in his new experimental-design bureau. With this one single decision, Mitkevich made an important contribution to rocket technology.[11]

We took difficult hikes along the rocky shore, got to know all the bays, and swam all the way to the *Zolotyye vorota*.[12] We played hotly contested croquet and volleyball matches. However, thoughts of our abandoned projects were always with us.

The "I" fighter, for which Isayev considered himself responsible, contained a multitude of original designs. His primary contribution was the pairing of two engines that used pusher rather than tractor propellers. The pilot's cockpit, armed with two guns and two large-caliber machine guns, was located in front of the paired pusher engines. Two beams replaced the fuselage. A remote-controlled machine gun was mounted between them behind the tail assembly. Isayev hoped that I would fulfill my promise for the remote control of this machine gun from the pilot's cockpit. I really had promised, but how was the pilot going to be able to aim at an enemy attacking from the rear? I could not come up with anything. And to top it off, my brain was busy with the heavy bomber B and alternating current.

We did not receive any newspapers and there was no radio in the boarding house. We got all our news from Chizhikov, who was spending his time with acquaintances at the neighboring military academy boarding house. One time he met us with a report that the Battle of Britain had begun. Soviet radio, citing German sources, broadcast news of vast fires and destruction in London.

Krymov was the first to give in. "Carefree swimming in the warm sea and basking in the sun when such things are going on—I can't do it. If Hitler makes short work of England, then we can't avoid war," he predicted.

We also gave in and left Koktebel three days early.

---

11. Aleksey Mikhaylovich Isayev (1908–71) later became the Chief Designer of OKB-2 (or KB Khimmash), the organization that designed rocket engines for many famous Soviet spacecraft (including Vostok, Voskhod, and Soyuz), anti-aircraft missiles, and submarine-launched ballistic missiles.

12. The *Zolotyye vorota* (Golden Gates) was a structure built in the eleventh century at the entrance to the city of Kiev (in present day Ukraine). Although the original structure no longers stands, a museum with a reproduction of the original was built during Soviet times.

Chapter 10

On the Eve of War

In the autumn of 1940, the enthusiasm of the Factory No. 293 collective, which had increased after our summer vacation, diminished noticeably. Our patron Bolkhovitinov, who had returned from sailing around large lakes and reservoirs, did not assign any new tasks that captured our imagination. When we dropped in on him to discuss routine matters, he shared his melancholy thoughts, but we did not receive any new directives.

After many years of anti-fascist agitation and propaganda, it was difficult to resign oneself to the stream of reports about the victorious Nazi *blitzkriegs* in Europe. The Germans had already captured Paris. Now we all feared for the fate of Britain. No one was indifferent. We were invariably and sincerely sympathetic to the British in our debates over reports from Europe about the air battle for Britain. We switched back and forth between discussing the problems of the air battle for Britain and analyzing the capabilities of our aviation. It was evident that the majority of our combat aircraft were inferior to those of the Germans in speed, engine power, weaponry, and equipment for night flights.

Now that the history of World War II has been thoroughly studied, one must give credit to the Germans for their efforts to mislead us regarding their true intentions. In 1940, Hitler had already planned Operation Barbarossa and had determined the dates for attacking the Soviet Union. At the same time, our delegations had been invited to Germany. They had visited military factories and concluded agreements for the sale of weapons prototypes, machine tools, and optical instruments. Among other things, the Germans allowed a commission of Soviet specialists to inspect their aircraft factories. The Soviet commission procured prototypes of combat aircraft. The aircraft were sent to the Air Force NII for flight tests and for industry specialists to familiarize themselves with them. All of the aircraft were fully equipped with weapons, state-of-the-art navigation equipment, and ultra short-wave and short-wave radio communications.

During the war it was precisely these types of aircraft that made up the Luftwaffe's primary strike force. The latest models were the Messerschmitt-109 and 110, the Junkers-88 and 52, the Dornier-215; the experimental Heinkel-100, which was not yet accepted as a standard armament; and several aerobatic training aircraft.

The Germans were completely convinced of their superiority. They calculated that if we got it into our heads to imitate their technical achievements, then we would waste several years mastering them and putting them into production, and during that time they would finish us off.

We conducted the inspection of the German technology as a group, without rushing. Above all I was interested in the electrical equipment, navigational instruments, radio system, bomb releases, and sights. The other equipment specialists and I were envious of the meticulousness and neatness of the interior layouts, the instrument panels, and the consoles. The electric bomb-release produced by the firm Siemens-Apparatebau had, as we would now say, a magnificent design—it was completely different from the one I had been laboring on for almost two years!

The bombers were equipped with electric autopilots. The first time we saw the Askania electrohydraulic control-surface actuators we could not even guess how they worked. Six years later, I would be faced with mastering the production of similar control-surface actuators—no longer for aircraft, but for long-range rockets. As soon as we switched on the onboard radio stations, we were convinced of the reliability of their air-to-air communications. The majority of our current fleet of aircraft had no radio communications whatsoever, neither air-to-air, nor air-to-ground.

Back home at the central airfield of the Air Force Scientific-Testing Institute, I ran into Larisa Dobrovolskaya, the former KOSTR *Komsomol* secretary at Factory No. 22, where she was now directing a large design team. On such an occasion, it would be impossible not to reminisce about our romantic *Komsomol* days.

Sharing her impressions of German technology, Larisa displayed her feminine powers of observation, "All the tiny details were so thoroughly thought out with German scrupulousness that we won't have time to reproduce them on mockups, much less in operating models. And look, everything is fitted so that there is nothing to get caught on in flight or during emergency evacuation of the aircraft when you need to jump with a parachute."

Dobrovolskaya continued, "But we have achievements. Now on every airplane we are installing tubes under the wings for rocket projectiles. Now military acceptance will not register a single aircraft without electric wiring and actuators to launch rocket projectiles. For the time being, evidently, the Germans don't have anything like our rocket projectiles."

At the far edge of the airfield stood a pair of dark green four-engine TB-3s. These were the same airplanes into whose production we had poured so much youthful enthusiasm six years ago. We sadly noted that these behemoths, with non-retracting landing gear reminiscent of the talons of a bird of prey, could become easy and safe spoils for the German airplanes we had so thoroughly inspected. Our first-hand knowledge of German technology showed that the Soviet Air Force, one of the most powerful in the world, was experiencing a crisis and was inferior to the German Luftwaffe.

Our collective was not the only one at a thematic crossroads. Many series-production factories were in the same situation. The production of obsolete aircraft developed during 1935–36 continued. New models were being put into production with tremendous difficulty after prolonged flight tests. Many factories were busy mastering new equipment procured abroad in large quantities. The directors paid more attention to special machine tools, multi-ton presses, trip hammers, and testing equipment than to the aircraft themselves. Pe-2 dive-bombers, Il-2 low-flying attack aircraft, and Yak-1, MiG-3, and LaGG-1 fighters had only just broken through into series production.

STALIN, WHO WAS VERY INTERESTED in aviation technology, decided after the series of repressions to once again strengthen the leadership of the aviation industry. In January 1940, Aleksey Ivanovich Shakhurin was appointed People's Commissar of Aviation Industry. Between 1933 and 1937, Shakhurin had worked at the N. E. Zhukovskiy Air Force Academy. He was secretary of the Academy's Science Council and Deputy Chairman of the Society of Inventors. Bolkhovitinov knew Shakhurin well and we expected that the new People's Commissar would give us the proper attention. Bolkhovitinov met repeatedly with Shakhurin. After these meetings he told us that Shakhurin reported to Stalin almost every day—but hardly at all about our work.

In the autumn of 1940, I often visited the People's Commissariat of Aviation Industry on Ulanskiy Lane. It reminded me of a stirred-up anthill. In 1939, the aviation industry released no more than twenty-five airplanes per day. Most were obsolete, but the industry had at least mastered their series production. Stalin demanded that Shakhurin increase production to seventy to eighty aircraft per day, primarily by introducing new models.

Increasing the capacity of the aviation industry to produce more than 20,000 combat aircraft per year was a task that specialists considered very difficult, but not impossible. The war had shown that even larger numbers were needed. Four years later in 1944, under considerably more difficult conditions than we could imagine, our industry manufactured 40,241 combat aircraft. For Russia in the year 2002 that number is a fantasy. At the air show in August 2002, the Russian Air Force Commander-in-Chief joyfully announced that in 2003 our Air Force would receive twenty new fighters! The aircraft of 2002 are of course many times more complex and expensive than their World War II predecessors—but not by a scale of one thousand!

The tough demands from the top and the cries of the factory representatives completely deprived the most hardened bureaucrats of their peace. Repeated calculations showed that the required production numbers could not be attained before 1943. No one considered that the war would start considerably sooner, or that they would be faced with solving the problem under conditions that were many times more difficult and immeasurably more severe.

The People's Commissar's concern about immediately finishing the new aircraft by Petlyakov, Ilyushin, Yakovlev, and Mikoyan and putting them into series production shoved our designs, which were exotic for those times, to the background. The idea of using alternating current, which I introduced for discussion at other design bureaus and at the offices of the People's Commissariat, met with sympathetic understanding; but the conversations usually ended thus: "We are not up to this now. Did you see the German technology? While they are busy with England, we should jump ahead of them. We have a maximum of one to two years for that. Your alternating current requires three to five. And then, where will it be used?"

ON ONE OF THOSE AUTUMN DAYS, Isayev and Bereznyak dropped into the room where my comrades and I were working surrounded by boxes full of alternating current machines. Apologizing for the intrusion, Isayev requested that my subordinates go to the design hall where his team was assembled.

This proposal was made in typical Isayev style: "Why don't you go on over and play with my kids for a while?" When my subordinates left, Bereznyak unrolled some Whatman paper. A drawing that had numerous erasures showed two views of a small glider with noble outlines. I was surprised at the dimensions of this glider. The wingspan was 6 meters, and the length from the pointed nose to the tail was only 4.5 meters. Four barrels were drawn above the forward fairing: two guns and two machine guns. Not having detected an engine with a propeller, I remarked that this was a strange glider. The area of the wings was very small.

"You're right. This isn't a glider at all," said Isayev. "OK, it's Sasha's turn to explain everything, and then I'll tell you why we are here. Swear that you won't tell anyone about this conversation. Even our patron mustn't know anything about this yet."

Bereznyak said that this would be a fundamentally new rocket-powered interceptor aircraft. It would have a liquid-propellant rocket engine installed in the tail. The diameter of the nozzle would be only 300 millimeters, and so the aircraft in the picture would be taken for a glider. Its entire mass would be no greater than 1,500 kilograms. This was half as much as a Messerschmitt-109. The maximum speed would be 1,000 kilometers/hour, and perhaps greater. That would all depend on the thrust of the engine. One had not yet been definitively selected. In the meantime, they had been promised an engine with a thrust of 1200 kgf. If it was boosted to 1500 kgf, then the thrust would be equal to the weight of the aircraft. This was almost a vertical takeoff. Its insane climbing capability would be the primary characteristic of this interceptor. As soon as an enemy bomber appeared at a distance of one to two flight minutes, the interceptor would take off and attack with lightning speed with at least twice the speed of its opponent. It would have only enough fuel for one attack. The engine would work for no longer than two to three minutes. It would then return to the airfield and land in glide mode. Given the bomber's speed of up to 600 kilometers/hour at an altitude of 5,000 meters, the interceptor would reach it one minute after takeoff if the pilot did not lose the target. The entire structure would be wood—plywood. The only duralumin permitted would be used for the assemblies and brackets. The landing gear would be lowered and retracted using compressed air.

"As an exception, the cylinders have to be made of metal," joked Bereznyak. Here Isayev interjected, "Sasha is dreaming of making even the guns out of plywood, but I won't allow him. We are mounting two ShVAKs on a wooden gun carriage. Such an aircraft doesn't need a factory. It will be made at all the furniture mills. Extraordinary low price and simplicity! Fifty furniture mills will put together twenty airplanes each in a year—that's already 1,000! Imagine, 1,000 interceptors around Moscow! Even if every other one misses, 500 bombers will still be brought down. That's what the Brits need now!"

Aleksey was carried away. He spoke so fervently and eloquently that I was taken by the prospect of plywood interceptors taking off at lightning speed from dozens of airfields against squadrons of heavy bombers.

"All of this is great, but why have you come here?" I asked. "There is simply nothing for me to do here. Provide a battery to light up the instruments and a firing actuator. That's a day's work for my drafters."

"That's not it at all," objected Isayev. "First, at night the interceptor pilot needs to understand where to fly in order to break out to the point of attack over the shortest route. Second, he can't make a mistake during his return. He won't have fuel to maneuver for a repeated landing approach if he misses or can't find his airfield. He might also get lost in the daytime in clouds. This aircraft is not capable of waiting in the air. His mission is to take off, attack, destroy, and immediately land."

I was supposed to devise an enemy detection and attack guidance system and navigation instruments for returning to the airfield. It was 1940, and they were proposing that I figure out the guidance system that would used in surface-to-air missiles about ten years later.

After several days, Bereznyak showed me his calculations with diagrams of the interception of enemy aircraft. The rocket-powered interceptor was supposed to take off as soon as an enemy aircraft entered its coverage zone—presumably 10–12 kilometers away. The attack would take place at any altitude up to 10 kilometers and continue for absolutely no more than two minutes. At that point the aircraft had not been designed to sustain an attack for even that much time.

"We will begin with just 90 seconds, and then we will increase it," said Bereznyak. "But we need to design the guidance system for pursuit, in case the enemy manages to break out of the coverage zone of our interceptors' base airfield."

The interception diagrams were beautiful and convincing. The flight trajectories of our interceptors aimed right at the burning bombers. After the attack, the interceptors were supposed to calmly return to their airfield, for which 5–8 minutes of their planned flight time had been set aside.

I asked for a copy of the top secret graphics to think over. My previous work had to do with developing equipment for bombers. My knowledge in the field of air defense systems was therefore quite superficial. Even when working on the "I" fighter, we had not devoted attention to target location and guidance issues. This was also the weak side of the military contractors' technical specifications requirements. At that time, the attention devoted to developing new navigation and guidance systems for blind flight was disproportionately less than that paid to the creation of the actual aircraft and engines.

I began my search among acquaintances at the Air Force NII. Chief of the Air Force NII special services group Sergey Alekseyevich Danilin—with whom I had been acquainted since the time of DB-A production and the N-209 flight—and his deputy, radio engineer G.A. Uger, listened to me with great interest. And then, in a very correct manner, they expressed their very skeptical thoughts concerning the rocket-powered interceptor's flight trajectories, which I had spread out before them. They did not call the aircraft's dynamic properties into question. Bolkhovitinov was highly regarded in military circles. They had already heard of rocket engines. But the existing resources of the air observation and aircraft detection and tracking services were, in their opinion, completely unsuitable for this interceptor. Optical, acoustic, and thermal means were immediately rejected. Only radio engineering could lead the interceptor into the zone of visual contact with the enemy. Subsequent action would depend on the pilot.

The idea of using radar facilities for Air Force aircraft had already been intensively developed in 1940. The lessons of the Battle of Britain had confirmed the need to provide aircraft with a means of navigation in night battle conditions. The well-known fixed-loop radio compasses that we had used on our long-range bombers, radio beacons, and the hyperbolic radio navigation systems that had been developed were not suitable for this purpose.

By that time, through the efforts of the scientists and engineers of the Leningrad Physics and Technical Institute, NII-9, NIISKA, and other organizations, experimental ground-based radar stations had been set up.[1] Danilin and Uger advised me to familiarize myself with this brand-new, ground-based radio engineering.

Every cloud has a silver lining. Studying the radar equipment that had only just appeared in our midst, I not only absorbed new ideas, but also met some very interesting people who were radar enthusiasts and fanatics. Several years later, I had the occasion to work with some of them again, this time in a new field dealing purely with rockets. In that sense, the meetings in 1940 proved to be very useful.

In radio engineering circles, there was a rivalry between proponents of pulse and continuous emission methods. Powerful transmitting tubes—magnetrons and Klystrons—were developed. Their names had only just been introduced into scientific terminology.

In 1940, the first experimental radar stations were established—*Redut* and RUS-2. Great changes in the country's air defense occurred because of the RUS-2 station. But how was all this to be linked with our interceptor? We did not know who would be capable of creating onboard systems.

It was precisely at this time that the Air Force NII first formulated the requirements for an onboard enemy aircraft detection station. At the radio industry NII, the

1. NIISKA stood for Scientific-Research Institute of Communications of the Red Army.

future NII-20, they told me the approximate mass of such a station. According to preliminary calculations, the complete set with power sources, cables, and antennas weighed 500 kilograms. Neither our interceptor nor any other fighter could bear such weight. At the institute, they assumed that testing of the station would begin on the Pe-2 aircraft. It was not simply a matter of kilograms. It was simply impossible for a fighter pilot to simultaneously pilot the airplane, control the detection station, and fire weapons. The onboard station required greater attention than the enemy!

IN THE SPRING OF 1941, Bereznyak, who had been following my research, said that he and Isayev had had several talks with the patron. Ultimately they had had a meeting of the minds and were now conducting their work openly and full steam ahead. I was expected to give a report regarding the development of a guidance and navigation system.

I do not remember whether it was March or April 1941 when I gave my report to Bolkhovitinov, Bereznyak, and Isayev on that subject. Having told them about the status of work conducted in that field, I concluded that in the next two to three years it would be unrealistic to create an onboard radio guidance system for such an interceptor given the limitations that we were applying in terms of mass and configuration. The maximum that could be done immediately would be to develop a radio system for landing that simplified the search for the home airfield. For the time being, we needed to design a compact radio station for guidance from the ground.

To Bereznyak and Isayev's displeasure, Bolkhovitinov reacted very calmly to this. He said that we had run ahead and there was nothing surprising in the fact that radio engineering was not yet ready to service such airplanes. We needed to start flying as soon as possible to demonstrate the absolute necessity of a fundamentally new intercept guidance and control system.

In the meantime, there was a more urgent problem for me. It was proposed that I set aside all my current work, study the startup and control system of the liquid-propellant rocket engine, automate everything that could be automated, and devise reliable methods for engine ignition and the monitoring of engine operation. Once I had received these instructions, I searched for Langemak and Glushko's book, *Rockets: Their Construction and Use*, which I had acquired back in 1936.[2]

During my earlier reading of this book, I had been interested in the section on solid-propellant rockets. In 1937, Factory No. 22's KOSTR developed drawings for the installation of rockets on the SB aircraft. The chapter "Construction of Solid-Propellant Rockets" was written by Langemak. Now I studied the chapter "Construction of a Liquid-Propellant Rocket Engine," written by Glushko. I found no other literature on liquid-propellant rocket engines in the libraries to which I had access, and nothing about automating their guidance.

---

2. G. E. Langemak and V. P. Glushko, *Rakety, Ikh Ustroystvo I Primeneniye* (Moscow: ONTI, 1935).

Bolkhovitinov advised, "Go to NII-3 and do some investigating there." He then sent me on my way after signing a letter addressed to institute chief Andrey Kostikov.

Thereby, I visited NII-3 for the first time in the early spring of 1941. Large letters on the wall of the main building facing Mikhalkovskoye Highway stated "Institute of Agricultural Machine Building." I felt no trembling or even respect for this institution. At that time, it was not for me to know that I was visiting for the first time a piece of real estate that would enter the history of cosmonautics. All of our rocket-space technology historians consider it obligatory to mention the founding role of RNII, the Reactive Scientific-Research Institute, in the origin of domestic cosmonautics. In order to avoid sending the reader off in search of various and contradictory publications, I will give a brief history of RNII—NII-3 in my own interpretation.[3]

On 21 September 1933, Mikhail Tukhachevskiy issued an order from the USSR *Revvoyensovet* on the organization of the Reactive Scientific-Research Institute. RNII was the first state scientific-research and experimental design organization in the world to combine different trends in the theoretical and practical development of problems of rocket technology. Only one-and-a-half months later, by resolution of the Council of Labor and Defense, RNII was transferred from the authority of the People's Commissariat for Military and Naval Affairs to the People's Commissariat of Heavy Industry.

RNII was established on the basis of two organizations: the Leningrad Gas-Dynamics Laboratory (GDL) and the Moscow Group for the Study of Reactive Motion. The domestic pioneers of rocket technology were associated with these small groups of early trailblazers.

GDL was created in Moscow in 1921 to develop the inventions of Nikolay Tikhomirov. After moving to Leningrad it was named the Gas-Dynamics Laboratory. Nikolay Tikhomirov had proposed using the reaction of gases obtained during the combustion of explosive substances for "self-propelled mines for the water and air." GDL activity was concentrated on creating smokeless solid-propellant projectiles and the technology needed to manufacture their explosive charges. Vladimir Artemyev was Tikhomirov's close collaborator and co-author in the development of the first solid-propellant rockets. He designed the first smokeless powder rocket and authored many inventions in the field of solid-propellant rockets.

In 1930, after the death of Tikhomirov, military engineer and artilleryman Boris Petropavlovskiy was appointed GDL director. Petropavlovskiy was also a professor at the Military Technical Academy, and he actively promoted the idea of rocket weaponry among his students. At his initiative, GDL developed launchers for the firing of rockets in the form of simple openwork tubes secured under the wings of an aircraft.

---

3. Between September 1933 and January 1937, the institute was known as RNII. In January 1937, it was renamed NII-3, a designation it retained until July 1942.

Petropavlovskiy became seriously ill in late 1932, and died in 1933. Ivan Kleymenov was appointed as the new GDL chief. Before coming to work at GDL, he had studied in the physics and mathematics department at Moscow University, and from there had been sent to the N. Ye. Zhukovskiy Air Force Academy. Upon graduation from the Academy, Kleymenov received the GDL appointment and took up the baton for the development of smokeless solid-propellant rockets for aircraft and multi-barreled rocket launchers.

Along with Kleymenov and Artemyev, one of the main leaders in the development of rockets at GDL was Georgiy Langemak. Like Petropavlovskiy and Kleymenov, he had volunteered for service in the Red Army during the civil war, and then was sent to study. After graduation from the Military Technical Academy, he selected internal ballistics as his specialty.

Valentin Glushko arrived at GDL in 1929 upon graduation from Leningrad University. One of the youngest employees at GDL, he had been captivated with ideas of cosmonautics since his youth. At GDL, Glushko set up a subsection for the development of electric and liquid-propellant rocket engines and liquid-propellant rockets. He developed a unique electro-thermal rocket engine and the first domestic liquid-propellant rocket engine using high-boiling propellant.[4] Glushko is now rightly considered by many to be the founder of the Russian school of liquid-propellant rocket engines.[5]

In 1930, GDL achieved its first practical results during the range testing of 82- and 132-millimeter rockets. In 1932, Mikhail Tukhachevskiy, *Revvoyensovet* Deputy Chairman and Red Army Chief of Armaments, was present when the first official in-air firings of RS-82 missiles from an I-4 aircraft armed with six launchers successfully took place. Solid-propellant reactive systems facilitating the takeoff of TB-1 and TB-3 aircraft were also being successfully developed. Vyacheslav Dudakov conducted this aircraft-related work at GDL.

By early 1933, around 200 individuals were working at GDL, which was directly subordinate to the Military Research Committee under the USSR *Revvoyensovet*.

It would be considerably later before Moscow's rocket technology enthusiasts were united. In the autumn of 1931, the Group for the Study of Reactive Motion (GIRD) was formed under the auspices of *Osoaviakhim*, a large social organization. GIRD's first director was Fridrikh Tsander—a scientist, inventor, and romantic who dreamed of interplanetary flight. Tsander was captivated by the problems entailed in flights to other planets, the movement of spacecraft in a planet's gravitational field, and determining the trajectories and duration of flights. He was also

4. High-boiling propellants, which were Glusko's specialty, are a subset of liquid propellants that do not require special cooling equipment to maintain them at normal temperatures; as such, they are also known as storable propellants.

5. Valentin Petrovich Glushko (1908-1989) founded and headed OKB-456 (later NPO Energomash), the developer of the most important and powerful rocket engines in the Soviet Union.

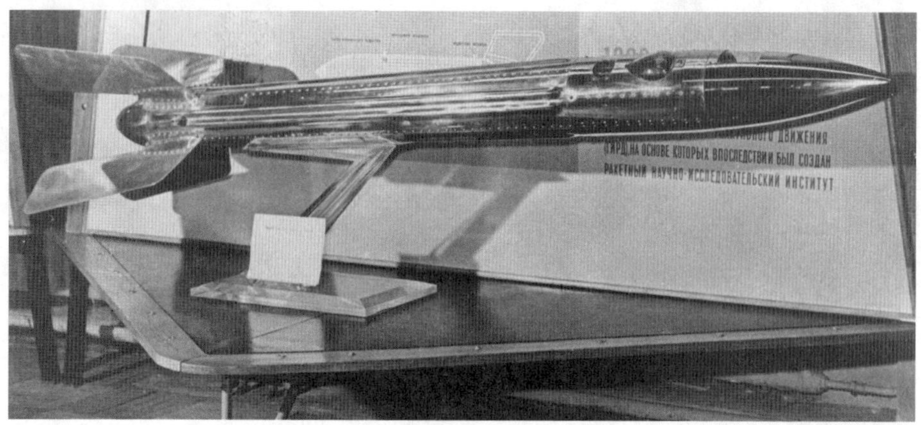

NASA History Office.

Above is the 09 rocket built by the Group for the Study of Reactive Motion (GIRD) in 1933. The 09, which was successfully fired in August 1933, was the first Soviet rocket to use liquid propellants. GIRD members included Fridrikh Tsander, Sergey Korolev, and Mikhail Tikhonravov, three important pioneers of Soviet rocketry and spaceflight.

involved in developing the theory and design layouts of different engines that did not need atmospheric oxygen. Tsander was a typical scientist—an enthusiast and a dreamer. He was also completely devoted to the ideas of interplanetary travel.

In 1932, Sergey Korolev replaced Tsander as director of GIRD. Tsander needed good therapy at a sanatorium. Korolev managed to get him a pass to Kislovodsk, which for those times was not a simple feat. Tsander became ill in Kislovodsk, however, and died suddenly in March 1933. Twenty-three years later, Korolev finally managed to track down Tsander's gravesite in Kislovodsk, and in 1957 a gravestone with a bust of Tsander was placed there on the occasion of what would have been his seventieth birthday.

Other talented engineers who experimented with the first Soviet liquid-propellant rockets at the Moscow GIRD included Yuriy Pobedonostsev, Mikhail Tikhonravov, Vladimir Vetchinkin, Yevgeniy Shchetinkov. Korolev developed designs for stratospheric aircraft with liquid-propellant rocket engines, and Tikhonravov directed a team that designed the first hybrid-propellant rockets and oxygen-gasoline propellant rockets. Pobedonostsev was involved with the problems of ramjet engines. In 1932, Andrey Kostikov came to work at GIRD, having graduated from the N. Ye. Zhukovskiy Air Force Academy. He joined in on the work of Pobedonostsev's and Tikhonravov's teams.

The Moscow GIRD, which had around sixty employees, was funded by the *Osoaviakhim* presidium and RKKA Directorate for Military Inventions. In 1932, meetings were held in Leningrad between GDL employees and GIRD leaders Korolev, Tsander, Tikhonravov, and Pobedonostsev. RKKA senior armament officials who were familiar with the work of GDL and GIRD were firmly convinced of the need to combine the two organizations and create a Reactive Scientific-Research Institute. In those years, organizational decisions aimed at strengthening defensive capability were made rapidly. Accordingly, they appropriated the grounds of the Agricultural Machine Building Institute in Moscow, found living accommodations for the Leningraders, and allocated sizeable resources for construction and to equip the laboratory of the new institute.

GDL DIRECTOR KLEYMENOV WAS APPOINTED CHIEF of the RNII, and Moscow GIRD chief Korolev was appointed his deputy. Because the young Korolev's technical interests diverged from those of the RNII leadership, Langemak was soon appointed instead to be the Institute's deputy chief. Langemak was also the chairman of a technical advisory board which was the official scientific leadership of RNII. The advisory board was comprised of Glushko, Korolev, Pobedonostsev, Tikhonravov, and Dudakov. Although many types of rocket technology were being developed, the preference for production was given to operational solid-propellant rockets and their launchers. Kleymenov, Langemak, and Artemyev, who had all transferred to RNII from GDL, already had a great deal of experience developing and producing such solid propellant rockets. By late 1937, RS-82 and RS-132 missiles had been developed under their leadership. The Air Force had accepted these missiles as standard armaments for I-16, I-15, I-153, and SB aircraft.

The RNII department under Glushko's leadership was involved with nitric acid liquid-propellant rocket engines. Glushko's collective continued to develop liquid-propellant rocket engines under the index ORM.[6] D. Shitov, V. Galkovskiy, and S. Rovinskiy joined Glushko's department. Among their developments were single-chamber and double-chamber engines with a thrust as high as 600 kgf using nitric acid/tetra nitro methane and nitric acid/kerosene propellant. Between 1934 and 1937, the ORM-65 nitric acid/kerosene engine was developed for Korolev's 212 cruise missile and RP-318-1 rocket glider. Korolev headed the winged-missile department. These were like airborne torpedoes designed to be launched from TB-3s. But Korolev's primary work at RNII was the design and construction of a rocket glider. Tikhonravov developed oxygen/alcohol liquid-propellant rocket engines and liquid-propellant rockets. Kostikov also worked in this department. Pobedonostsev first worked on ramjet engines, and then joined the work on solid-propellant rockets. RNII (named NII-3 after January 1937) made history with the legendary *Katyushas*, the combat artillery rocket launchers used during World War II. NII-3's central role in the development of this new type of weapon was indisputable. After RNII was established, its work was given the necessary framework, and joint operations began with the aviation sector and with the Main Artillery Directorate (GAU).

In the 1930s, only the Soviet and German governments were supporting work on a broad spectrum of rocket-related subjects. From 1932–1935, the Germans

6. ORM—*Opytnyy Raketnyy Motor* (Experimental Rocket Engine).

lagged behind us considerably, especially in the area of rockets. Beginning in 1935, however, the Germans started to catch up and then pass us in the development of liquid-propellant rocket engines, especially in those using oxygen/alcohol components.

During the tragic years of 1937–38, NII-3 lost its leadership. Kleymenov and Langemak were arrested in 1937 and shot to death in January 1938. Military engineer Boris Slonimer, who had returned from Spain, was appointed chief. Kostikov was appointed chief engineer and deputy chief.

In 1938, Glushko and then Korolev were arrested. In late 1939, they removed Slonimer from his post. Chief Engineer Kostikov became the sole director of NII-3. He was given free reign in all areas of research—the scientific technical advisory board no longer directed the institute. Repressions against the leadership had created a stifling psychological situation at the institute, crushing initiative and bold creative quests. But it was amazing to see the great intellectual potential of the RNII collective! People found in themselves the strength to throw off their depression and get back to their feverishly intense work. How many similar dramas were played out in those years!

Solid propellant rockets were accepted as a standard armament for aviation and set up for series production. The Air Force rated the weapons highly. This new field required the leadership's attention and reinforcement. Kostikov assigned Pobedonostsev to supervise the work. One must give credit to Yuriy Aleksandrovich. During that difficult time, he appreciated the promise that RS missiles held and did much to give them a second wind after Langemak's death. Pobedonostsev, as much as he could, also supported work on liquid-propellant rocket engines, assigning them to Leonid Dushkin. Kostikov assigned to Tikhonravov both Korolev's reactive aircraft and the continuing work on oxygen/kerosene liquid-propellant rocket engines and liquid-propellant rockets.

IN SPITE OF THE SUCCESS IN AVIATION, the development of ground-based solid propellant launchers for the infantry lagged behind. Vasiliy Aborenkov, senior military representative of GAU at RNII, played a large role in the development and final acceptance of the *Katyusha* rocket launchers as standard armaments for the infantry. Under strong pressure from Aborenkov, the scale of these operations expanded substantially. Military engineers Shvarts and Sorkin joined the work being conducted on projectiles, while Gvay, Pavlenko, Galkovskiy, and Popov began working on self-propelled launchers. In 1939, the first self-propelled launchers were manufactured using the ZIS-6 automobile.

In 1939, NII-3 was transferred to the People's Commissariat for Ammunition directed by Boris Vannikov.

Its departmental affiliation gave rocket projectile issues priority over liquid-propellant rocket engines, liquid-propellant rockets, and rocket-powered aircraft. Tikhonravov, the proponent of liquid-propellant rockets, did not have the opportunity to prove their relevance as weapons by that time. Work on oxygen liquid-propellant rocket engines, which Tikhonravov was trying to bring back, was not given the proper support. It should be mentioned that our intelligence services had no information concerning the scope of work being conducted on liquid-propellant engines and rockets in Germany. The Germans were building what was for that time a vast rocket center in Peenemünde. The primary mission of this center was to develop long-range, liquid-propellant guided rockets. The parameters of the liquid-propellant rocket engines being developed between 1935 and 1940 at NII-3 could not compare in any way with what the Germans were brandishing during that time at Peenemünde. Before the war we had ceded leadership in the development of liquid-propellant guided rockets.

Yet in the area of small, solid-propellant rocket projectiles we were far ahead of the Germans. This innovation, however, was not put into mass production for combat use until the onset of war.

High-ranking military leaders—the Deputy People's Commissars for Defense; the rapidly changing General Staff Chiefs, B. M. Shaposhnikov, K. A. Meretskov, and later also G. K. Zhukov; and the People's Commissar for Defense, S. K. Timoshenko—did not imagine the tactical capabilities of this new weapon and did not devise any plans for its use in the future war.

Usually all innovations in the field of weaponry were reported to Stalin. But Stalin had no information about infantry rocket projectiles before 1941. Artillery Marshal Kulik, as Deputy People's Commissar of Defense for Artillery, was responsible for the activity of the Main Artillery Directorate—for evaluating and accepting new rocket launcher systems as standard armament. It was his duty to report personally to Stalin, and if not to him, then to People's Commissar Timoshenko. But Kulik had underestimated this new weapon. Aborenkov was completely convinced of the effectiveness of rocket projectiles from his experience with their use in aviation. He went over the head of his superior, Marshal Kulik, and reported this development in a memorandum to Stalin. Aborenkov risked, if not his head, then certainly his career. One must give him credit for his daring.

In his memoirs, G. K. Zhukov recalls that soon after his appointment as chief of the General Staff, Stalin asked him if he was familiar with vehicle-mounted rocket launchers. Zhukov responded that he had only heard of them, but not seen them. Stalin said, "Well, then one of these days soon you'll have to go with Timoshenko, Kulik, and Aborenkov to the firing range and watch them fire."[7] At that time Aborenkov was a division chief at GAU. Someone of his rank was not supposed to have any dealings with Stalin, which makes it all the more interesting that Stalin knew about him.

I learned how Stalin's instructions were fulfilled from the stories and notes of the direct participants in the development and production of *Katyushas*. One of

---

7. G. K. Zhukov, *Vospominaniya i razmyshleniya* (Memoirs and Reflections) (Moscow: Voyenizdat, 1974), Vol. 1.

those who knew the subsequent events was Major General Pavel Trubachev, a Leningrad Mining Institute and Artillery Academy graduate who was sent to GAU after completing his studies in 1940. From his first days of service at GAU, he was directly involved in the testing, production, and acceptance of rocket launchers as standard armaments for the infantry. I made his acquaintance in 1945 in Germany. At that time he was a colonel-engineer.[8] Later, Trubachev became a military regional engineer at NII-88, at Korolev's OKB-1, and department chief at the Main Directorate of Rocket Armaments. I had the chance to interact unofficially with the retired Pavel Yefimovich at the Pirogovskiy Reservoir, where we both had garden plots. This is what he said about Stalin's assignment.

In March 1941, under Aborenkov's leadership, they had conducted successful range firings. There were no doubts as to reliability and effectiveness of the vehicle-mounted, multi-barreled rocket launchers. However, a demonstration of the rocket projectiles to the higher command was put off time after time. The demonstration was finally held on 15 June 1941.

Minister of Defense Timoshenko wanted a demonstration of new artillery armaments to be conducted at the firing range. The preparations for this undertaking took much longer than had been anticipated because the distinguished chief designers at the firing range were striving to present the new guns and mortars in the best light at the demonstration. The rocket launchers were mounted on ZIS-6 vehicle chassis at NII-3. The demonstration included a large number of artillery exhibits, and the plain outward appearance of the two trucks did not inspire much confidence. Standing by each artillery piece was its chief designer, sometimes along with the factory director and military representatives. They were all prepared to report to the People's Commissar of Defense not only the tactical advantages of their new model, but also its readiness for mass production.

Two BM-13 launchers, each with twenty-four projectiles, stood modestly to the side of the main exhibits. They were not accompanied by a chief designer or authorized industrial representative, since Colonel Aborenkov and the NII-3 engineers and workmen had taken the trouble completely on themselves to prepare the launchers for inspection.

The demonstration of the rocket weaponry was the last item on the schedule. The effect of the drumfire and the howling of forty-eight flying projectiles made a staggering impression on the marshals and generals. Clouds of dust rose up and flames raged in the area of the target. It seemed that nothing living could have withstood such an artillery strike.

Timoshenko addressed Kulik harshly, "Why have you kept silent and not reported on such a weapon?" Kulik excused himself by pointing out that the weapon had not been fully developed and troop tests had not yet been conducted.

8. Those officers in the Soviet armed forces who had received a higher engineering degree combined their rank with the title 'Engineer.'

In his memoirs, trying to explain the delay in accepting many new models as standard armaments, Zhukov was forced to admit that Marshal Kulik had not appreciated the capabilities of the rocket artillery before the war. However, a similar reproach could also have been leveled at Air Force Commander Rychagov, who did not appreciate the potential of Il-2 low-flying attack aircraft, and also at many other high-ranking military leaders, who in similar cases preferred to receive instructions personally from Stalin. According to the testimony of Galkovskiy, one of the test participants, Zhukov was not present at the demonstration of these firings. Perhaps that is why in his memoirs he did not revisit the events that followed the conversation with Stalin about the *Katyushas*.

After the demonstration, Timoshenko and Aborenkov demonstrated the necessary persistence and speed in putting the weapons into operation. Twenty-four hours before Nazi Germany attacked, a resolution was issued, signed by Stalin, for the series production of rockets and launchers.

NOW LET US RETURN ONCE AGAIN to events in which I was a direct participant.

When I entered the main building of NII-3 for the first time, I did not know its history and I had no gift of foresight to appreciate the historic role of the wide staircase leading to the offices of the institute directors. Fifty years later, millions of television viewers would see the building façade and this staircase in historic documentary films. This building was, at various times, the workplace of people whose names would not be revealed as part of the history of cosmonautics until decades later. Not knowing the past and not foreseeing the future, I calmly entered the office of the chief of the institute in March 1941.

Kostikov hospitably stood up and came out from behind his director's desk. He was wearing a military uniform with four bars on the gorget patches, which corresponded to the rank of a colonel-engineer. Smiling warmly, he said that he would keep no secrets from Bolkhovitinov and his representatives. At his command, Dushkin's deputy, engineer Vladimir Shtokolov, led me to his laboratory and showed me the liquid-propellant rocket engine test-firing bench.

From talking with Shtokolov, I understood that the engine that could be mounted on Bereznyak's airplane did not yet exist. Experimental work was underway, and I had shown up at the right time so that together we could work out startup, control, and in-flight guidance procedures for the future engine. During my subsequent visits to NII-3, we discussed the already defined tasks of designing the electric ignition, the remote control of the pressure in the combustion chamber, and the fuel and oxidant lines. I immediately rejected the idea of using a vacuum photocell device to signal the presence of the starter flame. While acceptable for laboratory experiments, in a combat aircraft this primitive, amateur device could only compromise the noble ideas of electronic automation.

Shtokolov let me in on a secret: the liquid-propellant rocket engine was much more likely to explode than a solid-propellant engine. If an excess of propellant components accumulates in the chamber before ignition begins, they will ignite with a violent bang or they will explode, destroying the chamber and spilling nitric acid over the nearby instruments. It would be a good thing to come up with a safety system to protect against such an eventuality.

Visiting NII-3, I became acquainted with L.S. Dushkin, the chief specialist on liquid-propellant rocket engines; A.V. Pallo, the chief tester; and other engineers. They were very calm about the explosive nature of the liquid-propellant rocket engines. I already considered myself an experienced aircraft engineer, and their attitude toward the rocket engine seemed, at the very least, to be strange. For them, the engine literally had a "mind of its own." Each test—if the engine started at all—brought so many surprises that it seemed impossible to predict the engine's behavior after the next modification.

Compared with "the song of the propeller" and the usual roar of multi-cylinder gasoline engines, the deafening, howling flame of a liquid-propellant rocket engine inspired no affection whatsoever. Each time the engine was activated, a rust-colored cloud of nitric acid vapor would explode from the nozzle. Our eyes teared our faces stung as if hit by an arctic blast, we continually felt like sneezing and coughing. It was obviously dangerous to inhale the rust-colored atmosphere, and I had the imprudence to allude to this. Pallo said that the vapor was nothing compared with the *explosion* of a rocket engine. That was when it was really harmful.

Pallo had previously worked as an aircraft flight mechanic. When I started talking about aircraft engines, his eyes got a look of nostalgia. Beside their other merits, propeller engines had a pleasant smell. We recalled the noble scents of aviation gasoline and hot engine oil, which at any airfield in any weather provided us no less pleasure than the fragrance of good *eau de cologne*.

As calmly as I could, I related to Bolkhovitinov my first impressions of the rocket engine. He said that Kostikov and Dushkin had promised him that they would modify the engine so that it was suitable for installation in an aircraft in three to four months. During that time, we ourselves still needed to fabricate the airplane and test fly it in glider mode.

Bereznyak and Isayev received my suggestions for automating the startup, but without enthusiasm. It turned out that Dushkin's designers had developed a rudimentary model of the turbo-pump assembly that fed fuel and oxidant into the combustion chamber. A new feed system needed to be devised, and consequently, a different automatic device needed to be developed.

The turbopump feed in Dushkin's engine permitted a thrust of no more than 600 kgf, and we needed at least 1,200 kgf. No electric automatic device was capable of doubling thrust! "I have an idea," said Isayev, "but we will have to redesign and redo the layout of the entire aircraft. Tomorrow, Sunday, I will work all day, and Monday we will go see the patron."

That was Sunday, 22 June 1941. A new time of reckoning was beginning.

Chapter 11

At the Beginning of the War

In the spring of 1941, there was no more hope for the series production of DB-A bombers. A sixteen-aircraft series was set up at the Kazan aircraft factory, but the DB-A did not survive the competition with the Tupolev TB-7 in terms of flight performance. Tupolev's "enemy of the people" status notwithstanding, the TB-7 (Pe-8) was still allowed into series production. By late 1940, after twelve aircraft had been released, work on our Annushka had been halted.

In spite of a series of modifications that experienced designers Zalman Itskovich and Ilya Flerov had made, the "S" aircraft did not stand up to competition with Petlyakov's Pe-2. After the Pe-2 multi-purpose dive-bomber was accepted as a standard armament, there was no more hope of series production for the "S."

The subsequent development of the two tandem engines led to the design of the "I" aircraft—a two-beam configuration dive-bomber with paired pushing engines. Isayev, who was selected by Bolkhovitinov to head this work along with Itskovich and Air Force Academy professor Dzyuba, introduced a multitude of innovations to the design, including the development of the first pilot ejection seat. I invented many unusual electric drive circuits for the remote control of weapons. After Isayev and Bereznyak got caught up in work with the rocket-powered interceptor, work on the "I" slowed down; after 22 June it was halted altogether.

Having restored the joint work with the Air Force Academy professoriate that was disrupted in 1937 by his transfer to Kazan, Bolkhovitinov revived his idea for a fast, high-altitude, long-range bomber. Provisionally, in place of "B," he assigned it the new designation "D." This design would be a four-engine, high-speed bomber with two paired engines with tractor and pusher propellers. The crew would be located in pressurized cockpits. The aircraft was to be controlled using electric drives.

It was for this aircraft that I worked on an AC electric equipment system; devised remote control circuits for the control surfaces, flaps, guns, and machine guns; and hoped to resuscitate the electronic bomb release.

By the summer of 1941, everything to do with the design of the "D" bomber was still on the drawing board only, except for my work on alternating current. Generators using permanent magnets, every imaginable type of AC electric drive, reduction gears, and voltage and frequency stabilizers gradually filled up every bit of free space in my special equipment department, prompting my neighbors in the cramped building to joke and gripe to Bolkhovitinov and myself, "All the resources allocated for the 'D' went towards alternating current, there is nothing left over for the aircraft itself."

THE LATEST DESIGN of the liquid-propellant rocket engine fighter-interceptor had been approved by Bolkhovitinov and included as "non-essential" in the schedule of operations in April 1941.

Bolkhovitinov allocated a separate room for the small team of developers for this preliminary design, which had been assigned the designation "BI." Colonel Engineer Volkov, the factory's chief engineer, who like many other "aviation wolves" had considered the work of Isayev and Bereznyak to be a childish escapade, set aside a 25-square-meter room for them.[1] There they began to work on the preliminary design. On the morning of 21 June 1941, I found Isayev and Bereznyak in a heated argument in that "secret" room. I wanted to coordinate the proposals for the automatic-startup control unit for the liquid-propellant rocket engine that had been developed after discussions with Shtokolov at RNII–NII-3. I also needed to know the latest version of the weaponry in order to design the electric circuit. Was it four machine guns or two ShVAK guns? Bereznyak informed me that all the machine guns needed to be thrown out and to design for two ShVAK guns with ninety rounds each. As far as the liquid-propellant rocket engine startup circuit, I could forget everything that I had agreed on with Shtokolov. Dushkin's turbopump assembly was in the most deplorable state and there was virtually no hope of getting it developed in the coming year.

Isayev announced that he would recalculate everything at home that day and the next day, Sunday. On Monday we would start over on a preliminary design that would not require a turbopump assembly. We then parted company, to meet again on Monday.

On Saturday evening I left for Udelnaya. Two-year-old Valentin felt more free at the home of his maternal grandmother in the fresh atmosphere of the pine forest than on the factory grounds where we lived with my parents.

The morning of Sunday, 22 June carried with it the promise of a hot, sunny day. Katya and I were about to hike to some nearby ponds for a swim and to give Valentin his first introduction to cold water. Just before we left, one of our neighbors shouted, "A government announcement is going to be broadcast at noon." We lingered. All the inhabitants of the nearby dachas gathered by the black "dish" of the broadcast network, the *Rekord* loudspeaker. Vyacheslav Mikhaylovich Molotov's short speech split time into two epochs: "before the war" and "during the war."

"Our cause is just. . . . The enemy will be defeated. . . . Victory will be ours." These words did not carry any specific instructions as to what each citizen was to

1. "Aviation wolves" refers to older, experienced aviators.

do. It was clear—our peaceful life was over. Now we needed to do something immediately to fulfill the promise, "Victory will be ours." On that summer day, we each had our personal joys, happiness, and sorrows. For all of us, everything personal was suddenly put off and immediately replaced by a single, common misfortune and concern: the war.

Unable to comprehend, Valyushka looked at his distraught mama and papa.[2] "When are we going to the ponds?"

"What swimming? . . .what ponds?!" I thought. I was convinced that the Germans would begin bombing Moscow that very night. We decided to dig slit trenches for shelter. After discussing the engineering required for such defensive measures, I set out for Khimki. The electric commuter trains were unusually crowded. Dacha residents and people returning from weekend recreation streamed into Moscow.

When I finally arrived at the factory, a crowd of familiar faces was already standing at the entrance. The people gathered there were discussing where the authorities were and what needed to be done now. It was clear to each of them that it no longer made sense to work on the "S" and "I", much less the "D."

We waited until evening for the arrival of our patron, Bolkhovitinov. A motorcycle with a popping engine rolled up to the gates. Isayev was driving, and clasping him from behind was the patron himself with his tunic unbuttoned. Without asking permission, everyone entered his spacious office. Bolkhovitinov then told us very calmly, in his usual quiet voice, that we needed to revise our plans immediately. The most important work now would be the liquid-propellant rocket engine interceptor. Beginning tomorrow, everyone who was not called up for the army would have to adapt to a barracks-like environment. He turned to his deputies—they needed to arrange for beds in the shops and KB, make sure that food was provided three times a day, equip reliable bomb shelters, and intensify security.

The following day, Isayev told his close comrades about his Sunday trip to get the patron. Isayev had labored almost all night over the aircraft's new layout. He decided to throw out the turbopump assembly and introduce a mechanism to feed propellant components to the engine using compressed air. He needed to determine the total amount of compressed air and the number of tanks, find space for them, recalculate the center of mass, replace the tanks with stronger ones, recheck the weight, and—by morning, when it seemed that everything would work out, he fell asleep.

Molotov's speech woke him up. When he heard the terrible news, Isayev filled up the motorcycle parked by his house and hurried off to Bolkhovitinov's dacha located on the bank of the Moscow-Volga Canal. When he arrived, he learned from the household that Viktor Fedorovich had left that morning in his yacht for the Pestovskoye, Yakhromskoye, or Klyazminskoye Reservoir. He had no radio receiver in his yacht. What to do? Isayev made himself comfortable on the reser-

---

2. Valyushka is a diminutive form of Valentin.

voir banks and continued to improve on the previous night's design while awaiting the arrival of the patron.

When the suntanned yachtsman finally arrived, Isayev overwhelmed him with the latest news. He suggested that they ride over to the People's Commissariat to see Commissar Shakhurin and propose that they immediately make rocket-powered interceptors. Bolkhovitinov changed his clothes in order to appear before the People's Commissar in full uniform, took his place on the back of the motorcycle, and tore off with Isayev to Moscow's Ulanskiy Lane.

The People's Commissariat was humming like a swarming beehive. Shakhurin found five minutes to receive Bolkhovitinov. The People's Commissar proposed that the preliminary design be completed in a week and then submitted to him for review. It was not until they were on the road from the People's Commissariat to Khimki that Isayev explained to the patron that we needed to start the preliminary design from scratch. If we did without the turbopump assembly, however, it would be a lighter machine, it would require less fuel, everything would be simpler, and the amount of time required would be reduced. All we needed from Kostikov and Dushkin was the "bottle."[3]

Beginning that Tuesday, everyone who did not live near the factory in Khimki really did switch over to a barracks-like regime. The preliminary design for the interceptor was completed in twelve days. According to the design, the aircraft had a wingspan of only 6.5 meters and a length of 6.4 meters. The landing gear was fully retractable using a pneumatic drive. The takeoff weight was 1,650 kilograms, of which 710 kilograms was nitric acid and kerosene.

Kostikov and Dushkin were not at all enthusiastic about doing away with the turbopump assembly and switching to a pressurized feed, but ultimately they agreed. It took two days to compose a letter to the People's Commissar listing all the advantages and the minimum steps required to build the aircraft over a period of three to four months. A month was set aside for state tests, and a decision on the startup of series production would be made in November. While agreeing on the dates, disputes arose as to whether such an aircraft would be needed in six months. By then the war would be over, "victory would be ours."

The letter was signed by Bolkhovitinov, Kostikov, Dushkin, Isayev, Bereznyak, and the military representatives of NII-3 and Factory No. 293.

On 9 July the letter was in Shakhurin's hands. Shakhurin personally reported the proposal to Stalin, and the following day Bolkhovitinov, Kostikov, Isayev, and Bereznyak were summoned to the Kremlin, where they drew up a draft resolution of the recently created State Committee of Defense (GKO). Stalin signed the resolution a day later. Shakhurin prepared a similar order, which allocated one month for the construction of the first aircraft for flight testing. This shocked us at the factory. As yet there was not a single drawing in production.

---

3. Here, Chertok is referring to the engine combustion chamber.

At RNII, the experimental engine had still not developed more than 600 kgf of thrust, while the preliminary design required 1100 kgf.

Bolkhovitinov, passing on the words of Shakhurin, said, "There's a war on. We need to have a different attitude toward deadlines." Shakhurin had nevertheless added five days in the final version of the order.

It was good that the aircraft had an all-wood design. Construction began without detailed drawings. The primary elements were traced actual size onto plywood. This was what they called template engineering. Joiners from the nearby furniture mill worked as though they had been building airplanes all their lives. They just needed to know what the designer wanted, and had no need whatsoever for drawings. But the steel tanks for the compressed air, the durable welded tanks for the nitric acid and kerosene, the pressure regulators, pipelines, valves, rudder control, landing gear, instruments, and basic electrical equipment all required deadlines for design and fabrication.

The war had flared up in a completely unexpected way. No one spoke any more about victory before the end of the year. Combat fighters appeared at the Khimki airfield in late June, and in July it became one of the air defense centers covering Moscow from the northwest.

On 22 July, I was given twenty-four hours leave to see my family. It was precisely that evening that German aviation carried out the first large air attack against Moscow. The air raid alarm caught me in Petrovskiy Park near the Dinamo metro station. I hurried on foot on Khoroshevskoye Highway and did not take cover in the metro. Dozens of searchlights crossed their beams on gleaming points. In the evening sky, hundreds of anti-aircraft projectiles were bursting in flames. The thunder of cannon fire was all around and the ground was strewn with shrapnel, but not a single bomb exploded in the vicinity. When I finally arrived in total darkness at my parents' home, the attack was still going on. I found my parents, together with the factory workers, in the trenches—an open shelter on the shore of a pond. Unseen anti-aircraft guns continued to thunder, searchlights swept the sky, and somewhere over Khodynka a glow had flared up; but as I pictured it, the German bombardment had not been successful. The Moscow air defense service had managed to ward off the first air attack.

My parents and their neighbors in the shelter were completely overwhelmed by the thunder of the anti-aircraft guns and the light effects of the first air attack. Not a single bomb fell in the vicinity, but the first glows in the area of Khodynka and Fili raised the question, what if these bombs fall on us? Observing the night-time battle of hundreds of anti-aircraft guns with an invisible enemy, I tried to imagine what our rocket-powered BI interceptors would be capable of doing under those conditions. No guidance system and no night-time airfield detection facilities for landing! Even if they were to give us a reliable engine and we were to "shape up" the entire aircraft, it would have very little chance of encountering the enemy after a night takeoff and even less chance of landing intact. We were ahead of the times with our rocket-powered aircraft.

On 1 September, five days late according to the deadline set by Shakhurin's order, the first BI was sent to LII to begin flight tests. Strictly speaking, it was not an airplane but a glider. It did not have an engine. Kostikov and Dushkin had not risked sending us the unmodified and frequently exploding assembly.

Test pilot Boris Kudrin was assigned to perform the testing of the BI glider aircraft. He had previously tested the paired engine aircraft, and had been the lead pilot for Bolkhovitinov's team. The tests began with takeoff runs on the airfield behind a towing aircraft.

DURING THESE FIRST DAYS OF SEPTEMBER, the entire factory work force, except for individuals who were directly involved with the beginning of flight tests, was mobilized to build anti-tank trenches near the village of Chernaya Gryaz along the Leningrad Highway. On both sides of our area, as far as the eyes could see, thousands of women and teenagers were digging the earth. The few men who had been granted a deferment from mobilization felt awkward. We worked there from sun up to sun down with two short breaks for food. During this time, the factory Party organization reported the latest news about the battle in Smolensk. The first successes of the Red Army, which had liberated Yelnya on 6 September, gave rise to an explosion of enthusiasm. Why put up fortifications around Moscow if the enemy has already been stopped? We believed that the fascists would now be driven to the west. But our joy was short-lived.

Having returned to the factory from the earthworks, I found out what had happened at the LII airfield. Our team, which had prepared the aircraft, did not find it in the hangar after arriving that morning. It turned out that the day before, Deputy People's Commissar for Experimental Aircraft Construction Aleksandr Sergeyevich Yakovlev had come. Behind his back the other chief designers called him ASYa.[4] When he had seen Bolkhovitinov's airplane, he expressed interest in the results of the wind tunnel tests. It turned out that the aircraft had not been in the TsAGI "tube." Yakovlev arranged for the single existing glider to be towed immediately to TsAGI for testing in the new wind tunnel. Fortunately, it was located quite near the airfield.

Bereznyak and Isayev were indignant, having seen in the incident ASYa's anti-Bolkhovitinov intrigues. For the sake of fairness, it must be said that Yakovlev's order proved to be beneficial. The wind tunnel tests were conducted under the direction of twenty-five-year-old lead engineer Byushgens. This was his first association with rocket technology. Forty-five years later Academician Georgiy Sergeyevich Byushgens, Hero of Socialist Labor, recipient of the Lenin Prize, Professor N.E. Zhukovskiy Prize laureate, and holder of many orders, presented his findings on the gas-dynamics and aerodynamics of the *Buran* orbital shuttle.

---

4. 'ASYa' is a play on the feminine name Asya, a diminutive form of both Anastasiya and Aleksandra.

On 15 November 1988, in a purely automatic unmanned mode, *Buran* was inserted into space by the *Energiya* rocket. Controlled by its onboard rocket engines, it made two orbits around the Earth. After 206 minutes of flight time, having splendidly executed all the difficult aspects of controlled descent into the atmosphere, it appeared precisely over the landing strip, touched down, and did a landing run of 1,620 meters with a lateral wind of 17 meters/second; the measurement deviated from the theoretical design point by only 3 meters laterally and 10 meters longitudinally!

Fourteen years after this flight, which brilliantly proved the enormous potential capabilities of the systemic combination of rocket, aircraft, and radio electronic technology, Academician Byushgens said in an interview for the newspaper *Izvestiya*, "The path to (flight) safety is through the complete automation of the flight during all phases— from movement along the takeoff strip and ascent to landing. The human being is needed only for monitoring purposes. The Buran spacecraft flew unmanned around the Earth and touched down precisely at the beginning of the strip, which was only 30 meters wide."

The BI wind tunnel tests showed insufficient directional stability. In flight, the aircraft could yaw off its heading. In spite of Isayev and Bereznyak's impatience, Bolkhovitinov gave instructions to build up the rudder along the trailing edge, put two circular plates on the horizontal fin, and enlarge the skid fairing. In historical photographs taken in 1942, one can clearly see the two vertical plates that appeared after the wind tunnel tests. Finally, Kudrin took off in the glider, without engine or guns, and towed by a Pe-2 bomber. At an altitude of 3,000 meters Kudrin disengaged and switched over to glider mode. He was an experienced glider pilot. Engineless flight was not a novelty to him.

In all, fifteen BI glider test flights were executed with the assistance of a towing aircraft. Kudrin and other pilots who flew the glider confirmed that the interceptor could return to its own or another nearby airfield in glider mode after its rocket engine was shut down at an altitude of 3,000–5,000 meters, as had been conceived. The flight tests coincided with a certain calmness on the central front. Muscovites had become accustomed to air raid sirens and obediently descended into the metro, the nearest bomb shelter, or trenches that had been dug in vacant lots.

In late September, having visited my parents, I learned that my older cousin Misha—who had volunteered to serve at the front, leaving the office of People's Commissar I. F. Tevosyan—had been wounded in the battle of Smolensk and was in the hospital. He had sent a letter, full of optimism and faith in our victory, but had also predicted that our next big war would be with China. I did not have the occasion to debate with him. A month later he returned to the front and died near Vyazma.

My parents took the news of Misha's death very hard. Before the conclusion of the "Molotov-Ribbentrop Pact," which allowed the Germans to begin World War II by invading Poland, my mother from time to time received news from her older

From the author's archives.

Participants in the development of the first BI rocket-propelled aircraft. Shown sitting (left to right) S. G. Chizhikov, V. A. Shtokolov, L. S Dushkin, K. D. Bushuyev, G. G. Goloventsova, B. Ye. Chertok, A. A. Tolstov. Shown standing: Z. M. Gvozdev, I. I. Raykov. A model of the BI rocket-plane can be seen in front of them. Among the men, Bushuyev and Chertok would go on to be extremely influential designers in the Soviet space program.

sister Fruma Borisovna and nephew Solomon who lived in Lódz.[5] "All of our relatives in Poland will be wiped out by the Germans," she predicted. "Misha is only the first of our relatives who has died at the front. You have worked all this time on airplanes, but where are they? Why is Moscow being bombed and not Berlin?" How could I answer her? Could I say that we were making a miracle—a rocket plane that would at least save Moscow? But I did not have enough confidence.

A new German offensive against Moscow began on 30 September. Holding our breath each morning, we listened attentively to the reports from the Soviet Information Bureau (*Sovinformbyuro*). An awful chill ran up my spine when they mentioned such nearby cities as Kaluga, Gzhatsk, Medyn, Mozhaysk, and Volokolamsk. We gathered around maps and interpreted the *Sovinformbyuro* reports our own way. Each day it became more clear that Moscow was under a direct frontal assault by the Germans, who were going around it from the north and the south. The Moscow-Volga Canal formed a water boundary covering the capital

5. The German-Soviet Nonaggression Pact of 23 August 1939 was negotiated by German Minister of Foreign Affairs Joachim von Ribbentrop and USSR People's Commissar of Foreign Affairs Vyacheslav Mikhaylovich Molotov. Among other things, the Pact laid the blueprint for the Soviet occupation of Latvia, Lithuania, and Estonia.

from the northwest. In the event of a swift breach by the Nazi tank armies, we in Khimki—along with our factory and all the work that had been done on the BI aircraft—would find ourselves on the "German side."

Increasingly alarming information was coming to us from Moscow about the evacuation of one military factory after another. After one of these routine distressing reports Isayev had a confidential conversation with me. He proposed that we create a guerrilla detachment. Isayev spoke with great enthusiasm, as if he were proposing that we take part in a recreational hike. This conversation took place on the eve of the Nazis' breach of the Mozhaysk and Volokolamsk lines.

On 15 October, the State Committee of Defense ordered the emergency evacuation of all central Party and State institutions from Moscow. The next day, 16 October, was the beginning of the mass evacuation that went down in the unofficial history of the war as "the Moscow panic." On that day, the People's Commissariats, the leaders of all central institutions, and all factory directors received a very strict order to evacuate east by any means available to new relocation sites for their institutions and enterprises.

Not having warned any of their subordinates, Bolkhovitinov and chief engineer Volkov disappeared on 16 October. Later we found out that they had not been cowards, but had been following an order by Shakhurin, who had summoned the directors and chief designers to the People's Commissariat and then ordered them to immediately leave Moscow without returning to their respective workplaces. Bolkhovitinov and Volkov had traveled to the population center specified by Shakhurin: Bilimbay, which was 60 kilometers west of Sverdlovsk in the Urals. Like the other directors, they would have to make arrangements with local authorities to begin receiving the evacuated enterprises so that operations could immediately be continued.

The next day the Moscow panic had reached Khimki, but we had not yet received an official order to evacuate the factory.

Blow up and destroy everything and join the guerrillas—that was my frame of mind as I set off on 17 October for Moscow, hoping that my wife Katya, who had been in Udelnaya, had come to the Golubkina Museum to stay with her cousin Vera, the museum director. Actually, having heard about the panic, Katya had tied all her things into a bundle, grabbed our son, and rushed to the station. One by one, overcrowded trains left Moscow and rushed on with no stops. They even coupled metro cars into the trains. But not one was traveling *to* Moscow! Finally, some overcrowded train headed for Moscow stopped in Udelnaya.

With the help of her mother and sister, Katya and Valentin squeezed into the jam-packed railroad car. Somehow Katya, carrying her enormous bundle and two-year-old son, managed to make her way from the train station to Bolshoy Levshinskiy Lane. That is where I found them. But Polya Zvereva was there before me. She was the former wife of Sergey Gorbunov. Several years after his death, she had married a well-known test pilot who had worked at LII. On the day of panic, she had remembered her fellow Zaraysk natives, had come to the

Golubkin family, and proposed that they evacuate on the LII and TsAGI special trains to Novosibirsk. Meanwhile, Vera announced that she would not abandon the museum; she was not ready for evacuation. Let Katya and her son take advantage of this opportunity.

I agreed with this proposal. In secret I told Katya that Isayev and I were joining the guerrillas and asked her to quickly trim my long lightweight overcoat so that it would be easier for me to maneuver. For some reason she did not fulfill my request.

Once I had returned to Khimki, I felt the need to inform the Party organizer about Isayev's guerrilla initiative and receive his approval to sign up with a volunteer detachment. Instead of support, I was promised a very severe Party reprimand for this unauthorized initiative. But, nevertheless, they took into consideration our "sincere patriotic impulse" and I received the assignment of convincing non-Party member Isayev to put any guerrilla thoughts out of his head. They had just been informed that we were getting a special train for evacuation. We were obliged not to burn or blow up anything, but to carefully dismantle and pack all valuable equipment, property, and documentation. All personnel and their families were to leave in the next few days for the village of Bilimbay in the Urals and continue to work there.

All those who had families in Moscow were ordered to quickly bring them to Khimki in preparation for evacuation. Instead of fleeing to the east, we were supposed to quickly bring our relatives toward the advancing Germans. By some miracle, I managed to intercept Katya and Valentin on the platform of the Kazanskiy train station. They were waiting for the special train to take them to Novosibirsk. Fortunately, the delivery of the railroad cars had been delayed. We returned once again to the Golubkina Museum. My parents had also come there. They had had to make their way there on a cart. "The closer to the front, the more orderly it is," I reassured them.

On the road to Khimki, we heard through a loudspeaker the appeal of Central Committee and Moscow Party Committee Secretary Shcherbakov to the capital's residents, "We will fight for Moscow tenaciously and bitterly until the last drop of blood."[6] Home guards armed with rifles moved along Leningrad Highway toward the front. There were no tanks or artillery in those columns. A train carrying wounded moved through the Khimki station, headed for Moscow. Would the fortifications that we had dug stop the Germans? And what if they didn't? Were the anti-tank hedgehogs on all of the roads leading into Moscow capable of holding off an avalanche of Nazi German troops?

After our "guerrilla plot" was uncovered, Isayev announced to the Party organizer that he would "blow his brains out" if, due to his stupidity, we ended up in the clutches of the Germans with no weapons.

---

6. Aleksandr Sergeyevich Shcherbakov (1901–45) was a candidate member of the Politburo in 1941–45.

Day and night we took down machine tools and lubricated, covered, and packed everything that we could into cases. We marked, inventoried, and loaded them. My comrades and I managed, in spite of the protests of the special train staff, to pack and load all of our AC machine models. The heated goods wagons were equipped with plank beds, iron stoves, and firewood.

Mama tried to abandon us all and run away to work at any hospital to care for the wounded. Father restrained her.

I managed to get my family to Khimki in time. On 20 October, Moscow and the areas adjacent entered a state of siege. Traffic on the streets was placed under strict control. A State Committee of Defense resolution stated: "Violators of the order shall immediately be called to account before a military tribunal, and agents provocateurs, spies, and other enemy agents calling for the violation of the order shall be shot on the spot."

We coupled an "air-defense flatcar" into our train. On it we mounted the same aircraft machine guns and guns that Isayev wanted to use for the guerrilla war. My family was in the heated goods wagon, together with the Mishin, Chizhikov, and Buzukov families. During the trip everyone was provided with white bread, butter, and groats. In "our" wagon, having pooled all the stores, we were convinced that we would not go hungry in the next two weeks.

Our special train was one of the last to evacuate the factories of the Moscow suburbs. On 25 October, the special train left Khimki and slowly chugged across the Moscow-Volga Canal bridge. The forward German units were located that day on the Kalinin-Yakhroma-Klin line and were approaching Tula and Kashira. Before our departure, I had conversations with the pilots from our combat airfield who had a better view from above. I got the impression that the Germans were played out and that the offensive was on the verge of being halted.

We moved eastward very slowly, always yielding the way to trains heading the other direction. Special trains carrying Red Army soldiers, well outfitted in light-colored knee-length sheepskin coats, were moving westward. "Where are you from?" we asked them at stations where we were getting boiling water. "Siberian troops!" the soldiers cheerfully responded. These were real warriors, unlike the Muscovite home guards.

The stations were full of factory personnel being evacuated to the east. Families were traveling with the bare necessities, while also carrying all the machine tools and supplies from the factories. We encountered many acquaintances on that trip. It seemed that all of Moscow would move to the east to colonize the Urals, Siberia, and Central Asia.

Streaming toward us heading west were flatcars filled with tanks and guns of various calibers. They gladdened us and gave us hope. Our special train made its way east during those rare windows of opportunity that opened in the stream of weapons and troops rushing westward on a green light to Moscow.

We did not make it to Kazan until 1 November. Here we finally heard the latest news about the course of the battle for Moscow. "It seems that the offensive against

Moscow has stopped," we told each other hopefully. Indeed, in the last days of October the Soviet troops' defense along the Western front had stabilized.

We arrived in Bilimbay on the morning of 7 November. A temperature of –20° C greeted us. In spite of the holiday—the twenty-fourth anniversary of the October Revolution—an all-hands rush job was announced to unload the special train. The local authority temporarily housed all the new arrivals in "God's Temple" (the church), right on the cold stone floor. While the women made arrangements for their children in the church and set up household, all the men began to haul the equipment to the iron foundry that had been placed at our disposal. According to the testimony of local residents, the factory had been built during the time of Catherine the Great. The work force of the Kamov-Mil helicopter design bureau and the Privalov assembly factory, which manufactured equipment for airborne troops, had arrived here several days before us.

"Ancient technology from the times of Peter the Great," said Volkov, who had already had a look around when we saw him. Fulfilling Shakhurin's order, he and Bolkhovitinov had left from Moscow in an "MK, " wearing the same summer uniforms they had arrived in. The journey by automobile over Russian roads at that time had lasted fifteen days. They had still managed to reach the site before the special train arrived, and had fought off others contending for the bulk of the space at the factory as well as prepared the church for people to settle there before they would be resettled in the wooden cottages of the local residents.

My mother got to work immediately. On the very first day, she found the local hospital and was taken on right then as the head nurse. From the hospital she brought improbable news about a military parade in Red Square on 7 November.

Having wound up on a team of "riggers" from our department, my comrades and I had the opportunity to see our future worksite while transferring the cases and machine tools. The factory had ceased to operate long before the war. Now, with the severe cold, the factory grounds sprinkled with the first snow made a depressing impression. The windows were knocked out and the frames broken open. We saw neither gates nor doors. The cupola furnaces and even some of the casting structures were crammed with "salamanders." In the courtyard and under the hole-riddled roof were heaps of slag, petrified in the frost, and tons of scrap metal of all sorts. Bolkhovitinov had found a builder somewhere, and the two had worked in a small office to draw up something like a remodeling design by the time we arrived. We were faced with converting this foundry graveyard into an aircraft factory.

Bilimbay was a large village which had developed at one time near the foundry on the bank of the rapid Chusovaya River. We needed to retrain the former metallurgists into aircraft builders. But only women and old folks remained in Bilimbay, behind the tightly closed, desolate gates of their homes. All the young people were in the army.

On one of the first days of the off-loading operations, Bereznyak suddenly appeared. But what a sight! His natty, lightweight overcoat and felt hat were covered with spots and his ears were bandaged. It turned out that on 16 October, he had received a critical assignment from Bolkhovitinov: he and a small group were to immediately travel out to the village of Bilimbay as our advance guard. First they rode on an open flatcar and got drenched in the rain. Beyond Kazan they were stricken by cold. Bereznyak's team, dressed in light clothing next to German tanks, were taken for prisoners of war. After many days of tribulation, they finally made it to Bilimbay, but it was already too late to fulfill the critical assignment.

We began with clearing and improving the grounds. We broke up into teams of loaders, riggers, carpenters, and glaziers. First of all we cleaned up the first floor for the installation of the machine tools. The second floor was set aside for the assembly shops, design bureau, and laboratories. We worked an average of twelve to fourteen hours each day. Our greatest enemy was the cold. By early December, the temperature had plummeted to -40°C. In such bitter cold one had to wear mittens, and it was impossible to do any work other than by using a crowbar and sledgehammer. By December the main dirty work had been completed.

Our Party secretary, Neyman, returned from a visit to the local Party regional committee, panting with excitement, and reported the latest news about our troops going on the offensive near Moscow. All of us in the depths of our souls had dreamed of such a miracle. Now that it had happened, the joy increased our strength and made us throw ourselves into our marathon project.

I was tasked with designing and assembling the factory's electrical equipment and lighting. Semyon Chizhikov headed the team of glaziers. The electricians and glaziers could not work in mittens. Every twenty minutes we ran to the scorching hot stoves, warmed our numbed fingers, and quickly returned to our work site.

Isayev headed one of the most physically grueling jobs, clearing the area for construction and delivering the machine tools to their installation site. Once, when we met up for our routine hand-warming by a makeshift stove, I did not miss the opportunity to comment, "Moscow got by without our guerrilla detachment."

"Yes, you're right! It seems Moscow was saved from the German tanks. Now we must save it from being destroyed by German bombers."

He pulled on his nicely warmed mittens, lifted the heavy crowbar, and commanded his team: "You've warmed yourselves by the fire. Let's go. We'll warm ourselves building our assembly shop." The team amicably threw themselves at the next heap of old metallurgical slag.

In December, our stores of provisions were used up. The primary source for sustaining essential activity was 600 grams of bread per person and hot *bilimbaikha*. *Bilimbaikha* was what we called brown noodles cooked in boiling water without any fat. A plate for our first course, a plate for our second course—we dined on this in a barracks structure that we called the "Great Urals Restaurant." Alcohol was a help. It was distributed in small doses among the workers. We saved it and would from time to time exchange it with the locals for milk or meat.

They soon managed to set up something akin to a kindergarten at the local school, giving the majority of the women the opportunity to work. Katya was included in a team that set up a "test station." This fancy name was given to the primitive stand they set up on the bank of an artificial lake formed by a factory dam on a Chusovaya tributary. The stand was a contraption welded from iron pipes and enclosed by plywood in which the wingless aircraft fuselage, including the pilot's cockpit, would be placed. The primary contents of the fuselage were the tanks of nitric acid, kerosene, and compressed air. The tail of the fuselage containing the engine was pointed toward the lake. The idea was that in the event of trouble during the firing tests, everything tainted with nitric acid would fall into the water. The water was covered with a thick layer of ice. For secrecy, the stand was screened from curious onlookers by a tall fence. They started by performed filling tests rather than firing tests. The snow cover around the bench took on a dirty brown tint.

Nitric acid fumes saturated the clothing of those who had worked at the stand. When Katya returned from work, her hole-riddled quilted jacket also filled the room of our cottage with that noble scent. My parents had taken up residence elsewhere. The local residents did not take in more than two to three evacuees per household.

During the very difficult last months of 1941, news about the crushing defeat of the Germans near Moscow was moral support. In the depths of our souls, each of us who had deserted Moscow during her most tragic days believed and waited. Now people started to ask, "Did we need to evacuate Moscow?"

Only when we returned to Moscow did we understand how likely it had been that the Khimki region would be captured. At the end of November, the fighting had already proceeded east of Kryukov and the Yakhromskoye Reservoir, where our patron had been sailing his yacht on the first day of the war. Only a good twenty-minute drive had separated the German tanks from Khimki.

But the miracle in which we so firmly believed came to pass.

Chapter 12

In the Urals

By New Year's we had restored the boiler house and heating system, and heat was flowing into all the workrooms. As a top priority, my team had finished the electrical equipment and run the lighting. The machine tools were humming on the first floor. On the second floor, the designers could work at their drawing boards free of quilted jackets and mittens. In the assembly shop they began fabricating three aircraft simultaneously in molding lofts.

I formed three groups from our small staff of personnel. Anatoliy Buzukov made the aircraft's general circuit diagram; Semyon Chizhikov designed and fabricated the pilot's instrument panel; Larisa Pervova converted the aircraft engine spark plugs into glow plugs for the ignition of the liquid-propellant rocket engine. The engineers not only designed and drew, but made what they had devised with their own hands.

Having been relieved of my worries about industrial electrical equipment, I was sent off "from Europe to Asia" on a temporary assignment. That's how we referred to my sixty-kilometer trip across the Ural Mountain Range. On the road to Sverdlovsk we crossed the boundary between the two continents, which was marked by a stone obelisk. Many scientific organizations had been evacuated from Moscow and Leningrad to Sverdlovsk. I tried to restore contact with the radio engineering organizations, but I achieved no concrete results.

With great chagrin, Bolkhovitinov reported that our factory test pilot, Boris Kudrin, was in the hospital. The Air Force NII Command, which was located at the Koltsovo Airfield near Sverdlovsk, had attached a new test pilot to us—Captain Grigoriy Yakovlevich Bakhchivandzhi. The pilot was soon introduced to us, and we all liked him immediately.

Each of the team leaders was tasked with personally familiarizing the pilot in detail about the aircraft's construction. Bereznyak and Isayev began first. It took them two whole days. Once he was free of them, Bakhchi, as Bereznyak christened him, landed with me. He did not doubt the reliability of our aircraft and showed an extremely respectful attitude toward us designers. My comrades and I did not sense in him any of the arrogance characteristic of many test pilots.

At Bakhchi's request, Chizhikov switched two instruments on the instrument panel, and Buzukov moved an instrument panel light to a different location.

Working together on the control stick, we selected and added the firing control actuator buttons, the toggle switches activating the engine start valves, and the ignition button.

Bakhchi astounded us by several times by flying in from Koltsovo in a light sport airplane, landing on the snow-covered ice of the lake, and taxiing right up to the test stand. In his black leather raglan coat, flight helmet, and chrome leather boots sunken into the snow, he seemed like an emissary from a distant world, from the warm airfields around Moscow. During my first days in his company, I was amazed by Bakhchi's confidence in our work. He seemed to be convincing us that we were creating the airplane of the future, instead of us convincing him.

Concentrated nitric acid came into contact with and mercilessly destroyed cables, parts of the electrical equipment, and the wooden structure of the aircraft. When the tanks were being filled for the bench tests, its reddish-brown vapor burned our lungs. Acid leaks occurred at the joints of the pipelines, in the valves, and at the engine inlet. Achieving reliable seals remained one of the most difficult problems in rocket technology for decades. But for Bakhchi back in 1942, it seemed that the suffocating scent of nitric acid fumes was more pleasant than *eau de cologne*.

Bakhchivandzhi was still young. He had served at the Air Force NII since 1938. They entrusted him with the high-altitude testing of aircraft. During the first days of the war, many Air Force NII test pilots became combat pilots. Bakhchi took part in air battles in a fighter squadron during the first months and shot down six enemy aircraft. After the losses of uniquely qualified Air Force NII test pilots became great, the Air Force Command changed its mind. In August 1941, they began to call back the surviving test pilots from the front. The testing of new aircraft technology—both domestic and some that was beginning to arrive from the allies—continued near Sverdlovsk, where the Air Force NII had been evacuated from the suburban Moscow area of Shchelkovo. Before we met him, Bakhchi had already flown in an American Cobra. He was touched by the BI's simplicity when compared to the complex, heavy Cobra. It was interesting to hear Bakhchi's unconventional musings about airplanes. We were won over by his intellect, innate simplicity, total lack of pretentiousness, and continuous internal visualization of flight situations. For him, test flights were not work but a way of life. Here was a pilot "by the grace of God."

While discussing routine BI control problems and the flight test program, Bakhchi contributed interesting ideas enriched by his combat experience. The lack of radio communications for air battle control was, in his words, one of the weakest points in our fighter aviation.

At the end of January, the stand by the frozen lake was put into operation and began to deafen the neighborhood with the characteristic roar of the liquid-propellant rocket engine. The engine creators sent Arvid Vladimirovich Pallo on temporary duty from Sverdlovsk to work at the stand in Bilimbay. I only saw Kostikov, Pobedonostsev, and Dushkin in Bilimbay one time. Bolkhovitinov and

Bereznyak, who had visited RNII in Sverdlovsk, were told that new assignments had come up and that improvement of the *Katyusha* had become their main priority. The interceptor's engine and the propulsion system required joint optimization during the testing and operational development process before a decision could be made as to its clearance for flight.

The firing test, conducted on the lake on 20 February 1942, almost deprived us of our test pilot on his birthday. According to the test program, Bakhchivandzhi himself was to have started up and shut down the stand-mounted engine while sitting in a makeshift pilot's seat. During Bakhchi's first combat training test session, conducted under Pallo's direction, the engine exploded. The nozzle went flying out over the frozen lake. The combustion chamber struck against the tanks, and nitric acid gushed from the ruptured pressurized pipelines. Bakhchi received a violent blow to the head—his leather jacket was covered with reddish-brown spots. Pallo's glasses saved his eyes, but the acid burned his face. Both were immediately taken to the hospital. Bakhchi recovered quickly. Traces of the severe burn remained on Arvid Pallo's face for the rest of his life.

DURING THE VERY DIFFICULT WINTER MONTHS of 1942, an attempt was made to begin work on yet another exotic rocket aircraft design. The engine design for this aircraft was fundamentally different than the liquid-propellant rocket engine.

Neither hunger nor cold could halt the initiatives springing up among the collective, even when we went out to cut timber. In the profoundly intense cold of January 1942, when the temperature dropped to –50°C, the most difficult problem was supplying wood. The local residents could not keep the evacuees living among them warm. The procurement of wood became as much of a compulsory activity as our primary work. Our workrooms, which had literally risen up out of the factory's ashes in three months, had heat. Childless bachelors preferred to spend the night at their work sites. At night, the design hall was converted into a spacious common bedroom.

We suddenly became severely overcrowded. After assembling the senior staff, Bolkhovitinov announced to us that the People's Commissar had decided to include the workforce of Arkhip Mikhaylovich Lyulka, director of Special Design Bureau 1 (SKB-1) attached to the S. M. Kirov Factory in Leningrad. The subject matter that Lyulka was involved with was similar to our new field of rocketry.

As early as 1935, Lyulka had developed the design for a turbojet engine (TRD).[1] "This is an alternative to the direction we are taking in the development of rocket aircraft," Bolkhovitinov told us, "We need to be very considerate to these new people who have evacuated from besieged Leningrad with

---

1. TRD—*Turboreaktivniy Dvigatel* (Turbo-Reactive Engine). In this particular sense, the phrase is more appropriately translated into English as 'turbojet' rather than 'turbo-reactive.'

great difficulty. Not only did they manage to get their people out, but they also saved the work they had done on an experimental engine with a thrust of 500 kilograms."

Lyulka theoretically designed a gas-turbine engine (GTD) in 1938 at the Kharkov Aviation Institute. The first RD-1 engine was developed in Leningrad at the Kirov Factory's SKB-1, which Lyulka directed. Bench tests were to have begun in 1941, but the war and subsequent blockade of Leningrad had dashed their plans.

Long before the war, Bolkhovitinov was interested in work being done on various designs of air-breathing jets (VRD).[2] He understood that the engine propeller unit of piston engines fundamentally limited flight speed and altitude. As soon as the opportunity presented itself, Bolkhovitinov had turned to Shakhurin with the request to evacuate Lyulka from Leningrad. Shakhurin arranged this with the Leningrad authorities and Lyulka and his surviving personnel were saved.

In Bilimbay we led a semi-starved existence. However, we heard from Lyulka's rescued Leningraders what real hunger was. They ate the brown *bilimbaikha* down to the last drop of the broth that we found barely edible. They did not drop a single crumb of bread.

Lyulka's workforce included several automatic control specialists. Bolkhovitinov suggested transferring them to me. Lyulka agreed under the condition that I would develop equipment to regulate and control the turbojet engine. I spent three days learning the principles of the turbojet engine. Lyulka personally explained to me the difference between the two classes of reactive engines—liquid-propellant rocket engines and turbojet engines. He did not disparage liquid-propellant rocket engines, which had to carry their own fuel and oxidizer. But with gentle humor, alternating Russian speech with melodious Ukrainian, of which he had a beautiful command, Arkhip Lyulka proved that everything had its place while telling me about the turbojet engine.

The air-breathing turbojet engine uses the oxygen of the air that enters through the aircraft's air intake from the atmosphere. The air is compressed by a compressor and then passes through combustion chambers where gasoline, or better still kerosene, is injected. The gas formed during combustion passes through the turbine, turning the compressor, and is ejected through the nozzle. This is not the dazzling bright flame of the liquid-propellant rocket engine, but hot gas that has already been depleted in the turbine and which is almost invisible in the daylight.

Once I had become familiar with the principles of turbojet engines and the ideas for regulating them, I came to the conclusion that the immediate problems related to the automatic regulation of the turbojet engine needed to be solved without any electrical devices, at least for the time being. I thought that we should instead use the capabilities of purely mechanical and pneumohydraulic automated

---

2. VRD—*Vozdushno-reaktivniy dvigatel* (Air-Reactive Engine). In Western technical vernacular, these types of engines are known as 'air-breathing jet engines' or simply 'jet engines.'

mechanisms. Lyulka did not agree, and the debate was carried over to Bolkhovitinov. Ultimately, I was relieved of the work on the regulation of turbojet engines, and Lyulka's engineers worked on this independently.

Lyulka spent eighteen months in Bilimbay, before transferring to Moscow in 1943. He soon thereafter received his own production base beside the Yauza River. The opening of new factories was a rare occurrence during the war. During the first postwar years, our engine-building industry reproduced captured German JUMO-004 and BMW-003 engines for the first Soviet jet aircraft. By 1948, Lyulka's work force was building domestic turbojet engines that were more powerful than the captured engines.[3]

One of the paradoxes of the history of technology is that liquid-propellant rocket engines using special oxidizers were developed and broadly applied much earlier than reactive (now referred to as jet) engines using a "free" oxidizer—the oxygen of the Earth's atmosphere.

DURING THAT DIFFICULT YEAR OF 1942, our collective's priority was to produce a reliable engine system for the BI aircraft. Lyulka's engines seemed to be for the distant future, while the exploding liquid-propellant rocket engine was in the here and now.

Bolkhovitinov and Isayev had flown to Kazan for several days. Our patron had, with difficulty, obtained permission to visit the NKVD's special prison attached to Factory No. 16 in Kazan. There they met with Valentin Petrovich Glushko. Upon their return, Isayev enthusiastically told us about their meeting. He said that their two days with the imprisoned Glushko and his associates had taught them more about the liquid-propellant rocket engine than the entire preceding period of dealing with the RNII.

"These *zeki* live better than we do," Isayev told us. "They have benches, laboratories, and production facilities that we wouldn't even dream about. They are under guard, so they weren't able to talk openly about their life. On the other hand, they are fed better than we are, who are free. The most important thing is that their engines work much more reliably."

This first encounter with Glushko in Kazan in the winter of 1942 determined Isayev's subsequent fate. To the end of his days, he remained true to the decision made at that time regarding the development of reliable liquid-propellant rocket engines.

Upon his return from evacuation, Isayev set up a special engine-design bureau. A postwar temporary assignment to Germany profoundly influenced Isayev's engineering career. He became a leader in the design of liquid-propellant rocket engines for air defense, missile defense, submarines, spacecraft, and many other applications. The collective that he founded inherited Isayev's marvelous enthusi-

---

3. Lyulka served as Chief (and later General) Designer of OKB-165 (or KB Saturn) from 1946-84. During that time, he oversaw the development of many generations of Soviet jet engines for civil and military aviation.

asm. The notion of a school of rocket engine construction is indelibly linked with his name.

THE WINTER OF 1941–42 IN BILIMBAY was the most severe of all the prewar and wartime winters. First there was the strenuous physical labor to build the factory in temperatures of –50°C (-58°F). Then there was the construction of the airplanes and the operational development and firing tests on the engines. All of this took place under conditions of food rationing that put us on the edge of survival. It is amazing that under such arduous conditions no one complained of illnesses that were normal for peacetime.

But illnesses inherent to wars did occur. Somewhere near Sverdlovsk, Polish troop units had formed, and an epidemic of typhus erupted. Mama volunteered for the team that went out to fight the epidemic. She could not protect herself and ten days later they brought her back to Bilimbay with a fever of over 40°C (104°F). By the time I ran over to the hospital she no longer recognized me or my father. The female physician who was attending her told us that more than one typhus patient lying in the cold barracks owed their life to my mother. But she had completely disregarded her own safety. When she realized that she had been infected, she requested that she be taken quickly to Bilimbay so that she could say goodbye to her husband and son. "But we didn't make it in time," the doctor wept, "If only we'd had just a couple more hours."

Mama died on 27 March 1942. At her funeral, the head physician said that medical personnel were performing feats not only on the front but also in the rear. "The selfless labor of Sofiya Borisovna is a living example of that." Her death was a terrible blow to my father and me.

A week after my mother's funeral, my father and I found out about other painful losses among our relatives. In Sverdlovsk we found our cousin Menasiy Altshuler, whom our family called Nasik. He was a railroad engineer who was my age. Nasik and his wife, both only half alive, had managed to get out of Leningrad by traveling over the ice of Lake Ladoga. He told us how his father, a mathematics professor, his mother, my father's sister, and his younger brother, my namesake, had died of hunger in Leningrad.

Chapter 13

15 May 1942

In April, we sensed that spring might even be coming to the Urals. We sent our first aircraft on two trucks to the Air Force NII air base in Koltsovo. They assembled it and placed it in the hangar, enclosed it behind a tarpaulin, and posted a sentry. Only those involved in its flight preparation passed behind the tarpaulin.

Even before our evacuation, Bolkhovitinov had appointed the calm, very business-like Aleksey Yakovlevich Roslyakov to be the lead engineer from our factory supervising the preparation of the BI flight tests. Mikhail Ivanovich Tarakanovskiy was appointed lead engineer from the Air Force NII. He already had a wealth of experience as an aircraft engineer and tester.

When the tests started, all those involved moved into the Air Force NII dormitory in Koltsovo. They began hauling the aircraft behind a truck by towrope around the airfield for a "shaking" test. One of the landing gear struts broke. They repaired it. They performed a filling test and found an acid leak in a valve. They drained off the acid and repaired the valve. There were a few minor negative remarks.

On 25 April 1942, an order was issued to create a State Testing Commission. Air Force Academy professor General Vladimir Sergeyevich Pyshnov was appointed chairman of the commission. The State Commission approved the program, which called for testing to begin with takeoff runs over the airfield and liftoffs to an altitude of 1–2 meters.

On 30 April, Bakhchi performed the engine's first fire tests at the airfield. This was a sensation for the military pilots who had not understood why such a large number of engineers and high-ranking military officials were bustling about such a tiny little bird of a glider.

On 2 May, Bakhchi executed the first takeoff run of the BI with a working engine rather than being towed behind a truck. By this time it was evening. Darkness had fallen, and the bright shaft of fire and the roar exploding from the tail of the tiny aircraft made an extraordinary impression. During the takeoff run, Bakhchi made sure that the tail would lift up when the speed was increased, then revved the engine—the aircraft became airborne. It flew at an altitude of 1 meter for around 50 meters and landed smoothly. During inspection, a steaming gush of acid was found. Once again repair was required. Bakhchi climbed out of the cockpit after the liftoff flight, then announced, "The airplane can be cleared for flight."

After the takeoff runs and liftoffs there were repairs, adjustments, and repeated leak checks. They tuned the altitude recorder, air speed recorder, acceleration load recorder, and control position recorder, then weighed and balanced the aircraft.

The first flight was scheduled for 12 May. They were aiming for good visibility throughout the flight from takeoff to landing. But we had no luck with the weather—it rained.

The aircraft had no radio equipment, not to mention the fact that "black boxes" did not yet exist. Nor was there a single radar at the airfield. We relied completely on visual control from the ground, a post-landing report from the pilot, and the processing of the recordings made by the delicate recorders installed on the aircraft—if, God willing, they remained intact.

State Commission Chairman Pyshnov conducted a session during which they reviewed the bench test results, the reports on the takeoff runs and liftoffs, and the remarks of lead engineers Roslyakov and Tarakanovskiy. They reviewed the flight diagram drawn on a sheet of Whatman paper. Bakhchi reported that he was ready for the flight.

Finally, on the advice of the meteorological service, they scheduled the flight for 12:00 p.m. on 15 May. But by that time, low clouds had once again obscured the entire sky. We cursed the weather and empathized with the hundreds of curious onlookers who, in spite of the secrecy, had climbed onto the roofs of hangars and houses and up into trees, hoping to find out why the airfield had fallen silent since morning. All other flights had been canceled. For the first time, combat aircraft that used Koltsovo as a stopover airfield on flights from Siberian factories to the front were prohibited to land.

After many days of intense work, the wait for a break in the weather created a nervous atmosphere. Bakhchi was also nervous. At 4:00 p.m., Air Force NII chief P. I. Fedorov, who was no less nervous than the rest of us, suggested that Bakhchivandzhi take off in a training airplane to determine the airfield visibility and reference points from the air. Twenty minutes later Bakhchi returned from the flight and reported, "The weather is breaking up. I can fly!"

The day was fading away. It was now or never. They inspected the plane yet again. Arvid Pallo sniffed for acid fumes more carefully than anyone else. Everyone reported his readiness status according to procedure. Bolkhovitinov embraced Bakhchi and gave him the traditional Russian kiss on both cheeks.

The pilot adroitly climbed into the cockpit and began to settle himself. He ran his hand over the throttle one more time, rocked the control stick forward and backward, and moved the pedals. Everyone dispersed from the airplane except Pallo, who wanted to check one last time that there were no leaks. Everything was dry on the exterior. Bakhchi calmly said, "Away from the tail," and closed the cockpit canopy; he then switched on the propellant component feed and ignition.

We all crowded together about fifty meters from the aircraft. Each of us had seen the engine work numerous times on the stand and in takeoff runs here at the airfield, but everyone flinched when the blinding flame exploded from the tail of the tiny airplane. Apparently the nervous tension from the prolonged wait was having its effect.

The roar of the engine over the quiet airfield and the bright flame heralded the beginning of a new era. On 15 May 1942, hundreds of people observed as the airplane began its rapid run down the takeoff strip. It easily became airborne and gained altitude very quickly. With its engine firing, the aircraft made a 90-degree turn to one side, then the other, just managing to make the transition from a steep climb to horizontal flight—then the flame disappeared.

Roslyakov, standing beside me, glanced at the stopped chronometer: "65 seconds. The fuel was used up."

The BI landed, having approached the ground headlong with its engine no longer running. This was Bakhchi's first landing in that mode. It was a rough one. One landing gear strut broke and a wheel came off and went rolling over the airfield. Bakhchi managed to get out of the cockpit and climb down from the airplane before Fedorov and Bolkhovitinov drove up with a fire truck and ambulance. Bakhchi was very upset about the bad landing. Good grief, how awful—the landing gear broke. In spite of his protests, the crowd that had run up to him immediately started to lift him up and carry him in celebration.

Later that evening, we all gathered for a celebration dinner at the spacious Air Force NII dining hall. Fedorov, Bolkhovitinov, and Pyshnov sat at the head of the table together with Bakhchivandzhi. Opening what was for those times a luxurious banquet, Fedorov congratulated Bolkhovitinov, Bakhchi, and all of us on our tremendous success. The results of all the flight recorders had already been processed. The entire flight had lasted 3 minutes 9 seconds. The aircraft had reached an altitude of 840 meters in 60 seconds with a maximum speed of 400 kilometers/hour and a maximum rate of climb of 23 meters/second.

NASA History Office.

BI-1 rocketplane.

A day later there was a ceremonial reception and meeting in the assembly shop of our factory in Bilimbay. A banner hung over the presidium table: "Greetings Captain Bakhchivandzhi, a pilot who has made a flight into the future!" Fedorov and Pyshnov sent an optimistic report to the Air Force Command and People's Commissar Shakhurin. A GKO decision followed calling for the construction of a series of twenty BI aircraft, with the correction of all known defects and full armament.

Our team was elated, even more so because of summer's arrival. According to the signs observed by the locals, it would be a warm one. But I was not destined to spend the summer in the Urals. Having basically solved the liquid-propellant rocket engine's ignition and control problems, Bolkhovitinov returned to the idea of radio-controlled guidance for the interceptor. He wanted to solve that problem for the next series, and I was ordered to return to Moscow. The People's Commissariats and scientific organizations were gradually reassembling after their evacuation.

My family, orphaned after the death of Mama, remained in Bilimbay. It took me two days to get to Moscow flying on military airplanes with stopovers. I did not return to Bilimbay and did not take part in the subsequent BI flight tests. But this is how the events developed there.

Bolkhovitinov freed Isayev from his involvement with the engineering follow-up on the interceptor and made him responsible for the entire propulsion system. Taking advantage of his consultations with Glushko, who worked in the special prison in Kazan, Isayev was to try to make a complete break from the Dushkin-Kostikov engines. Our patron considered this all the more important because Kostikov was preparing proposals for his own aircraft with a liquid-propellant rocket engine and was not interested in transferring a series of reliable engines to us.

Bakhchivandzhi's second flight was not executed until 10 January 1943 in the second model of the BI aircraft. It was winter and the wheels had been replaced with skis. The liquid-propellant rocket engine had been regulated for a thrust of 800 kilograms. The aircraft reached an altitude of 1,100 meters in 63 seconds at a speed of 400 kilometers/hour. The landing on skis went well. Suddenly Bakhchivandzhi was summoned to Moscow to review our rivals' design and mockup—an experimental interceptor with the designation "302." It had been proposed that this interceptor contain two Dushkin engines, plus a ramjet engine to increase range.[1]

The BI's third flight was assigned to Lieutenant Colonel Konstantin Afanasyevich Gruzdev. During this flight the engine had a maximum thrust of 1,100 kilograms and the aircraft reached an altitude of 2,190 meters in 58

1. Unlike the BI, the '302' experimental rocket-aircraft was never successfully flown. The project was eventually cancelled in 1944.

seconds, having reached a speed in excess of 675 kilometers/hour. During the takeoff, the aircraft's left ski tore off. There was no radio on the airplane, and it was not possible to communicate to the pilot that landing posed the threat of an accident. Gruzdev went in for a landing completely unaware of what had happened. Nevertheless, the experienced pilot made contact with the ground so smoothly, that having touched it with only one ski, he understood immediately that there was a problem with the other one. When the aircraft had slowed down it gently turned and rested on its wing.

On 11 and 14 March, having returned from Moscow, Bakhchi executed the fourth and fifth flights. The aircraft could now reach an altitude of 4,000 meters after the engine had run for 80 seconds, with a maximum climb rate of 82 meters/second. During the sixth flight on 21 March, a third model of the BI flew into the air for the first time. The tests were conducted with the aircraft carrying the full battle scale of ammunition at maximum thrust.

The objective of the seventh flight on 27 March was to reach maximum speed in horizontal flight. Fedorov, Bolkhovitinov, Bereznyak, and Isayev were present, as they were for each flight. When we met in Moscow, Isayev told me that everyone expected a world speed record to be set during the flight. The abrupt, steep takeoff with a transition to horizontal flight lasted 78 seconds. The characteristic reddish-brown vapor let them know that the engine had shut down. A small cumulus cloud impaired visibility for a couple seconds. And then something totally inexplicable happened—the aircraft flew out of the cloud with its nose down and plunged to the ground in a steep, steady dive.

I did not witness this flight. According to others' accounts, there was no cloud and the aircraft shifted from horizontal flight at maximum speed into a dive at an angle of around 45 degrees and literally stabbed into the ground 6 kilometers from the airfield. I learned about Bakhchivandzhi's death when I was at the People's Commissariat of the Aircraft Industry. They had immediately received the news from Koltsovo.

For me, 27 March became a black day. It was the date of my mother's death. A year later, to the day, Bakhchivandzhi died. And twenty-five years later on that same day, Gagarin died. I have always considered myself a diehard atheist and materialist who does not believe in any omens or inauspicious dates. But after these coincidences, I have an internal fear with the approach of every 27 March that I will receive bad news.

After that accident, BI flights were discontinued. Gruzdev, the second pilot to have flown in a BI, also died soon thereafter during a routine flight in an Aerocobra fighter received from the Americans.

Experienced test pilots Bakhchivandzhi and Gruzdev were called from the aviation front to the deep rear to perform their testing work. They had survived dozens of air battles, always emerging as victors. Gruzdev's death resulted from the failure of American technology. The Cobra was pretty, but they had failed to perform the proper engineering follow-up required to make it a reliably operating fighter.

Bakhchivandzhi died having executed a "flight into the future." There was much in this "future" ahead of us that was unknown and dangerous.

The special commission that investigated the BI catastrophe was not able to determine its true cause. They estimated that the aircraft had reached a speed close to 800 kilometers/hour. The true air speed was never documented; the flight recorders were destroyed in the crash and there were no precise ground measurements. It is worth remembering that at that time the official world speed record was 709.2 kilometers/hour. The commission determined that the aircraft did not break up in the air. They could only hypothesize that during high speeds in flight new phenomena occur which affect controllability and loads on the controls. Four years later, tests conducted in the new TsAGI wind tunnels confirmed the possibility of an aircraft being dragged into a dive at speeds of around 800–1,000 kilometers/hour.

Bakhchivandzhi was the first Soviet man to take off directly from the Earth's surface using the thrust of a liquid-propellant rocket engine. After the war, when we were in Germany, we discovered that Willy Messerschmitt had built the Me-163 fighter at about that same time with a liquid-propellant rocket engine produced by the Walther company. Only a small number of these airplanes were manufactured, and they did not take part in any air battles. But the Me-163 was not the first. The very first aircraft to take off aided only by a rocket engine was the He-176 aircraft produced by Ernst Heinkel's company.[2]

For a long time, Bakhchivandzhi was remembered only within the narrow circle of those who took part in the events of those distant war years. His test pilot friends have done much to preserve his memory. In 1973, thirty years after his death, Bakhchivandzhi was awarded the title Hero of the Soviet Union. A bronze bust of the pilot was placed at Sverdlovsk Airport—the former Koltsovo military airfield. In 1984, a monument to Bakhchivandzhi was erected in his hometown, the Cossack village of Brinkovskaya on the Kuban River.

The history of rocket technology has linked Bakhchivandzhi and Gagarin. Both men were propelled into the air by a rocket engine. Both died in airplane crashes at the age of thirty-four—on 27 March. In both incidents, separated by twenty-five years, the investigating commissions could not determine the true causes of the pilots' deaths. In the case of Bakhchivandzhi there was no radio. Gagarin and Seregin's airplane had a modern aircraft radio, but the real causes of their death remain unsolved.

I knew both pilots and can affirm that they were extraordinary people. Gagarin was never in combat; he never had the occasion to bring down the enemy in mortal battle. But it fell to his lot to be the Earth's first cosmonaut and then to endure the test of fame, which few pass. Gagarin passed. In our dealings with

---

2. An He-176 aircraft completed the world's first rocket-powered flight in 1939. A Soviet rocket-glider designed by the young Sergey Korolev, the RP-318-1, performed the first Soviet rocket-powered flight in 1940, although it did not take off under its own power. In 1942, Bolkhovitinov's BI rocket-plane became the first Soviet rocket-plane to take off using its own engine.

Bakhchivandzhi, he also never overplayed his role. He did not consider himself a celebrity pilot of higher standing than the aircraft designers. He did not have a trace of conceit. This was also true of Gagarin, in spite of the fact that his worldwide fame gave him license to conduct himself otherwise.

To what extent were the creators of new technology guilty of these catastrophes? The death of test pilots is just as possible as the death of pilots in air battles. But their opponent was the unknown—that very "flight into the future" about which we knew very little. Bakhchi died during the war in an air battle against an enemy that was still unknown to science—transonic speed. Such air battles continued for many years after the final victory on the war fronts. The creators of the new technology were guilty of what they did not understand. This is the law of historical progress.

Chapter 14

Back in Moscow

In the summer of 1942, Moscow lived by the laws of a city near the front lines. A strict curfew was observed. Red Army soldiers patrolled the streets. Passes were affixed to the windshields of cars. Strips of paper crisscrossed the windows of houses; the blackout was compulsory. In the evening, hundreds of barrage balloons floated over the city.[1] The use of ration card coupons was strictly enforced and required for purchases in the stores and dining halls. Our existence was far from luxurious, but none of the dozens of people that I met were going hungry. At any rate, it was much worse in Bilimbay.

The streets, especially in the center of the city, were kept clean. One did not see the trash, rubble, or traces of fires typical for cities subjected to bombing. The main form of crime was the pickpocketing of ration cards, mainly by hungry teenagers. But one never heard of armed attacks for the purpose of seizing ration cards. The 20 November 1941 order declaring a state of siege did the job.

The Germans had been pushed back 150–200 kilometers from Moscow. Air-raid sirens often sounded as darkness fell, but there were no signs of panic. The operation of the Metro and above-ground transport was interrupted only during air raid warnings. Metro stations were used as bomb shelters. During the first month of the war, all private radio receivers were turned over to special government storehouses. To make up for this, wired radio relays operated twenty-four hours a day without interruption. The black dishes were found in every apartment. Powerful loudspeakers hung in the streets and squares. The daily *Sovinformbyuro* reports provided no comfort: the Leningrad blockade continued in the north, the battle of Stalingrad was raging in the south, and Europe was under Hitler's control. But the British were staunchly holding their ground and had successfully repelled air attacks. America was aiding us with airplanes, canned meat, and powdered eggs. We really sensed this, and were comforted that we were not alone in the world. By all objective indicators it was a very difficult time. But surprisingly, in this very

---

1. Barrage balloons were huge balloons raised up low over the city to discourage low flights and precision bombing by attacking aircraft. Each balloon was normally hooked to a wagon with a cable that was strong enough to damage an aircraft upon impact.

difficult situation, our certainty about our ultimate victory and the security of Moscow was simply understood. This mood distinguished the hungry Muscovites of the summer of 1942 from the satisfied Muscovites in October 1941.

On 16 October 1941, all the People's Commissars and senior staff of the government institutions had abandoned Moscow in a terrible hurry. In the summer of 1942 they tried to forget about that. Once again all branches of the defense industry and the restructuring of all non-military industries was being directed from Moscow under the slogan All for the Front, All for Victory. At the People's Commissariat of the Aircraft Industry I could see that all the offices were occupied. There were typewriters chattering in the reception rooms. The hallways were once again filled with clerks, hurrying to deliver reports, and messengers sent from distant factories, gray-faced from the fatigue of travel.

WITHOUT ANY RED TAPE, I was provided with a night pass and ration cards then ordered to go, without delay, to our abandoned factory in Khimki and set up the production of remote-controlled onboard radio stations and suitable control knobs. I was to work out ways to protect the onboard receivers against interference from spark ignition. They disregarded my objections that this was not my area of expertise and that I had flown here specifically with the assignment to come up with a radio control system for the BI interceptor. "A war is going on, every day is precious. Your BI is still to come, but we need radio communications in battle not tomorrow, but today—yesterday even. We already have fighters that are as good as the Germans', but our radio communications are deplorable!"

At the Scientific Institute of Aircraft Equipment (NISO), which was tasked with helping me, I met many old acquaintances who had already returned from a brief evacuation.[2] Among those who worked there were Sergey Losyakov, my old school chum; his chief, Veniamin Smirnov; engineers and avionics specialists Nikolay Chistyakov and Viktor Milshteyn; and radio communications systems developer Yuriy Bykov. They were all subordinate to the extraordinarily dynamic and energetic chief engineer Nikolay Ryazantsev. I received help in the form of advice, drawings, and individual pieces of equipment from allies and captured materials.

By late summer, at the same Factory No. 293 we had abandoned in October 1941, a special equipment department (OSO) was already in operation. Jack-of-all-trades Sokolov and his team of two mechanics and one lathe operator were the production division. Three designers and two drafters drew from early morning until late in the evening, and after my inspection they redid the greater part of their labors. Two radio engineers darted about Moscow and combat airfields trying to equip the laboratory with instruments and models of real radio stations. I considered the five-man "special purpose" team our main accomplishment. This team,

---

2. NISO—*Nauchnyy institut samoletnogo oboroduvaniya.*

which consisted of an engineer, two technicians, and two mechanics with "golden hands," was tasked with making forays to frontline airfields for the practical application and commissioning of our achievements. This did not require long trips. Aviation units taking part directly in the air war were located 30–50 kilometers from Moscow. I went along to one of these airfields near Kalinin so that I could hear with my own ears the effectiveness of our measures. The test was very simple; it involved two squadrons of Yaks.[3] All of the aircraft in one squadron were equipped with ignition-shielded wires, noise filters, and remote-controlled radio stations. The fighters in the other squadron had the same radio stations, but had no noise protection. After both squadrons had taken off, the commander gave heading change instructions from the control post. He transmitted codes, asked how they were understood, and requested a response.

The squadron with the optimized radio communications executed all the commands concisely, answered questions the first time they were asked, and went in for a landing precisely upon command. The second squadron, by contrast, responded, "Did not copy that; say again." Their maneuvers did not always correspond to the commands, and toward the end, the commander took the microphone from the communications officer and let loose the kind of Russian expressions that overcome any radio noise; the squadron returned for its landing.

Bolstered by frontline flight rations, our team worked around the clock after this test to optimize the equipment of the regiment's airplanes. One of the tasks of this aviation unit was to escort and cover the low-flying attack aircraft that were regularly attacking the enemy's nearby rear areas. After returning from a raid with the upgraded equipment, the escort pilots reported, "It's a whole new ballgame! We've even stopped cursing now, since we can understand everything." For the first time they returned without losses.

When all the work was finished, we were officially thanked in front of the ranks. We then returned to our Khimki airfield, having been provided with both "dry" and "wet" rations.

In November 1942, a landing force arrived in Khimki from Bilimbay to prepare for their return from evacuation. The victory at Stalingrad had for the time being overshadowed rumors of failures on other fronts. Design bureaus, NIIs, factories, and people who had fled from Moscow in the panic of October 1941 were trying to return. Under the leadership of Chief Engineer Nikolay Volkov, restoration work was being conducted throughout the factory.

Lyulka's deputy, Eduard Eduardovich Luss, headed one of the first groups returning to Moscow from Bilimbay. My family also returned with this group in February 1943. Worn out more from the return trip in 1943 than she had been by the evacuation in 1941, Katya arrived with Valentin and my father, who had aged terribly. The problem of accommodations became very acute. While alone, I had

3. Yak aircraft were built by the Yakovlev design bureau.

From the author's archives.

Engineers from the special equipment department of Factory No. 293. Standing: third from right B. Chertok—department chief. Moscow. Khimki, 1943.

preferred to live at work using a folding cot. With great difficulty we managed—semi-legally—to settle temporarily in separate rooms on Novoslobodskaya Street. Our four-year-old son, Valentin, was once again transferred to the care of his grandmother in Udelnaya.

Katya became involved in the feverish activity of the "external expeditions" department, which is what we called the team that traveled out to frontline aviation units. Father went to work at our factory as a transport department shipping clerk. He quickly lost his strength, however. He became ill and died on 19 March, eight days before the anniversary of my mother's death.

After the death of Bakhchivandzhi, our patron Bolkhovitinov returned to Khimki from Bilimbay. Having familiarized himself with my activity in the field of combat aviation, he said that the work was useful, but it wasn't what we needed. He repeated the parting request he had made when he sent me off to Moscow. "First, set up work on guiding the interceptor toward the enemy. Second, we must have the most state-of-the-art measurement equipment for the new test-firing stands for the liquid-propellant rocket engine that we will build here soon. As far as the aircraft is concerned, we will redo it. We will increase the engine's thrust. It's time to do away with the high-pressure tanks and return to the turbopump feed engine. Isayev should be getting back in touch with Glushko and use his Kazan experience. RNII has obviously botched its aircraft design. As soon as we get started with our firing tests, Shakhurin is going to visit us and we will discuss the idea of combining RNII with our factory, with new management of course. We will not be under Kostikov!"

Isayev returned to Khimki full of enthusiasm about creating our own production and testing base for liquid-propellant rocket engines. He could not conceal from his friends the secrets revealed to him by the higher-ups in confidential conversations. He and Bereznyak had taken part in Bolkhovitinov's meeting with Shakhurin. In Isayev's words, the patron assured him that the fate of RNII had been decided beforehand. The institute would switch to aviation, and it would have new management—from aviation. Shakhurin frequently met with Stalin and would select a suitable moment to raise the issue. It was almost certain that he would receive Stalin's approval for the reorganization of RNII–NII-3.

INSPIRED BY SUCH A PROSPECT, I once again began to work with NISO. Smirnov and Losyakov drove me to the Central Airfield. A Pe-2 aircraft shrouded by a tarpaulin was standing at the edge of the flight field where the United State Political Directorate (OGPU) separate special purpose division (ODON) camps had once been located.[4] When, after much effort, we managed to find a crew, dismiss the guard, and uncover the aircraft, I saw that it was covered with the most random assortment of antennas sticking up all over it. Having sworn me to secrecy in my dealings with radio specialists from other organizations, they told me that they were testing a decimeter-range onboard radar developed at NISO based on the ideas of Gerts Aronovich Levin.

Meanwhile, there were two problems. The first was weight. If the radar was installed on the aircraft, the ammunition reserves would have to be reduced by 500 kilograms. The second problem was the "key effect." When the radar was activated, if someone came near the aircraft with a bunch of keys in his pocket, then the displays started to flicker!

"Well, that's good. It means your radar is highly sensitive," I hastily complimented the inventors.

"It's too sensitive," explained Smirnov. "The thing is, if you put your hand in your pocket and start to jingle your keys, then it's impossible to understand what's going on with the displays. That's the 'key effect'!"

After spending the entire day on my boyhood home turf of Khodynka, I realized that this development would not yield a system for our BI in the foreseeable future. Sensing my pessimism, Sergey Losyakov cheered me up by saying that he

---

4. The OGPU—*Obyedinenoye Gosudarstvennoye Politicheskoye Upravleniye* (United State Political Directorate)—existed between 1924-34, and then mutated into several intermediate agencies before becoming the KGB in 1954.

had recently met with our mutual classmate Abo Kadyshevich. Abo had graduated from the physics department at Moscow State University and had been kept on by the department. He had some interesting thoughts and designs.

I tracked down Kadyshevich—who in his time had been the chess champion of our School No. 70—to get his take on the radar problems. He contributed some very original thoughts. "You don't need to put a heavy, complex radar set onboard the aircraft. You need to shift the guidance tasks to the ground." He had studied the American SCR-584 radar sets, which we had recently received for air defense through the lend-lease program.[5] He considered this an excellent set for gun laying. The Americans had passed us by, and had probably passed the Germans as well. If this radar set were optimized, it could simultaneously track an interceptor and a target. The set could also be used to guide a fighter to a landing airfield after an attack.

Kadyshevich was a talented physicist. From the standpoint of pure physics, he suggested that the weight of the aircraft equipment for such a setup would be no more than 10 kilograms.

"This is not fantasy. My friend, radio engineer Roman Popov, is already working on real layouts and equipment."

A month later, Roman Popov and Abo Kadyshevich were working at Factory No. 293 in the OSO department. A special radio laboratory had been created for them within the OSO structure. We started to prepare lists for Bolkhovitinov of radio engineers that were to be sent to us to develop the aircraft coordinate radio locator, or ROKS, as we called our new idea.

The idea interested many radio specialists. Even the all-powerful (at that time) Aksel Ivanovich Berg, who had heard about our idea, came to Khimki to meet with Bolkhovitinov.[6] He was interested not so much in the radio engineering as in the aircraft that was the reason we had cooked up all this soup. He was a very decisive individual, a lover of new ideas, who without hesitation supported us. With his help we received the American SCR station. We had some trouble accommodating the military personnel that accompanied the station. Three officers and five Red Army soldiers had to be housed and fed according to army frontline standards. All of the organizational problems were quickly straightened out. Once he had received five specialists, Roman Popov, who proved to be a talented engineer and good organizer, swore that it would take no longer than six months before he would be able demonstrate the new idea. If we did not have a rocket-powered airplane by that time, we could test the principles using a conventional fighter.

---

5. The Lend-Lease Act, passed by U.S. Congress in March 1941, allowed the U.S. government to sell, exchange, and provide equipment to any country fighting against the Axis powers. The Soviet Union was one of the principal beneficiaries of the Lend-Lease Act.

6. Academician Aksel Ivanovich Berg (1893–1979), later a deputy minister of defense, was the most influential radar expert in the Soviet Union.

AN ORDER WAS ISSUED to convert NII-3 into NII-1, and soon afterward I was "removed" from Khimki and transferred to Likhobory, to the post of avionics department chief. It was a shame to part with my cozy special equipment department at Factory No. 293 and the work at combat airfields. In Khimki, I left behind the branch under Roman Popov developing the ROKS system and the group expanded by Isayev that was servicing measurement equipment and electrical ignition systems.[7]

In 1944, Pobedonostsev helped me establish good relationships with the NII-3 staff. He first introduced me to Tikhonravov, Artemyev, and to the few original creators of the *Katyusha* still remaining at the institute. They opened up for me certain pages of the pre-history of our rocket technology. But the names of Korolev and Glushko were still not mentioned at NII-1 during this time.

At NII-1, during my first year of activity, I successfully combined the avionics teams of Factory No. 293 and NII-3. My new department received two men who are now firmly ensconced in the history of our great rocket technology and cosmonautics. Nikolay Pilyugin, who had worked on instruments for autopilots, transferred from LII to NII-1 immediately after it was formed.[8] Leonid Voskresenskiy from the Nitrogen Institute transferred to NII-3 even before it was reorganized. At the beginning of the war, he developed anti-tank incendiary grenades at the Nitrogen Institute. Upon coming to NII-3, Voskresenskiy began work on pneumohydraulic systems for the liquid-propellant rocket engine.[9]

One of the first creative achievements of the new department was developing a unified avionics system for liquid-propellant rocket engines for aircraft. After flying off to Germany in April 1945, I found out that Voskresenskiy and I had been awarded combat Orders of the Red Star for this work.

The medals were presented to us in the Kremlin only after our return from Germany in 1947. Immediately after receiving them, we headed to Tverskaya Street—in those days Gorky Street—to the Aragvi restaurant. Voskresenskiy went up to the director and asked whether he knew Comrade Chertok, who had just been at the Kremlin. The restaurant director was very sorry that he did not know him but was happy to make his acquaintance. After that introduction, we were afforded the honor of having a banquet set up for us in a separate dining hall. I also recall this evening because it seems it was the last time that all of us who had just returned from Germany—Korolev, Pobedonostsev, Tikhonravov, Pilyugin,

---

7. In February 1944, the former NII-3 (at Likhobory) was completely reorganized. The institute was briefly known as NIRA before becoming NII-1. As part of this reorganization, Bolkhovitinov's design bureau, the OKB-293 (at Khimki) where Chertok worked, was attached to NII-1 as its Branch No. 1. This 1944 reorganization of NII-3 into NII-1 focused the resources of the Soviet aviation industry toward the development of jet and rocket engines.

8. Nikolay Alekseyevich Pilyugin (1908-1982) later became an Academician and chief designer of Soviet missile guidance systems.

9. Leonid Aleksandrovich Voskresenskiy (1913-1965) later became deputy chief designer under Sergey Korolev at OKB-1.

Voskresenskiy, Mishin, and I—were in military uniform. We had all come from NII-1, where in July 1944 a real rocket euphoria had begun after information about the Germans' work on long-range ballistic missiles was received.

SINCE ABOUT MID-SUMMER 1944, Pobedonostsev, Tikhonravov, Pilyugin, Mishin, and I, plus five other specialists in various fields attached to us, devoted much time to studying captured German technology. From fragments brought from the German firing range in Poland, and based on prisoner of war testimony and intelligence materials, we reconstructed the appearance of the rocket that was to play an important role in the initial history of rocket technology.

Germany's secret "vengeance weapon," the V-2, inflicted a very damaging blow against our institute in Moscow before the Germans first began launching them against London. General Fedorov wanted to take part personally in the search for remains of the German technology in Poland. The airplane that was flying him to Poland crashed near Kiev. Twelve people died along with Fedorov.[10] Among them was Roman Popov. Popov's death practically killed our ROKS development. Ten years later, I found that the ideas Popov was developing then anticipated to a certain extent the principles used to create the Moscow air defense radar systems in the 1950s. The chief conceptualizer of these systems was Aleksandr Raspletin.[11]

My activity at NII-1 virtually ended on 23 April 1945 when Bolkhovitinov sent me "briefly" to Germany as part of General Petrov's "special purpose commission." I did not return to NII-1.

Thus ended nearly a decade of work, with brief interruptions, in Viktor Fedorovich Bolkhovitinov's collective. Returning to those times, I invariably recall Bolkhovitinov, whom we always referred to as "patron", with a mixed sense of vexation and gratitude. I recall with gratitude that he instilled a camaraderie among his creative team, regardless of their ranks. Chief Designer Bolkhovitinov was never an overbearing, threatening chief. He was our elder comrade, whom the authorities had vested with the necessary rights and duties for those times. We did not fear him—we simply loved him. The sense of vexation is associated with Fate's obvious unfairness toward scientist-designers such as Bolkhovitinov. In order to be a full-fledged "chief" or "general," in addition to having a certain amount of class, one had to have more of what is commonly called "exceptional organizational capabilities."

Bolkhovitinov loved talented people, and in contrast with others, was not afraid that his pupils were capable of surpassing their teacher and eclipsing his glory. Isayev, Bereznyak, Mishin, and Pilyugin—Bolkhovitinov's former compatriots and

10. Petr Ivanovich Fedorov (1898–1945) was Director of the reorganized NII-1 in 1944–45. He was killed with twelve others in a plane crash on 7 February 1945.

11. Academician Aleksandr Andreyevich Raspletin (1908–67) was the Chief/General Designer of the famous KB-1 from 1955–1967 and oversaw the development of the first Soviet air defense systems.

subordinates—became "chiefs" themselves. To a great extent they owe their successes to the Bolkhovitinov school.

One of the qualities that was cultivated in that school was the ability to ponder—and during work's daily hustle and bustle—to synthesize individual facts and events, striving to understand the forces that drove the development of science and technology. However, none of my friends from the war years could imagine the scientific revolution that was about to take place with our direct participation. Although the most unusual ideas were encouraged in our team, the fantastic idea of creating a combat missile capable of flying across the ocean never occurred to anyone.

Chapter 15

Moscow—Poznan—Berlin

In early April 1945, military actions raged across a significant portion of Germany. Soviet troops were on a forced march from the east and the allied troops were advancing from the west.

In the central sector of the Soviet-German front, the troops of the First Byelorussian Army Group under the command of Marshal G. K. Zhukov were conducting combat operations on the left bank of the Oder. The primary concentration of this army group was already located 60–70 kilometers from Berlin. The troops of the Second Byelorussian Army Group under the command of Marshal K. K. Rokossovskiy were delivering the main thrust from the area of Stettin (Szczecin) with a subsequent breakout to the shores of Pomeranian Bay on the Baltic Sea.[1]

The atmosphere at NII-1 of the People's Commissariat of the Aircraft Industry in Likhobory where I worked was extremely exciting. It was not just the euphoria of imminent victory that had seized the entire population and the sensation of stepping out into the bright light after four agonizing years of war. We now had our own special interests in Germany.

By studying materials that had been recovered from the German rocket firing range in Poland near Debica, as well as intelligence data, the scant reports from our British allies, and testimonies and accounts provided by a few prisoners of war who had information, we had been able in general terms to get an idea of the scope of the work that the Germans had conducted on a new type of weapon—the long-range guided missile.

Subsequent events showed that we had been very close to putting together a fundamentally correct description of the V-1 and V-2 "vengeance weapons." It was evident that neither we nor our allies had developed similar weapons, either in terms of the parameters achieved or the production scale. We were extremely interested in the problems of guidance technology, the instrument design, the actual parameters and control systems for powerful rocket engines, and the role of radio-control. All of our specialists—the rocket specialists, engine specialists, and

1. Szczecin is located in northwest Poland where the Oder river meets the Baltic Sea.

guidance specialists—who had only reached the initial stages of rocket technology in developing the *Katyusha*, the BI-1 rocket-powered airplane, and the small rockets of the RNII were stirred by the questions: What was their experimental base? and, How did they manage to create such a powerful liquid-propellant rocket engine?

We already knew for certain that the primary German center for the development of rocket weaponry was located on the Baltic Sea coast, on Usedom Island. Rokossovskiy's armies were rushing there. But we had to get there before the possible destruction of this center by our "fellow Slavs," who had no clue how valuable these instruments, benches, laboratories, and papers could be to specialists. Indeed, all of this could simply be blown up, burned, or destroyed if the Germans themselves did not destroy everything before the Red Army arrived.

On this matter I appealed more than once to our immediate supervisor, Professor and General Viktor Fedorovich Bolkhovitinov. Aleksey Mikhaylovich Isayev approached him with these same questions. We all attacked the new NII-1 chief, Air Force Lieutenant General Yakov Lvovich Bibikov.[2] He was a sufficiently competent engineer to understand how important a mission it was to obtain the captured material, not only in the form of machine tools, which our production workers cared about most of all, but also in the form of intellectual output. However, somewhere in the multi-runged hierarchy of the bureaucratic ladder of our People's Commissariat of the Aircraft Industry, the military intelligence of the People's Commissariat of Defense, the State Committee of Defense (GKO), military headquarters, and God only knows where else, something didn't work.

The troops of the Western allies had already forced their way across the Rhine; they had routed the enemy's grouping in the Ruhr Valley and broken out to the Elbe. Who knew—if the Germans showed them no resistance and threw all their forces only at the Eastern front, then perhaps all the captured rocket materials, including those at Peenemünde and the laboratories and factories in Berlin, would end up in the hands of the Americans and Brits. But at our People's Commissariat, and among aircraft specialists in general, there was no particular interest in unmanned guided missiles or "guided projectiles", as artillerymen called them. But jet aircraft—that was a different matter! Our pilots had already encountered the Me-262 with two turbojet engines in air battles. This aircraft was considered a real technological achievement that we needed to capture immediately and thoroughly study, especially with regard to the JUMO engine. After A. S. Yakovlev's article in *Pravda*, which caused a lot of stir and described German work in the field of jet aviation as the death throes of Nazi engineering thought, there had been a sobering process. Being Deputy People's Commissar of the Aviation Industry and a man

2. Bibikov had been appointed to head NII-1 after the previous director, Petr Ivanovich Fedorov, died in a plane crash in February 1945.

close to Stalin, Yakovlev wanted this article to answer the question, "Why don't we have such engines and aircraft?," especially since he was clearly hostile toward both our work on the BI and Arkhip Mikhaylovich Lyulka's work on the first domestic version of the turbojet engine.

We needed to quickly find a roundabout way to get to the front and be the first to seize the intellectual war spoils of rocket technology. We understood that the future of our program depended on what we would see, find, and then be able to test back home. Using our connections at the "friendly" institutes of our aircraft agency, I decided to act without waiting for a resolution on the matter of the affiliation and departmental jurisdiction of the "guided projectiles."[3]

I had close ties with the Scientific Institute of Aircraft Equipment (NISO) from my work during the previous war years. My old school chum Sergey Nikolayevich Losyakov, future professor and prominent radio receiver specialist, worked there. I was closely acquainted with the leading engineers—all talented and extremely likeable—in the new fields of aviation radar, radio communications, remote measurements, and avionics: Veniamin Ivanovich Smirnov, Nikolay Iosifovich Chistyakov, Viktor Naumovich Milshteyn, and Yuriy Sergeyevich Bykov. They would all become professors in charge of departments. Alas, many of them are no longer with us. During those war years I was very indebted to them for fresh technical ideas, engineering optimism, and moral support in the face of the many difficult technical problems.

In 1944, the NISO director was Gerts Aronovich Levin, one of the country's leading radio specialists and a pioneer of radio communications theory. His scientific authority was indisputable. But his ethnicity clearly did not suit one of the high-ranking directors, and he was therefore replaced by Air Force General Nikolay Ivanovich Petrov.[4] The general was well received both by People's Commissar of the Aircraft Industry Shakhurin and by Air Force Directorate Chief Novikov.

With the help of my friends, it was not particularly difficult to explain to General Petrov how important it was to be the first to seize the captured materials, to prevent them from being trampled by the advancing armies or split up and hauled away to the quarters of various agencies. Experienced with such problems, he immediately grasped how vital it was not to pass up this chance, even at the price of a certain risk.

And so, on 16 or 17 April, Bibikov and Bolkhovitinov summoned me and announced that I would be part of NISO chief General Petrov's group that by GKO decision was being granted authority to inspect, study, and when necessary seize German aircraft radar and instrument prototypes and materials.

---

3. The Soviet government was in the midst of a debate over assigning the development of long-range guided missiles to one among several competing ministries (or people's commissariats).

4. Levin was Jewish.

This group consisted of eight or ten individuals, including Smirnov and Chistyakov. We formed a troika that was given the special assignment of studying German aircraft instruments, autopilots, special equipment, weaponry, radar, radio navigation, and communications. The range of issues was very broad, but it was exceptionally interesting for each of us.

On 20 April, I was invited to the Sokolniki regional military commissariat. Here, after checking my name on a secret list, the officer explained that I should immediately obtain a uniform, including shoulder boards with two bars and one major's star. In the record of service I was listed as "untrained private" and now suddenly I was already a major! "We're dressing a lot of you civilians now, and right off the bat you get high ranks," was the somewhat pained remark of the colonel, who was decorated with medals and stripes for the wounds he had received.

Without any red tape, I received a full officer's uniform, including an officer's belt, map case, an excellent overcoat, and a TT pistol with two clips. Our light industry was still functioning surprisingly well for those difficult times! Forty years later my officer's belt is a special treasure for my grandson. And the cutoff skirt of my overcoat still serves to keep my automobile engine warm when it's parked in the winter.

In the early morning of 23 April, our group took off from the M.V. Frunze Central Airfield, the same old Khodynka that I had known so well since my childhood and subsequent work there. We departed on a Douglas C-47 cargo plane. At that time it was the most popular transport plane. We were flying to the First Byelorussian Army Group! Our movement orders stated: "To perform a special assignment for the GKO."

An hour later we became distracted from our conversations and thoughts about the mission ahead of us and were pressed up against the windows. Soon we were over Minsk. From an altitude of about 3,000 meters, the interiors of the houses were quite visible—almost all of the homes were without roofs. Viewed from above in an airplane, destroyed cities look entirely different than from the ground when you are in the midst of the rubble. As surprising as it is, the view from above has a much greater impact and is far more depressing. Perhaps this is because from up high you can immediately grasp the scale of the catastrophe—the destruction of a large city.

Two hours later we were over Warsaw—a picture even more horrible than Minsk. Perhaps it was because of the black soot at the sites where fires had raged. Beyond Warsaw we saw the intricate patterns of tank tracks on the untilled fields.

We landed in Poznan to refuel. Here at the airfield we observed a meeting of the Polish government delegation that had flown in from London. After Poznan we did not tear ourselves away from the windows. Fortunately, the weather was excellent. We passed over forests, farms, and white village cottages with red tile roofs. Surprisingly, in the big cities almost all of the houses were roofless, while from above, the villages and various farms seemed untouched. And if columns of all sorts of vehicles had not been crawling over the light-colored roads, if there had not been a thick network of tank tracks on the ground, you would not have realized that a moving artillery barrage had just rolled through here, one of the last operations of World War II.

THE ASSAULT ON BERLIN was in full swing when we crossed the German border. Before that trip, I had flown a great deal from Moscow over various routes, and I have flown even more since then. But it is difficult, especially now after so many years, to describe the feeling that I experienced during that flight. For me, in terms of emotion, this flight was unique.

I was thirty-three years old—about the same age as everyone on that flight, except for General Petrov, who was ten years our senior. We flew out of Moscow, where I had lived since I was two years old, and where my father had quite recently died of dystrophy. We flew over Smolensk. Somewhere down there my older cousin Misha, the darling of the entire family, had been killed. He had worked for People's Commissar Tevosyan and therefore had a deferment. He also had two sons, and had still volunteered. He was wounded, and after he was released from the hospital, he returned to the front and died near Smolensk. We flew over Poland where I was born. I didn't feel any particular closeness to Poland although I knew from my parents' stories about my escapades there up through the age of two. But somewhere here below us, in Auschwitz or the Warsaw ghetto, my other cousin, Solomon Zlatin, and his entire family had been annihilated. He had left Moscow for Poland to join his mother, my aunt Fruma Borisovna, back in 1921, immediately after the end of the war with Poland. The last time I saw them was on the platform of the Belorusskiy (at that time it was still the Aleksandrovskiy) train station, when my parents took me along to see them off. Fruma Borisovna, my mother's older sister, had helped my parents a great deal when they had settled in Lódz and were enduring financial difficulties after wandering around Europe as emigrants. Then her only son, Solomon; his beautiful wife, Lyuba; and their five-year-old son, Yasha, returned home. Yes, and I was also linked to Poland by Sigismund Levanevskiy, to whose tragic flight I devoted more than a year of intense labor. And how much more effort was spent preparing the expeditions to search for the airplane that had become known worldwide by its number, N-209.

And now we were flying over Germany, who had shattered all of our pre-war plans, hopes, and way of thinking.

Later on, I directly participated in what would be historical events for humanity, events that changed the history of the world: the launch of the first artificial satellite; the first launch of a human being, Yuriy Gagarin, into space; the creation and launch of the first spacecraft to reach the surface of the moon; and the creation of the first intercontinental missiles capable of carrying a warhead with a yield of tens of megatons to America. But never have I had such a feeling of oneness with human history than on the day of that flight at the end of the war. Perhaps it was because I had been burdened with enormous worries and responsibilities during the previous and subsequent events of historical significance. I always had to be

From the author's archives.

Boris Ye. Chertok, photographed in Germany, 1945.

taking care of business, keeping track of something, taking readings, mulling over results as they occurred, and deciding what to do in the next few minutes, hours, and days. Here in the airplane flying to Germany on 23 April 1945, neither my fellow travelers nor I were troubled by anything concrete or urgent. We understood that we would encounter something completely new, unusual, and unprecedented even after four years of war. But it would be somewhere over there, in a different country, where we were flying without any visas or passports, by right of victory.

Toward evening we landed at a temporary airfield near Strausberg. Once we had jumped out of our airplane there, the atmosphere of a combat airfield at war immediately surrounded us. Low-flying Il-2 attack aircraft were continuously taking off and landing. I had previously visited combat airfields and had observed air battles more than once during the war, but this was the first time I had seen such an assembly line process: takeoffs, landings, refueling, the racking of bombs and rockets—and all of this done quickly, continuously, in a very business-like manner. We saw no fighter cover in the air. They were all "on business" over Berlin.

While the general was tracking down the officers in charge of the army group's logistical unit, we were mingling with the crews and asking them about their encounters with the new "Messers."

Night fell. In the west, the glow of fires was bright and the continuous rumble of the war, muffled by the distance, was approaching—or so it seemed. Finally, General Petrov appeared. He distributed us among three jeeps, and we set out for Strausberg. We reached the officer's mess hall first. There we found unusual cleanliness, bright light, and waitresses in snow-white aprons and hats.

Where were we? Could mortal combat really be going on just twenty kilometers from here? They fed us a tasty meal according to the standards for frontline army officers. The sergeant major then showed us our accommodations. Chistyakov, Smirnov, and I asked to be together. He took us to a two-story cottage and said that beds for three had just been prepared on the second floor.

And so the three Soviet engineer majors climbed to the second floor of a German house whose owners had fled. We were immediately overwhelmed by the comfort. It was hardly necessary for the rear service to care particularly about providing beds and toilets for officers. The entire household was unharmed, and as we determined, "top notch." We cast lots. Chistyakov and Smirnov got the bedroom with the wide double bed and adjoining bathroom and toilet. For my part, I got the study with a sofa made up with crisp, clean sheets. In the study, a portrait of Hitler still hung on the wall, and on the desk was a photograph of an officer of undeterminable rank with his wife clinging to him.

Before going to bed, wanting to jot down my first impressions, I sat at the writing desk and pulled open the top drawer. Good grief! The former proprietor of the study had left his Iron Cross (German medal for meritorious combat service) and Walther pistol with a full ammunition clip in the drawer. With these spoils in hand, I immediately paid a visit to my comrades, who were wallowing in their snow-white feather beds. For them it was quite unusual to use feather beds instead of ordinary blankets.

"Give the pistol to our general," advised Smirnov, "and hand over the medal to Smersh."[5]

In the morning, the general briefed us about the plan of action. Our first task was to perform a thorough inspection of "DVL," the German *Luftwaffe* research center in Adlershof.[6] But Adlershof had not yet been taken. In the meantime we had to find maps, and we conversed with the army group's intelligence representatives.

The general was pleased with my gift. The Walther pistol was much lighter than our TT. The Smersh representative showed no enthusiasm when he took the German medal. "We have quite enough of these," he commented.

The army group's intelligence service officers did not brief us so much as try to understand what we were interested in and what they needed to look for. They were competent combat officers, but clearly were not in the habit of dealing with scientific-technical intelligence matters. We were one of the first official echelons within the military for scientific-technical intelligence, which previously had been completely uncoordinated within the combat armies. Subsequently, many detachments of civilian specialists were sent to the occupied zones of Germany and called "trade union" or "civilian" officers.

The Smersh military intelligence representative posed the following question to us at a meeting: "The Germans have issued leaflets that say that we will not take Berlin, and that we will receive such a blow that there won't be any bones to gather. The Führer has a secret weapon in store so that the Red Army will be completely annihilated on German soil. What could that weapon be?"

Indeed, what could it be? If it were the V-2, then no matter how many of them Hitler had "in store," this weapon would not bother the Red Army. This much was clear to us. Chemical weapons? using them in any form on German soil would now be more dangerous to the Germans than to us.

We decided that this was pure propaganda. And it turned out we were right. In the United States, Germany, and the USSR, a new, top-secret weapon really was

---

5. The organization *SMERSH—Smert Shpionam* (Death to Spies)—was the armed forces counter-intelligence agency whose primary task from 1943 to 1946 was to uncover spies and saboteurs in the military, screen all liberated Soviet POWs, and protect military factories

6. DVL—*Deutsche Versuchsanstalt für Luftfahrt* (German Aviation Research Institute).

being developed—an atomic weapon. But even we, who had access to top-secret materials, knew virtually nothing about it until 6 August 1945, when the atomic bomb was dropped on Hiroshima.

Back then, we did not know that just a stone's throw away, a group of specialists from Kurchatov's team was already preparing to search for German atomic secrets. This team had the highest authority, for at that time the main chief of our atomic operations was Lavrentiy Beriya himself, and other related special committees were being directed by powerful organizers such as Vannikov and Malyshev. We did not know that in addition to the allied armies heading toward our troops from the west, there were also special missions being sent out to seize German rocket technology and rocket specialists, search for nuclear physicists, and seize everything that had been done in Germany on the new scientific frontiers—first and foremost in the fields of guided missiles, nuclear fission, and radar.

We received "directives" and instructions thought up by God knows who, such as: "While inspecting German factories and laboratories, don't get carried away with intellectual achievements, but first and foremost compile a list of the types and number of machine tools, industrial engineering equipment, and instruments. As far as documentation and specialists are concerned, that is a matter of our judgement, and initiative will not be prohibited."

FROM 24 THROUGH 26 APRIL, the troops of the First Byelorussian Army Group were breaking through the perimeter of the Berlin defensive region, practically uniting with the troops of the First Ukrainian Army Group; they had surrounded the entire Berlin grouping. On 25 April, we heard about the meeting of Soviet and American troops on the Elbe in the vicinity of Torgau.

We intensively studied maps and routes for two days, and gathered the addresses of factories and companies of interest to us in the greater Berlin area. Finally, on 28 April we began our sorties along the roads to Adlershof in Berlin.

In those days, the roads in Germany east of Berlin were overflowing with two streams coming from opposite directions. Heading westward, toward Berlin, were Studebakers filled with soldiers and goods, all sorts of trucks, and columns of tired, but joyous and animated soldiers on foot. "To Berlin" was emblazoned on tanks, trucks, and artillery pieces in all colors of paint. Eastward bound were trucks and horse-drawn vans with red crosses evacuating the wounded to the nearest hospitals. Along the road were many medical battalions and army hospitals.

Crowds of the liberated of all nationalities streamed in disarray toward the oncoming army ranks. We received many shouts of greeting. French, Czechs, and Belgians stood out in particular when they saw our jeeps carrying Soviet officers. A woman with a French flag jumped down from a carriage being pulled by three young lads and practically threw herself under the wheels of our car just to shake our hands, and overflowing with the joy of freedom, she shouted, "Vive la russe!"

Our "herdsmen" were driving pedigreed black and white cows straight down the roads. How were they going to get across Poland? There were many horse-drawn carriages carrying luggage of all kinds. Each amateur conveyance was traveling under its national flag. Sullenly, slowly, and silently, gray-green columns of prisoners of war plodded toward the east. Blankets, brushes, and briefcases on ropes hung all around them, and sometimes a loaf of bread dangled in a string bag. We were surprised that there were only four or five of our soldiers guarding several hundred German prisoners who had just come from battle.

Adlershof, located in southeast Berlin, is part of Köpenick. We remembered from magazine articles that this was an industrial workers' area where German communists and "Red Front" military detachments had been prominent. The streets and houses had remained intact. Everything appeared to be quite homey. But many buildings already housed Soviet military institutions: "General Petrov's unit"—arrows pointing to the military commandant. Troop units also took up positions without any identifying marks. There were directional arrows on posts that read "Berlin Center," "Buchow," "to Kostrzyn," "to Frankfurt." These were used both for advancing and evacuating to the rear. Posted in the intersections were our female traffic controllers, who were surprisingly attractive during those spring days, wearing the shoulder boards of lieutenants junior grade, white gloves, and beautifully fitting uniforms. They were waving traffic on or stopping it, smiling, and giving directions. As we soon found out, they even knew German.

There were so many vehicles—lots of Studebakers and Dodges, damaged tanks and self-propelled guns. Here the Germans had already lived through the worst. There wasn't any firing in their area; bombs and shells weren't exploding. The smoke and thunder were elsewhere, closer to the city center. German women had gathered in groups around soldiers, having grown bolder, and were bombarding them with questions.

Our placards alternated with the German ones: "A beaten German is a bootlicker and a scoundrel. Don't trust him, soldier," "Kapitulieren? Niemals!" (Capitulate? Never!), "Red Army soldiers do not wage war with the civilian population—this degrades the warrior's honor," "Berlin bleibt deutsch!" (Berlin will remain German).

An excerpt from Stalin's orders: "History shows that Hitlers come and go, but states and peoples remain."

Two days later, we were moved out of the Strausberg area, which was jam-packed with rear services and frontline aviation, to a location close to the surprisingly comfortable Buchow. With five officers of our "spoils" team, we settled into what was, by our Soviet standards, a quite magnificent villa.

In spite of its proximity to Berlin—Adlershof was twenty-five kilometers away—Buchow was completely preserved. This was yet another inexplicable phenomenon of the war. When the Germans went to Moscow in 1941, all of the surrounding towns and villages in their path were destroyed. The well-tended appearance of this idyllic resort did not match our programmed conceptions about war and about the condition of the Nazi "lair" after four years of war. Here were comfortable hotels and pristine, though mostly empty, stores and cafes—and all of this on the slopes of wooded hills on the shore of a lake. It was particularly difficult for us to understand why, three days after German units had withdrawn and Red Army tank units had entered the town without a fight, Germans had appeared and were cleaning up and washing the streets! Lilacs were blooming in the gardens of the numerous villas and the returning residents were tending their lawns and rose bushes right where the Soviet officers were staying.

Later we would see many such "oases" throughout the entire territory of Germany, and significantly more of these than destroyed cities. Sometimes it seemed that all of Germany was a chain of excellent autobahns connecting well-scrubbed, very comfortable towns and villages, or "Dörfer." But even in the villages, almost every house had a sewer line, hot and cold water, and electric stoves in the kitchen.

A couple of months later, we had almost forgotten that in Moscow each of us lived in a communal apartment with one tiny bathroom for eight to ten people, a wood stove in the kitchen, no bathtub, and a single sink with cold water for everyone. And here, among these "Fritzes," only workers from the east, prisoners of war, and convicts in correction camps understood what barracks were!

FINALLY WE BEGAN our thorough inspection of the buildings of Adlershof. The fighting here had not been very intense—all of the buildings were intact. In the entryway beyond the spacious bicycle stand, on large wall panels, were keys, keys, and more keys—all numbered with German attention to detail. Everything was intact!

DIARY ENTRY. 29–30 April 1945.
We are inspecting DVL. Administrative building. Archives, papers, personal documents—in safes. How do you open a safe? The sergeant and soldier detailed to us from the BAO already have experience.[7] The soldier holds a large chisel against the safe doors. The sergeant, a "non-combatant" well over forty, delivers a precise, powerful blow with a heavy sledgehammer. Usually it opens the first time. Sometimes if the safe is especially "difficult," it takes three or four blows. The safes are full of reports with a red stripe, "Geheim!" (Secret) or "Streng Geheim!" (Top Secret). We leaf through the pages—reports, reports about all kinds of tests.

"DVL"—this is, after all, the equivalent of our TsAGI, LII, and Air Force NII all rolled into one. There is neither the time, nor the physical capability to read and study them.

The general has ordered us to list everything, load it into boxes, and send it by plane to Moscow. But where are we going to get as many boxes as we

7. BAO—*Batalyon Aerodromnogo Obsluzhivaniya* (Aerodrome Maintenance Battalion).

need? It turns out that the rear services and BAO do have them, and can organize everything! But there isn't even time to make a list of the reports.

Now let me confess: I sinned. I concealed one report and have kept it to this day. This was the work of Dr. Magnus on the damping gyroscope—an angular velocity sensor. But I'll talk about that later. Now I will continue to cite from my diary:

> Laboratory building. The aeronavigation laboratory is filled with benches for testing onboard instruments. The photochemical laboratory, the laboratory where materials are tested for strength and fatigue, vibration benches. A bombing and firing sights laboratory, accelerometer calibration units. And what magnificent drafting and designing equipment! I am envious of the German designers' workstations. Aside from the nice Kuhlman drafting unit, the swivel chair, and comfortable desk with lots of drawers, it is full of details, and everything has its place. Oh, this German love for details and this exactness, which has engrained such top-notch work into the culture.
>
> The thing that every laboratory needs the most and that is in the shortest supply is the Siemens four-mirror oscillograph. There we found various models: two-, four-, and six-mirror models. Without them, conducting research on rapidly occurring dynamic processes is impossible. This is a new epoch in the technology of measurements and engineering research. In Moscow, at NII-1 we had only one six-mirror oscillograph for the entire institute. And these Germans had so many! No, we no longer felt the hatred or the thirst for vengeance that had boiled in each of us earlier. Now it was even a pity to break open these high quality steel laboratory doors and to entrust these diligent but not very careful soldiers with packing priceless precision instruments into boxes.
>
> But faster, faster—all of Berlin is waiting for us! I am stepping over the dead body of a young German *panzerfaust* operator that has not yet been cleared away.[8] I am on my way with my detachment from the BAO to open the next safe.
>
> My middle-aged sergeant has lagged behind. It turns out that he has pulled the body of the German soldier out of the passageway, laid him on his back next to the wall, and placed his unused *panzerfaust* beside him.
>
> "He's just a kid," he said as if trying to justify himself, "like my youngest. If he had stayed home with his mama, he would still be alive."
>
> "And do you still have an elder son?" I asked.
>
> "I had one, comrade major. A fighter pilot. He died in 1943 in the Battle of Kursk."

8. A *panzerfaust* ("armor fist") was a German recoilless anti-tank rocket launcher.

With no further discussion, we sat down on the boxes, took off our garrison caps, and lit up a cigarette.

The electric instrument laboratory was fantastic! There were so many unique (for us) instruments of all types and ranges from the world-renowned German firms Siemens, Siemens und Halske, Rohde & Schwarz, and the Dutch firms Philips, Hartmann-Braun, and Lorentz! And again—photographic enlargers, slide projectors, movie projectors, chemicals, bulky stationary cameras, cine-theodolites, phototheodolites, and optics of incomprehensible purpose…

We christened a separate building the electrophysics building because of its contents. Electronic low- and high-frequency frequency meters, wave meters, precision noise meters, octave filters, harmonic analyzers, nonlinear distortion factor meters, motor generators and dynamotors for various voltages, even the scarce cathode ray (now called electronic) oscillographs. The richest building of all was the one containing radio and acoustical measurement equipment.

We are writing the addresses of our firms on the boxes: "P.O. Box" so-and-so.[9] But what will really happen to this stuff? Who will meet the airplanes in Moscow?"

After much time had passed, I indeed never found a single report from that mass of secret and top-secret reports that I had sent from Adlershof. They were dispersed over LII, TsAGI, NISO, and other aircraft industry institutions. Only about one-tenth of the instruments that we had sent ended up at NII-1, provoking a justifiable reaction from my immediate chief. At NII-1 in Likhobory, they began preparing the next, independent expedition to Germany. This time, on their own airplane.

9. In the Soviet era, enterprises engaged in sensitive activities were frequently identified simply by post office box numbers.

Chapter 16

May Days in Berlin

Well here it was, 1 May. How could we pore over reports in Adlershof? In spite of strict warnings not to poke our noses into the city center, the three majors and a driver with an automatic weapon decided to make their way to the Reichstag under the pretext of looking for the Askania factory in Friedrichshafen.

We had left the outskirts and now saw heaps of ruins. It was an apotheosis of destruction—piles of broken stone, brick, and stucco—very broad streets and extremely narrow thoroughfares. This was primarily the work of Allied aviation.

The destruction in the center of Berlin had been going on for over two years, and the Germans had arranged for equipment to clear the streets so that the city did not suffocate in obstructing ruins. The closer one was to the center, the more fires and damaged and burned tanks we saw. Here were two tanks side by side—one Russian, one German, leaning up against one another, both black and burned out. Somewhere nearby you could hear the explosions of grenades. And over there were German men and women digging in the ruins and carrying things away. Chistyakov wanted to orient himself using a compass because there were no street names. We shot ahead onto a straight section, having already lost our bearings. Suddenly a middle-aged German almost threw himself under our vehicle. "There, up ahead, the bridge was blown up." We thanked him. Perhaps he saved these Soviet officers' lives. What force ripped him away from his handcart in the ruins and threw him toward our car?

Suddenly there was a crowd—a line of people waiting for water and bread, our field kitchens. Finally we made it to the Tiergarten (zoo). All around were mangled anti-aircraft guns and dead bodies that no one had managed to clear away.

The Reichstag! A red flag flapped in the breeze in the midst of a sculpture group. Smoke rose from the Reichstag. We didn't have a camera. I had gotten a Leica camera in Adlershof, but there weren't any film cartridges. We were silent. But it really made our flight from Moscow worthwhile to see a red flag flying over the smoking Reichstag!

Suddenly there was a deafening burst nearby from a large-caliber anti-aircraft gun. We numbly looked and saw that a German gun crew had fired a round toward the Reichstag. Surprisingly, the Germans were so busy firing on our tanks that they did not notice our Jeep some 100 meters away.

From the author's archives.

Brandenburg Gate in Berlin, photographed in the summer of 1946. Mercedes shown is being driven by Boris Ye. Chertok.

Our driver Vasiliy was a lot more experienced than these eccentrics to whom he had been attached. In spite of his rank, he shouted, "Comrade Majors, why should we die for nothing? Quickly, let's get out of here!" So that's what we did—drive right back into the chaos of destruction, past the Brandenburg Gate, without understanding very well where the front lines were. On 1 May, there were essentially no boundaries. The shattered garrison units that had defended Berlin were resisting bitterly. They fought desperately in spite of the obvious futility.

A dust-covered lieutenant colonel with a bandaged head and a group of men wielding automatic weapons stopped us.

"Where are you coming from?"

Without batting an eye, Smirnov said, "From the Tiergarten."

"But the Germans are there!"

"But not many of them."

The combat officer evidently took us for intelligence officers and with a gesture of vexation waved us away. Having commanded his soldiers, "Let's go," he strode off to the sounds of the nearby fighting. Returning that evening through Karlshorst, we stopped at the headquarters of a random unit to get something to eat and some gasoline.

Here we heard the following communiqué: "Today, 1 May, by the end of the day, Nazi units consisting of more than 1,500 men surrendered, having lost the battle at the Reichstag building. But individual groups of SS officers, who have withdrawn to sections in the Reichstag basement, continue to resist."

Late that evening we made our way to "our" Buchow. Here we were able to dine well—we were even allowed 100 grams of vodka—and before we went to bed we could take a bath! "This is out of this world!" I said.

"No, it's not—it's in Europe!" someone corrected me. We agreed to get up a little earlier than usual and go to the Reichstag again the next day.

On 2 May,.having stocked up with dry rations, we once again "dashed" from Buchow to Berlin, this time along a reconnoitered road.

We entered the Reichstag, but it was more like a stream of soldiers and officers had carried us in. Bursts from automatic weapons still resounded from somewhere inside. Grubby soldiers who had made it out of the lower floors warned, "They're still down there in the basement." But upstairs, along the staircase decorated with smashed sculptures, a frenzied event was underway—they were autographing the walls of the Reichstag. And there were so many signatures! We could barely find free space. We had to help each other and stand on each other's shoulders so that, after we had found something to write with, we could sign the following: "Majors Smirnov, Chertok, and Chistyakov from Moscow." I don't remember now whether we put down our initials. But we signed and then we circled this memorial inscription twice for good measure. I remember an inscription below our autographs: "Ivan the Russian has brought order to Germany. Ivan Kochetov."

On the steps of the Reichstag and in the square, soldiers were taking group photographs everywhere. What a photo we missed!

While we were at the Reichstag, very close by at the imperial chancellery the Nazi leaders were rejecting the demand for unconditional surrender. We didn't know this, of course, and when we were getting ready for the trip back to Buchow,

From the author's archives.

Reichstag. Berlin, 1946.

we sensed something was wrong. A cannonade roared with unusual force all around—it was difficult even to determine a direction that was quieter. They were executing the order to wipe out the Berlin resistance as quickly as possible.

Our driver, a soldier who had gone through the war from the very beginning in Moscow, complained for the umpteenth time, "With you, Comrade Majors, stupidity will get us killed a day before total peace." But we nevertheless made our way to the familiar avenue and finally there, we once again saw the traffic controllers and "our" resting tanks. We then found the already familiar road to Buchow.

That evening we decided that we still needed to cover our debt from the day before and make our way to the Askania factory, since our visit had been hindered by the ongoing battles and events at the Reichstag.

We had heard a lot about Askania when we were still in Moscow. We had also found traces of the company's multifaceted activity at DVL. And here we were at the factory, though, not surprisingly, it was difficult to find. The regional commandant, who had just been appointed on 1 May, was a frontline soldier, a combat infantry lieutenant colonel. But he was already an active mayor—on his right arm was a band inscribed, "Bürgermeister." Having heard us out, he quickly pulled out a map of the area and clearly explained where to look for the factory. It seemed that he knew the area well. But he was surprised: "But that's a very small factory. It's only one of Askania's divisions."

Indeed, the entire factory was located inside a small two-story brick building and two wooden barracks-like buildings. The administration that had remained in place explained that they had just moved there about a year-and-a-half previously. Nevertheless, the factory was very interesting. It manufactured altitude gyros and heading gyro assemblies for the V-1, and had also begun to master the new remote-controlled gyromagnetic compasses based on American models.

"What is this—an exact copy of the Sperry?" we asked.

"Yes, we're studying American technology we gather from airplanes that have been shot down. We have to admit that they are far ahead of us when it comes to remote-controlled instruments."

Per our assignment, we inspected and listed the machinery. We particularly liked the precision drilling machine with the broad range of speeds from 500 to 15,000 RPMs and very smooth control.

The night of 2 May, we slept very peacefully at our cozy villa in Buchow. At some point after midnight, chaotic firing caused me to jump up. I looked out the window. Searchlights were sweeping the sky and rockets were being fired. I saw the trails of tracer shells and heard bursts of automatic weapons.

What was this? I grabbed my pistol. My comrades had already run out into the courtyard. We found Red Army soldiers simply firing into the air. "What's going on?"

"What do you mean, what's going on? The war is over!"

Well, how could we not join in on such a salute? For the first time I used my TT pistol, emptying the entire magazine into the air.

"It wouldn't be bad to toast such an occasion, but we don't have anything," we complained to each other.

But the ubiquitous soldiers!

"Comrade Majors, what do you mean you don't have anything?! Hand me a glass."

Instantly they hauled out glasses and poured each of us 50 grams of pure alcohol. In celebration we gulped it down— it almost took our breath away. It's a good thing that the soldiers immediately splashed water into our glasses. They had thought of everything.

And that is how we celebrated the end of the war, even before the official end, along with the entire Berlin garrison.

On 3 May, we received information to the effect that we must not postpone our inspection of the western part of Berlin because this portion might be handed over to the three Allies by the end of the month and then we would have no access or restricted access there. At this point there was not a single allied soldier there.

In the morning, having again noted down the Spandau-Tremen route, we set out in our Jeep through the center of Berlin, which we had already mastered.

Again the already familiar picture—up until the point where we entered the city center there were lots of people with handcarts and all sorts of luggage. There were tanks and shifting troop units. Once again we entered the ruins of the city center. The Germans had set up chains of people to sort through the rubble; they were passing stones from one person to another among the formless shells of buildings. Clouds of dust were everywhere. This was the road to Nauen. At one time a ham radio operator's greatest dream had been to pick up Nauen—one of the strongest long-wave radio broadcast stations in Europe—on his homemade receiver.

Nauen is about twenty-five kilometers northwest of Berlin, but we did not get there. On a country highway, almost at our destination, our military driver suddenly slammed on his brakes, grabbed an automatic weapon, and shouted, "Germans!"

What Germans? It's peacetime now—capitulation. But our Vasiliy was right. We jumped into a ditch and watched as a gray-green column of Germans, armed and in full gear, stretched along the road intersecting our autobahn. The soldiers moved quickly. From time to time they raised their automatic weapons and fired a burst, aiming who knows where—they seemed to be aiming at us. We didn't understand what was going on until tanks began to thunder along the highway behind us, nearly crushing our Jeep. They were T-34s, so familiar to us. Belching fire, they moved to cut across the path of the gray-green column. And behind them—standing upright and firing on the march—came Red Army soldiers. A captain with a pistol jumped down into the ditch with us. "Where are you from?" We tried to explain, but he only threw up his hands, "Don't you see? the Germans are fleeing Berlin. They could finish you off just for the hell of it. Now, go to the colonel. Over there on the highway, in the inn—that's our headquarters. They'll figure out what to do with you there!"

We had ended up in the crossfire of one of the tragic and last fatal skirmishes of the war. The Germans had wanted to break through to the West—it would have been better if they hadn't tried. It was terrible to watch our tanks and infantry come from behind and gun them down almost point blank. They were surprisingly submissive and hardly returned fire. Along the highway where we were standing, another column in Studebakers moved calmly toward Berlin as if nothing had happened. Before their eyes a battle was going on, but they were traveling their own road. This battle had nothing to do with them.

Out of nowhere, the captain appeared again, waving his pistol excitedly and swearing a blue streak. He jumped into our Jeep and commanded, "Drive to the division headquarters."

We drove to the "inn." It turned out to be a roadside restaurant named Breakfast at the Tavern. The colonel, evidently the division commander, was sitting at a large table set with all kinds of bottles and hors d'oeuvres. Another ten or so military men, who were clearly off-duty, sat at the table.

The captain approached, saluted, and reported, "During our operation we found these guys in a Jeep." We presented our documents to the colonel and explained who we were and why we were there in Tremen looking for a factory where they made instruments for the V-2. The colonel gave our documents a cursory glance, then broke out laughing. "You're alive, and thank God. And I'm not going to let you look for any factory now! You see what kind of skirmishes there can be! Sit down and have a drink and a bite to eat with us in honor of the victory." I can't say that we protested very much.

We returned to our quarters after dark without having found any Siemens factory in Tremen. The next day, we found a beautiful Siemens factory in Spandau. This was the Siemens *Apparatebau* (apparatus construction) multi-story modern building that had suffered absolutely no bombing damage. It was an aircraft equipment factory. The entrance was open and there was no one at the checkpoint. All of the doors to the shops on the various floors were open. We walked around—there wasn't a soul anywhere. Everything was spread out on the workbenches and machine tools. Everyone, without taking or hiding anything, had left their workstations and run away without taking or hiding anything. We began to feel so uneasy that we walked around the completely empty shops with our pistols drawn.

Suddenly we heard a shout from somewhere below, "*Hände hoch!*" (hands up), followed by a burst of automatic gunfire. We ran down the staircase from the third floor and saw an officer wearing captain's shoulder boards facing two trembling Germans in civilian clothes.

"What happened?"

"I just arrested them. They probably wanted to blow up the factory."

Chistyakov took it upon himself to talk with the detainees. They explained that the *Bürgermeister* had sent them to guard the factory until the occupying authorities showed up.

From the author's archives. Photo by V. I. Smirnov.

N. I. Chistyakov (right) and B. Ye. Chertok at the bank of the Spree River on the outskirts of Berlin, May 1945.

"They're lying. I have my eye on them. Let's go down to the basement. I'll let them have it with my automatic and we'll be done with it." The captain spoke so clearly that the Germans understood without interpretation.

"Wait, Captain, why do that? They don't have any weapons or explosives. Come on, let's take them to the commandant's office."

"I don't have time to mess with them now. And I don't advise you to walk around like that. I can take your weapons away from each of you! I'm a Smersh representative!"

But we persuaded him. He finally threw up his hands and went back to his captured vehicle, in which we saw what was probably a "captured" woman.

The Germans whom we had saved turned out to be workers at the factory, and we made them guide us through the shops. We saw the most interesting thing on the first floor, in the factory's accounting and cashier's office—the entire floor was strewn with a thick layer of Reichsmarks scattered helter-skelter. How many thousands or millions were there! We kicked them about with our boots like autumn leaves, and without picking anything up, hurried off to look around the factory. I was doggedly searching for traces of the production of V-2 gyro instruments which had been manufactured by Siemens, the remains of which had been found in Poland as recently as 1944. We found nothing, however, that was not purely related to aircraft production.

On 6 May, we had to leave our cozy but remote Buchow and resettle into semi-residential, barracks-type buildings in the immediate area of Adlershof. Here it was not nearly as comfortable. But the main thing was that our officers' mess was not at all the same. The large mess hall, it seems, had been set up especially for "trade union" officers. It was staffed not by dazzling Muscovite waitresses, but by young women who had been liberated from various camps and were awaiting their hour of repatriation. At that time, neither they nor we knew what awaited them.

A kind of semi-starvation existence began. During our expeditions around Berlin, we therefore had no qualms about stopping for a visit at units where they fed "starving" Muscovites pretty well.

In the ensuing days, we continued to investigate the Askania facility. The factory had a broad spectrum of interesting items and competed with Siemens for our interest. We discovered another large factory and design bureau in Mariendorf. Here I finally found intact actuators for the V-2, as well as similar actuators for aircraft autopilots. Sets of autopilot equipment intended for delivery were assembled on the test stands. With astonishment, we discovered shops with submarine periscopes, periscope range finders, bombsights, and anti-aircraft fire control equipment (PUAZO).[1] There were special cockpits equipped for crew training and testing where blind flight conditions were simulated. There was a rather large shop that was involved with purely optical production. Optical glass polishing machines stood next to the finished products—virtual mountains of lenses of various diameters up to 50 centimeters! The test laboratories were excellently equipped. They contained pressure chambers, thermal vacuum chambers, vibration stands, and rainfall simulators. Every area was equipped with all-purpose and special-purpose measurement instruments and also with our dream instrument: the Siemens multi-mirror oscillograph!

On 8 May we inspected another Askania factory in Friedenau. Here we met with the firm's technical director. He drew a diagram (since there was no ready-made diagram available) of the polarized relay for the V-2 actuator and told me that their company had the most advanced instrumentation and machinery in Europe. He boasted in particular about their unique set of jig-boring machines and optical benches.

During later journeys, interagency squabbles began. The first took place that same day, 8 May, when we approached the *Kreiselgeräte* factory.[2] There was a sign out front—Saburov's unit. What unit? Was Saburov a *Gosplan* representative?[3] At the entry checkpoint there were two men wielding automatic weapons. "Comrade officers, we cannot allow you to enter." After some wrangling, one of the men left and brought back a lieutenant colonel—a "trade union" officer like us. We introduced ourselves. He apologized that he could not admit aviation representatives since the factory had been transferred to the shipbuilding industry. He went off somewhere, then returned with permission for us to enter.

This was Zinoviy Moiseyevich Tsetsior. We soon made friends with him and for many years he remained our compatriot in the development of gyroscopic

---

1. PUAZO—*Pribory Upravleniya Artilleriyskim Zenitnym Ognem.*

2. *Kreiselgeräte* is the German word for 'gyroscope.'

3. Maksim Zakharovich Saburov (1900–77) was the Chairman of *Gosplan*, the State Planning Commission, in 1949-55.

instruments for rockets. Viktor Ivanovich Kuznetsov was in command at the factory. He permitted us to inspect the gyro-stabilized platforms that had already been put into series production. According to the German Kreiselgerät specialists, they had received an order two years ago from Peenemünde for guided projectiles. Only the senior officials, who had defected to the West, knew precisely which ones were used there. Viktor Ivanovich, the future chief designer of gyroscopic instruments for rockets and spacecraft, academician, and two-time Hero of Socialist Labor, was at that time a lanky colonel in a soldier's tunic that was obviously too short for him. He enthusiastically told us about the design of the gyro platform and especially about the transverse and longitudinal acceleration integrators mounted on it. Kuznetsov explained, "Yes, this is a state-of-the-art product. We already make pretty good instruments for ships—but for rockets? And with these dimensions?!"

DIARY ENTRY. 9 May 1945.

Our visit to the Telefunken factory in Zehlendorf was very interesting.

Originally it was a radio tube factory, but in the last few years it has switched almost completely to radar. In contrast to many other enterprises here we found almost all of the personnel, including chief engineer Wilki and his immediate staff. Chistyakov and I already spoke German rather briskly. For that reason we did not need an interpreter. Wilki and the production chief showed us the factory and laboratory. Wilki directed research in the field of centimetric waves. His laboratory, which is not located within this factory, has been conducting a thorough study of American and British radar installed on aircraft as well as radar sights for bombing and reconnaissance.

According to the assessments of the German specialists, the Americans and Brits have been very successful in the field of radar—especially in submarine warfare. Their aircraft detect periscopes from tens of kilometers away. In this regard, they have also worked a great deal on instruments to alert submarine crews that they have been illuminated by aircraft radar.

At the factory, they were involved in the series production of aircraft radar using American and British experience. The radar manufacturing shops were well equipped with electronic monitoring instruments. The factory turned out to be relatively new. They finished building it in 1939. In all, counting the Ostarbeiter, some 6,000–7,000 people worked there.[4] Of that number, 3,000 were engineers and technicians. They experienced no shortage of materials or supplies.

The Lorentz and Blaupunkt companies provided large television screens for the radar and receivers. "But you didn't study Soviet radar?"

---

4. *Ostarbeiter*, literally "East workers," referred to workers, including prisoners of war, from Eastern Europe who were used as slave labor in Germany.

According to information from our military, they did not find radar on a single one of your airplanes. And among the captured materials that they were able to provide us during our troops' offensive, there was also nothing of interest. We decided that the Russians had safeguarded this technology so well that it did not fall into the hands of our military."

I think that he spoke of "safeguarding" to be polite. In actual fact, they had surmised that during the war we had virtually no aircraft radar and radar sights. Wilki said that last year they had been fed very poorly. They got a total of 250 grams of bread per day and 200 grams of meat in the dining hall. Very little sugar and fat. At the same time, as a rule, workers were fed better than engineers. Foreign workers worked at the factory, including Russians and French. Supposedly (although we did not believe it) the Russians were fed the same as the Germans. It is true, they did not have the right to live in private apartments, and they spent the night in a camp. He felt it necessary to add, "In my opinion, all the atrocities are the result of the SS. They are animals, not people."

"Did you hear anything about the extermination camps—Maidanek, Treblinka, Auschwitz, Buchenwald? About the extermination of 6 million Jews?

"No, I don't know anything about that."

"Do you know what a 'gas wagon' is?"

"No, I've never heard of it."

As best we could, we explained to the German specialists the design and purpose of the gas chambers and gas wagons. On their faces we could detect neither surprise, nor any other emotions. They listened very attentively. Once again the comment was, "It's all the SS and Gestapo."

We wore them out with questions about other firms and studies. Like all radio and electronics specialists, they were well informed about related firms and developments and told us that Telefunken and Lorentz were involved primarily in radar technology for air defense purposes, while Askania and Siemens were involved in remote control. Over the past six months, many directors along with staff and laboratories had been moved to Thuringia and Westphalia. They knew that the secret weapon, the "vengeance rocket," was being made in Peenemünde. None of them had ever been there—it was very secret. But other divisions of Telefunken were building ground-based radars and stations for the radio control of rockets.

The tube shops were excellently equipped. Here they were making magnetron tubes with a pulse power of up to 100 kilowatts!

When asked who was considered the most prominent among the specialists in the field of vacuum tubes, Wilki responded, "Germany is proud of Professor Manfred von Ardenne. This is a man with big ideas. He was a great engineer and visionary."

"Why do you say 'was'?"

> "The past two years he was working on some new idea—a new secret weapon. We don't know anything about it. I think it is at the Postal Ministry or the Kaiser Wilhelm Institute.

We were very familiar with the name Manfred von Ardenne from prewar literature on vacuum tubes. Much later we found out that he had been in Dahlem collaborating with German physicists who were working on the atomic bomb. The U.S., British, and Soviet intelligence services were hunting for a lot more information about Ardenne and his activities. When the Americans took prisoner practically the entire German elite involved in work on the atomic bomb, von Ardenne was not among them. He turned up in the Soviet Union and worked very productively for many years at the Sukhumi Institute of the Ministry of Medium Machine-Building.[5] He was treated respectfully and high government awards were conferred on him.

That is how we first heard about the Kaiser Wilhelm Institute in Dahlem. Later, while exchanging impressions about everything that we had seen at Telefunken and then at Lorentz, we mused how, in spite of the strictest secrecy, scientific knowledge and its progress are ultimately shared between countries. Thoughts are transmitted between scientists over some sort of telepathic channel. Not only did we all toil separately, but we believed, and quite rightly, that the Germans were our mortal enemies. Our allies, out of consideration for secrecy, hardly shared their work with us. Nevertheless, with the exception of small lapses, parallel developments were taking place in the scientific fields of radar, nuclear energy, and rocket technology.

> In front of the entryway, a large crowd of workers, primarily women, surrounded us. I must say that after G. K. Zhukov's very strict orders concerning discipline and prohibiting offensive behavior toward the population, especially women (Zhukov promised a military tribunal for rape, and his were not empty words), the Germans had grown bolder. What is more, there was a rumor that these three majors, after spending almost the whole day with the factory management, had even shared their rations. That means you can safely talk with them.
>
> "Gentlemen officers, we would like to know—what's waiting for us? Are they going to send us off to Siberia?"
>
> "Nobody is getting ready to arrest you, nor take you prisoner. As for active Nazis, your *Bürgermeister* is going to sort them out."
>
> "No, you misunderstood us. When are you going to give us work? And who will pay us now? You mean you don't need the equipment that we know how to make?"

---

5. Sukhumi is located on the Black Sea in the Abkhazian region of Georgia. The Ministry of Medium Machine-Building (*Minsredmash*) was the predecessor to the present day Ministry of Atomic Energy.

Yes, these were difficult questions to answer just five days after taking Berlin.

Of course we promised that everything would be looked into; they had a good factory and so they would not remain without work.

However, the Telefunken factory in Zehlendorf soon ended up in the American zone of West Berlin, so we never found out if the women who attacked us on 9 May 1945 were provided with work.

DIARY ENTRY. 10 May 1945.

We barely managed to make our way to the Lorentz company in Tempelhof. There was a large crowd of women and children. They stood silently looking through a high lattice fence separating the façade of the building standing in the interior of the courtyard, away from the street. Our soldiers with automatic weapons guarded the entryway and in the courtyard were several Jeeps with officers scurrying about between them loading cardboard packages. When we had managed to pass through, after long negotiations, we determined that it was not secret radar instruments that had been hidden in the Lorentz basement, but over a hundred barrels of fruit preserved in alcohol—some sort of base product for all kinds of liqueurs. To be on the safe side, the first wave of the assaulting Red Army soldiers who had discovered this basement had filled the darkness with bursts from their automatic weapons before investigating its contents. A curious liquid began flowing from the perforated barrels. When subsequent echelons of attackers realized what was in the basement, the liquid was already ankle-deep. Nevertheless, they feverishly filled all sorts of military-issue containers with the precious nectar.

The garrison authorities found out about the disorder only after the emboldened women and youth had started to "help" the Red Army soldiers, showing them that, in addition to liqueur and wine, Lorentz also had a store of chocolate products made by the renowned Sarotti company. And in fact, it was Sarotti that had given up its basement to Lorentz and not the other way around. It was this, and not Lorentz's output, that was tempting the local residents. Given their half-starved existence, the possibility of obtaining the finest quality chocolate overcame any fear of the enemy, who were right in the basement drinking a mixture of wines ladled out of broken barrels. Until the arrival of the garrison patrol, they had quite a party.

Now order had been established and chocolate and wine was being distributed only with the permission of the commandant's office and the Bürgermeister. But the crowd standing in the street was counting on the kind-hearted nature of the officers who were carrying out the boxes and packets of chocolate, powdered milk, and bags of powdered sugar. I must say that they were convinced of Russian generosity. To begin with, we also went down to the basement, which was illuminated with flashlights. After making sure that our imitation leather boots were watertight, and on the advice of the duty officer from the commandant's office, we sampled the beverage from a barrel that contained either oranges or peaches preserved in alcohol. Of course, we immediately admitted that we had never drunk such a beverage. It was truly the "drink of the gods". But we didn't have anything in which to take some with us. We were limited to packets of powdered sugar, cocoa, and powdered milk to enrich our breakfasts in hungry Adlershof.

Even before our visit, the factory itself had been taken over by the "trade union" officers of Moscow's radio factories. They had also appropriated from the basement, but did not interfere with our inspection. We talked with the German specialists for about two hours. They showed us transmitters for 3- and 9-centimeter range radars. It was interesting that the laboratory, which specialized in the development of television receivers, was quickly reoriented for instruments with large radar observation cathode ray tubes.

The factory produced ground-based radio stations with large rotating antennas for guiding aircraft to their airfields. We ascertained that in practice these radar stations were also used to control air battles in the direct coverage zone. We were surprised by the number of circular scanning stations with large screens that made it possible to see hostile aircraft and distinguish them from one's own. The Germans claimed that they had already produced around one hundred of these stations. It was hard to believe, considering the exceptional complexity and labor intensity of the system. Development of the Freya radar began as early as 1938. It enabled the detection of an aircraft at a distance of up to 120 kilometers. The Würzburg radar with a spherical antenna was developed to control anti-aircraft fire. Night fighters homed in on the target using the powerful "Würzburg Giant" radar station. At the beginning of the war, all of the German radar technology was in the decimeter range. The German engineers advised us, "Our war with the Brits was fought not only on the battlefield and in the air, but also in laboratories. They had already achieved great success as early as 1942 thanks to their daring switch to the centimeter range. At that time we did not have the same tube technology."

After spending quite a while talking with the German radio specialists, as we were leaving the "drunken" Lorentz radio factory, we stopped in to report to the colonel who had given us permission to conduct an inspection. Our formal presentation turned into a prolonged conversation and exchange of impressions. It turned out that the colonel was a "trade union" officer just as we were.

He was Aleksandr Ivanovich Shokin, representing the GKO Council on Radar. At that time I had no way of knowing that I had met with the future deputy minister of the radio engineering industry, who would become the minister of the electronic industry. I would have the occasion to meet with him more than once in the latter period of his hypostases, almost up until his demise in 1986.

At that time in Berlin he said bitterly that, in spite of serious scientific achievements, our radio engineering and electronics industry was undoubtedly poorly developed compared with what we were seeing here. On this visit, as during all of our visits to German factories and laboratories, we were stunned by the abundance of instruments—both universal and special-purpose, especially in comparison to their scarcity at home. Vacuum-tube voltmeters, oscillographs, audio-signal generators, filters of all kinds, standard amplifiers, wave meters, frequency meters, etc., etc.—and all of it was high quality. Instrument models that we had considered precious before the war were continually showing up here. Not one of our institutes, factories, or laboratories could even imagine such abundance.

But indeed the war of the laboratories was not only a war of pure intellects. Each "intellect" had to be armed with the most advanced instruments for scientific research. This required a well-developed instrumentation industry.

Alas, even today, fifty years after the war, we do not fully appreciate the strength of the research scientist's laboratory weaponry, much less that of the engineer. Incidentally, one of the burning topics for the past ten years, our scandalous lag behind in the field of personal computers, has not only economic but also ideological roots: indifference to the specific needs of the human being as an individual, since, in the opinion of the country's senior leadership, above all, we had to be ahead of the "entire planet" in the smelting of steel and iron, in coal mining, oil production, and in the number of tractors and machine tools produced.

These garish indicators got through to the dullest bureaucrats at the highest levels of the Party-State hierarchy, but for the longest time they did not comprehend why it was necessary to lead or at least be on the level of an average capitalistic country in terms of providing measurement technology, not to mention expensive computers. And when it suddenly occurred to us, it turned out that we were one of the most backward countries in the world in that field.

Well, these are modern issues, but back then in Berlin and its surrounding areas we continued to collect worthwhile literature and send it to Moscow. I also insisted that we send back a wide variety of measurement technology.

Measurement equipment was my weakness during our collection of "spoils." I carefully prepared the cases containing instruments that had been retrieved by Red Army soldiers from the aerodrome maintenance battalion, then and waited for "my" airplane to deliver them to "my" institute.

By the middle of May, our *troika*—reinforced by several more specialists from NISO and LII, including Professor Sergey Nikolayevich Losyakov, had already put together a more or less clear picture of the instrument and radio industry in Greater Berlin. Our list contained more than thirty enterprises, each of which surpassed our own in terms of technology and production. The most interesting were the laboratories and factories of Askania, Telefunken, Lorentz, Siemens, AEG, Blaupunkt, and Loewe Radio.

For us it was a novelty that the company List, which specialized only in the development and mass-production of multi-pin plug connectors, existed and flourished among the Germans. They had produced hundreds of thousands of connectors for German aircraft and rockets. The concept was very simple, but the engineering and production involved were fundamentally new to us. This innovation developed in response to the extreme complexity of the electrical circuits used in flying vehicles. The connectors enhanced rapid assembly and allowed electrical components to be connected and disconnected reliably during the repair and testing of individual compartments.

The very term *shteker*, or plug connector, made its way into the Russian language from the Germans after the war. Throughout history much has been transferred to the victors from the vanquished. Only after the war did we come to appreciate what a tremendous technical role such a seemingly simple device as the plug and socket connector was destined to play in aircraft and rocket technology! The Germans spent years developing reliable connectors, and introduced into aircraft and rocket technology the standard List *shteker*, which had from two to thirty pins. We needed three years to reproduce connectors that were as reliable. However, during our first years of mastering rocket technology they gave us a lot of trouble.

Now our industry produces connectors—both tiny and enormous, airtight, onboard and ground-based—to connect and remotely disconnect more than one hundred electric circuits. Despite all of these achievements, the problem of connector technology remains one of the most complex in the entire world. This is why there are booths at every international aerospace exhibition advertising hundreds of modifications of quick and reliable cable connectors. Dozens of powerful companies in many countries produce them by the millions.

We were interested not only in individual factories, but also in the organization and structure of the instrument and radar industry. German companies worked on a lot of technical problems on their own initiative, without waiting for instructions "from the top." They did not need the decisions of the *Gosplan* or People's Commissariats, without which not a single factory of ours could produce anything. Before the war, the electric measuring technology, instrument, and radio industries had developed rapidly to conquer the entire European market, and their products had successfully competed with those of the United States. The companies Hartmann-Braun, Telefunken, Anschütz, Siemens, Lorentz, AEG, Rohde & Schwarz, Askania, and Karl Zeiss enjoyed worldwide fame long before World War II. This created a solid technological base, which we simply did not have in these industries on the required scale at the beginning of the war.

Our general-purpose electrical instrument industry, our aircraft industry, and also our nautical instrumentation industry were all housed in just a few buildings in Moscow and Leningrad (Elektropribor, Teplopribor, and Svetlana in Leningrad; Aviapribor, the Lepse Factory, Elektrozavod, and Manometr in Moscow).

It is revealing that when we began to reproduce technology for the V-2 rocket after the war and develop our own new rockets, we found out that in our country there was only one factory, Krasnaya Zarya in Leningrad, that was able to manufacture such a mundane device as the multi-contact electrical relay. In Germany, Telefunken had three such factories and Siemens had at least two. This is one of the reasons that German weapons production did not drop, but continuously increased until mid-1944, despite the continuous bombing that Allied aviation inflicted on German cities.

Chapter 17

What Is Peenemünde?

I flew into Peenemünde on 1 June 1945. The wealth of measurement instruments that I had gathered in Berlin and then delivered to Moscow had prevented me from visiting this legendary rocket center immediately after the troops of the Second Byelorussian Army Group had entered there. But every dark cloud has a silver lining. We managed to organize a special flight of our B-25 Boston from Berlin to Peenemünde. My traveling companion was Veniamin Smirnov, with whom I had flown to Germany. In Berlin we had inspected many enterprises and together we had hunkered down in a ditch, ready to defend ourselves against the German troops forcing their way out of Berlin.

What fascinating collisions history was suddenly producing! Soviet specialists with officers' ranks were leaving Berlin, where only twenty days earlier the document stipulating the unconditional surrender of Nazi Germany had been signed, to fly to Peenemünde on an American bomber piloted by a man who tested the first Soviet rocket-powered aircraft, the BI-1. At that time, I still did not realize that I was flying to the site on the shore of the Baltic Sea that would become the launching pad for the beginning of the twentieth century's great missile race. Dozens of nations from all continents would be drawn into this race, and by the end of the century almost every army in the world would acquire missile weaponry in one form or another.

We are proud of our fellow countrymen Tsiolkovskiy, Kondratyuk, and Tsander for their work on the theory of rocket flight. The Americans posthumously recognized the theoretical and experimental work of prominent scientist and inventor Robert Goddard. The Germans and Austrians are proud of the works of Hermann Oberth and Eugen Sänger. These lone pioneers who dreamed of interplanetary flight were the inspiration for those who took on the official responsibility for the practical development of long-range ballistic missiles. However, the first authentic, operational long-range guided missiles were developed in Peenemünde not for interplanetary journeys, but first and foremost to destroy London.

These days, reports about "missile attacks" on the frontlines of small, localized wars surprises no one. Missile attacks are now common even in numerous interethnic clashes. I don't believe I will be mistaken in predicting that lightweight, "portable" guided missiles will be as accessible as *Kalashnikovs* to terrorists, saboteurs, and guerrillas at the beginning of the twenty-first century.

But in those days we could not imagine the prospects of such a historical turn in weapons technology. We were drawn to Peenemünde purely by our professional engineering curiosity and a sense of duty to our country. I had been somewhat prepared for what I might see from the detailed impressions of Aleksey Isayev and Arvid Pallo, who had flown into Berlin from Peenemünde just the week before. But when, at our request, the airplane flew over the entire island, I was so delighted by everything that I saw that even a half century later, I still see the broad beaches, the white caps of the surging surf, and the forested hills in my memory. I didn't want to tear my eyes away from the sights of this wonderful natural preserve. The landscape contrasted sharply with the ruins of Berlin which we had grown accustomed to over the past month. Among the pines at Peenemünde we could see the outlines of buildings and the enormous iron structures of raised drawbridges, along with some other structures that I couldn't make out from the air; they were clearly production facilities. A network of roads, practically concealed by the shade of the pines, connected the entire island. To the right, the forests reached into the distance and one could see patches of lakes. To the left was the gray sea. We flew again over the developed area of the island, and saw attractive white, pink, and multicolored villas and hotels peeked out from under the green of the pine trees. Simply put, it looked like a resort.

From the air, we saw no traces of the brutal bombings that the Brits had reported to us. The airfield that received our Boston proved fully suitable. It had been designed for the landing of high-speed bombers. Our arrival had been expected and we were driven immediately to the Schwabes Hotel after we landed.

Seeing the areas surrounding Peenemünde, everybody's first impression had less to do with rocket technology than the natural beauty of the Baltic coast. This is where the elite of the German rocket specialists had lived, worked, and played. Now the best of the hotels—the Schwabes Hotel—housed the headquarters for Peenemünde research headed by Major General Andrey Illarionovich Sokolov, who during the war had been the deputy commander of the Guards Mortar Units. While they were looking for people in Moscow who would supervise German rocket technology, the Main Artillery Directorate entrusted Sokolov with studying and safeguarding Peenemünde. One has to give him credit—he organized this work very well.[1]

Virtually no competent German specialists remained on Usedom Island. General Sokolov's group had gathered several ill-informed specialists from among the local residents. With their help and with the speculations of Soviet engineers, they compiled a report on Peenemünde's former state before our army arrived—

1. Andrey Illarionovich Sokolov (1910-1976) later headed NII-4, the most influential ballistic missile research institute within the Soviet military.

"former" being the key word. Allied aviation had damaged most of the buildings and laboratories, but they had not been totally destroyed. The firing rigs were bigger than anything we could have imagined.

Near the rigs, the Germans had built bunkers which had remained in good condition. From the bunkers they controlled and observed the testing of engines and rockets. The facilities in total occupied several tens of hectares which were connected by excellent roads. Tens of kilometers of power, measurement, and signal cables had been run in cable ducts that the Germans had not had time to dismantle.

All of the equipment down to the last instrument—even the machine tools in the large factory whose building was almost untouched—had been dismantled and removed. What they did not manage to evacuate before the arrival of Marshal Rokossovskiy's troops had been rendered useless by SS *Sonderkommandos*.[2]

To a significant degree, General Sokolov had managed to restore the old order in the residential area of the *Zinnowitz* resort. Arvid Pallo had already prepared me for this. Back in Berlin he had warned me that Peenemünde was on par with fashionable prewar resorts. It was as if there had been no war with all its horrors.

At the Schwabes Hotel restaurant they served a *table d'hôte* for the entire officer staff. It was covered with a snow-white tablecloth and each place was set with numerous utensils, clearly exceeding the number of dishes. The deft waiters presented the plates with a modest appetizer so that the hotel's logo was positioned in front of us just so.

None of the officers dared sit down at the *table d'hôte* until the general had entered. Then began a ceremony known to us only from the movies. A line of waiters in black suits, white shirts, and bow ties, led by the *maître d'hôtel*, marched in solemn procession around the table, beginning with the general, and then proceeded around by rank. In this process the first waiter ladled soup, the second placed a potato, the third showered the plates with greens, the fourth drizzled on piquant gravy, and finally the fifth trickled about 30 grams of alcohol into one of the numerous goblets. Each person diluted his drink with water according to his own taste. To some extent this entire spectacle revived the protocol that had reigned at the Schwabes Hotel when distinguished guests had visited Peenemünde. According to the *maître d'hôtel*, almost all of the top brass of the Nazi Reich had stayed there, except for Hitler himself. "But, of course," added the *maître d'hôtel*, "at that time I served the best wines. When Dornberger and von Braun evacuated Peenemünde, they took all the food and wine stores with them." We were introduced to the general in the dining hall and, with honor, upheld the rules of etiquette prescribed for "high society", in spite of the provocative smiles and remarks of the old residents.

---

2. *Sonderkommandos* were special detachments of prisoners at Nazi death camps that the SS organized to aid the killing process.

I need to write about Peenemünde from a historical standpoint. For that reason I have singled out only the basic results of the work in Peenemünde. In so doing, I have used "sources" that we found in Germany and information from memoirs published later by individuals who participated directly in the work in Peenemünde.

The inspection of Peenemünde in May and June 1945 showed that the actual scope of work on rocket technology in Germany was far superior to what we had imagined. We Soviet specialists needed to investigate the entire volume of work that had been done in Germany in the field of rocket technology. But it was just as important to obtain information on the history of these developments and the methods used by German scientists and engineers to solve many difficult problems, such as those involved with the development of long-range guided ballistic missiles.

Before 1945, neither we, the Americans, nor the Brits had been able to develop liquid-propellant rocket engines with a thrust greater than 1.5 metric tons. Those that had been developed were not very reliable, had not gone into series production, and were not used to develop any new type of weapon. By that time, however, the Germans had successfully developed and mastered a liquid-propellant rocket engine with a thrust of up to 27 metric tons—more than eighteen times greater! What is more, they had produced these engines in large-scale series production by the thousands! And the automatic guidance system! It was one thing to fundamentally and theoretically show that for the given level of technology it was possible to control a missile's flight and consequently the engine mode in flight at a range of 300 kilometers; it was a quite another thing to put this into practice and bring the entire system up to a level suitable for acceptance as an operational armament!

As a result of World War II, at least three new scientific and technical achievements emerged, which to a great extent revolutionized previous ideas on strategy and tactics for possible future wars. These achievements were automatically guided missiles, radar technology, and nuclear weapons. The first two did not require the discovery of any new laws of physics. Nuclear technology was another story. Its development was spurred by the discovery of new laws of nature and new scientific methods involving penetration into the atomic world and the nature of the fundamental principles of matter.

These three new forces, like genies, were released from laboratories during World War II. Historically, the obvious and undoubtedly objective fact that military technology is the strongest stimulus for the development of a society's industrial power and the progress of civilization as a whole seems paradoxical, even shocking. The fact that the foundation of modern cosmonautics was the scientific and technical creativity of the military-industrial complex of the Soviet Union and the United States has also been obvious for a long time.

What role did German scientists and Germany's military-industrial complex during World War II play in the future development of cosmonautics? We obtained much of the information needed to answer that question during our first months of postwar work in Germany. Our sources were the facilities that we inspected, the onsite study of examples of rocket technology, the accounts of German specialists, and the historical reports that they drew up at our tasking.

The history of German rocket technology does not start with Peenemünde. Practical work to implement pioneering theories about interplanetary travel began with the development of primitive rocket engines as early as 1929. The evolution of amateurish research by lone enthusiasts into large-scale activity to develop rocket-powered flying vehicles began after military agencies had assessed the outlook for missile weaponry.

According to the Treaty of Versailles, Germany had been forbidden to develop and produce new types of aircraft, artillery, and other weapons technology that had been known during World War I as offensive systems. Missile weaponry had not been included on the list of banned weapons. The authors of the Treaty of Versailles had not even considered the possibility of using missiles as offensive weapons. The top army leaders in Germany, however, believed that the work started by the romantics of interplanetary travel could have real applications in warfare. One of the first German generals to render real support to the amateur work of rocket enthusiasts was Karl Emil Becker, author of the widely known textbook *Lehrbuch der Ballistik* (Ballistics Textbook) and head of the directorate for infantry artillery weaponry. Becker formed a special subdivision for rocket technology, which in 1934 was headed by Walter Dornberger, a thirty-nine-year-old artillery captain. One of Dornberger's first subordinates was the twenty-two-year-old engineer Wernher von Braun, who was well known for his scientific study *Konstruktive, theoretische und experimentelle Beitraege zu dem Problem der Fluessigkeitsrakete* (Constructive, Theoretical, and Experimental Contributions to the Problem of the Liquid Rocket), published in 1931.[3] Dornberger and von Braun's focused work to develop long-range, liquid-propellant ballistic missiles began at an artillery range in Kummersdorf near Berlin. The first experimental missiles were assigned the ordinal designations, A-1 and A-2.

In 1933, with Hitler's rise to power, all restrictions placed on the armaments of Germany by the Treaty of Versailles were discarded. The Nazis openly encouraged industry to work energetically to restore Germany's military might. In 1935, von Braun and Dornberger approached the Directorate of Armaments with a design for a liquid-propellant missile capable of achieving a firing range of 50 kilometers. To optimize the guidance system, the Germans developed an experimental missile under the designation A-3. A launch pad for the missile was set up on the small island of Greifswalder-Oie in the Baltic Sea.

---

3. Wernher von Braun, *Konstruktive, theoretische und experimentelle Beitraege zu dem Problem der Fluessigkeitsrakete*. Ph. D. dissertation, Friedrich-Wilhelm-Universitaet of Berlin, 16 April 1934, subsequently reprinted in *Raketentechnik und Raumfahrtforschung*, Sonderheft 1. Frankfurt: Umschau Verlag, n.d.

For the first time in the history of rocket technology, a gyro-stabilized platform was tested on a missile. All four A-3 flight tests proved unsuccessful due to failures in the control system. But the tests did produce a fundamental milestone: the first vertical guided launch of a freestanding missile on a launch platform. Dornberger and von Braun understood that the experimental base in Kummersdorf was not suitable for work on the scale that they had conceived, and they began to look for a new site for the construction of a large scientific-research center combined with a test range.

In early 1936, Dornberger, now a colonel, was appointed director of missile development for the German armed forces. Dornberger and von Braun proposed to the senior army leadership that a missile with a range of more than 200 kilometers be developed under the designation A-4. They would need to create a powerful scientific-research and testing center to develop and test the missile.

In August 1936, the Germans decided to begin construction of the military testing center on Usedom Island on the Baltic coast, next to the resort of Zinnowitz and near the fishing village of Peenemünde. Dornberger was named chief of the entire test range, which consisted of two parts: Peenemünde-Ost (East), the unit under the command of the infantry forces; and Peenemünde-Westen (West), which was transferred to the *Luftwaffe* (Air Force). Twenty-four-year-old von Braun was appointed to the high post of technical director for Peenemünde-Ost. The rapid construction of the center began in the fall of 1936. In 1937, the first ninety employees moved to Peenemünde to the infantry's research center. Development, research, and testing were conducted simultaneously with the construction, which was more or less complete after three years. The company that carried out the primary construction operations in Peenemünde was headed by the future President of the Federal Republic of Germany, Heinrich Lübke.[4] Even in 1945, after the brutal bombardments of the war, we had no reason to complain about the quality of the construction.

In Europe, preparations for war were underway, but none of the Allied intelligence services, including the celebrated British intelligence, had imagined the objectives and scale of work being conducted at the Peenemünde center.

On 1 September 1939, World War II began. In November 1939, an unknown anti-fascist dropped a detailed report concerning the scope, objectives, and tasks of the Peenemünde center into the mailbox of the British Embassy in Oslo. By that time, more than 3,000 individuals were working in Peenemünde, not including the construction workers. It is astounding that it took four years for the British to react!

In March 1940, the first firing tests began on chambers of the 25-metric-ton thrust, liquid-propellant rocket engine intended for the A-4. This was the first time the Germans used a turbopump feed to deliver alcohol and liquid oxygen into the

---

4. Lübke served as *Bundespräsident* (Federal President) of the Federal Republic of Germany (West Germany) between 1959 and 1969.

combustion chamber instead of the pressurized system using compressed nitrogen that was used on the first engines in Kummersdorf.

With the onset of World War II, the scope of the research and development activity at Peenemünde rapidly expanded. New services were created and a large number of new specialists needed to be recruited and quickly mobilized to the center, many being summoned from the army. The schools of higher education in Darmstadt, Berlin, Dresden, Hannover, and later Vienna and Prague were also enlisted to support the research activity at Peenemünde and conducted their work on assignment from there.

From 1937 through 1940, the Germans invested more than 550 million Reichsmarks into the construction of the Peenemünde center, an enormous sum for that time. All of Germany's leading electrical and radio engineering firms provided the center with special testing equipment and the latest measurement instruments. Despite our anti-fascist views, we had to credit the directors of the operations, first and foremost Dornberger and von Braun, for the energy and confidence with which they acted.

As a matter of fact, it wasn't just the enthusiasm and organizational capabilities of the Peenemünde directors. They understood perfectly well that the enthusiasm and brilliant capabilities of lone scientists were far from sufficient. What was needed was a clear conception of the scale of operations needed to achieve set objectives, and the determination to create the strongest state-supported scientific-technical, production, and military-testing infrastructure possible. All of this was conceived prior to the war, then refined and implemented during the hostilities under the conditions of Hitler's totalitarian regime, which spared no expense to develop the proposed secret weapon of mass destruction. There was no need to account for the project before parliamentarians. To a significant degree, this contributed to the success of the new endeavor.

On 22 June 1941, Germany attacked the Soviet Union. While certain of a rapid victory on the eastern front, Hitler was troubled by Britain's tenacity. On 20 August 1941, at his "Wolf's Den" headquarters in East Prussia, he received Colonel Doctor Dornberger, Doctor von Braun, and the lead specialist for the development of missile guidance systems, Doctor Steinhoff. There was no need for the new missile weaponry for the Blitzkrieg against the East, since the Germans would be in Leningrad within a month and Moscow no later than October. Why then such attention to missile specialists? Dornberger and von Braun familiarized Hitler with the current status of their work on the A-4 missile, which had a range of up to 300 kilometers. They also discussed the potential for new missiles that would have intercontinental ranges. As a result of this meeting, Hitler gave the programs at Peenemünde the highest priority.

In autumn 1941, roused by this support from the highest level, the work force at Peenemünde began to speed up the design process for two-stage and perhaps even three-stage A-9, A-10, and A-11 systems. But the basis was to have been the A-4 design.

The infantry command believed it had a monopoly on missile weaponry, but the Air Force also found a place at Peenemünde. The western part of Peenemünde-Westen became the center for the flight optimization of the Fiesler Fi-103 flying bombs with pulsejet engines. Equipped with a primitive autopilot, these winged missiles gathered speed on launching rails and took off horizontally, continuing to fly aided by their engine at a maximum altitude of 3,000 meters and a maximum speed of 300 kilometers/hour. The Fi-103, which later gained fame as the V-1, reached a range of 250 kilometers. Its warhead, which had a maximum yield of 500 kilograms of TNT, was intended to hit targets such as large cities. Their dispersion was tens of kilometers.

After a series of failures, the first completely successful firing rig test of the A-4 rocket engine took place on 18 May 1942. The engine generated a thrust of 25–26 metric tons over a period of 60 seconds. I will remind the reader that three days before that event, on 15 May 1942, the pilot Bakhchivandzhi flew in our first rocket-powered aircraft, which had the most powerful Soviet liquid-propellant rocket engine at that time—with a thrust of 1200 kgf. Our liquid-propellant rocket engines did not exceed 1200 kgf until 1948! We could soothe our wounded pride only with the fact that the United States' own liquid-propellant rocket engines hadn't even reached that amount of thrust yet.

The successful test-firing of the A-4 was followed by rocket explosions during the first attempts at flight optimization. Finally, on 3 October 1942, the A-4 completed its first successful flight. With a launch mass of 13 metric tons and an equivalent "payload" of 980 kilograms, the rocket flew 190 kilometers, reaching an altitude of 85.5 kilometers at the apogee of its trajectory. After this first success of the A-4 program, the first tests of the winged Fi-103, the future V-1, were also successful.

In April 1943, the A-4 was successfully tested at a maximum range of 330 kilometers. Having later been given the name V-2 (*Vergeltungswaffe Zwei*—Vengeance Weapon Two), the A-4 rocket program and the V-1 were competing. Meanwhile, on the Eastern Front there was no *Blitzkrieg*. The rout of the Germans near Stalingrad was followed by another defeat at the Battle of Kursk. The German Air Force no longer had overwhelming air superiority on the Eastern Front and proved incapable of withstanding the air raids by British and American bombers against German cities.

In addition to their exceptional efforts on the A-4 project, they started work at Peenemünde on the design of an automatically guided air defense missile, under the codename *Wasserfall*. This liquid-propellant rocket was part of a radar detection and guidance system targeting enemy aircraft. The onboard control system used homing devices during the final flight segment. The A-4 was used to test several control principles for the *Wasserfall*.

In 1943, there were more than 15,000 dedicated personnel working at Peenemünde. New firing rigs made it possible to conduct firing tests on engines with thrusts from 100 kilograms to 100 metric tons. Peenemünde aerodynamics specialists prided themselves on having the largest wind tunnel in Europe, created over a period of just eighteen months; the largest factory for the production of liquid oxygen; and spacious and excellently equipped design halls. Launching areas for rockets and launch control bunkers were provided during the earliest days of construction on Usedom Island. Correspondingly, the entire path of possible launches along a line running north-to-northeast was equipped with monitoring and observation facilities.

By late 1943, the struggle had once again heated up in military circles over the priorities for the development of new types of armaments. Almost all of the senior military leaders supported the *Luftwaffe*'s Ju-88 program, which called for the mass production of medium bombers to support frontline operations by striking enemy cities and strategically important targets. The *Luftwaffe*'s leadership understood very well that the organization of large-scale series production of the A-4 could interfere with the fulfillment of the numerous industrial orders for the Ju-88 program. The *Luftwaffe* undoubtedly had weighty claims to the highest priority because the new bombers were sent directly to the front, to active aviation units. The A-4 rockets would also be used to attack Britain, but it was bombers that were the real weak point in the German Air Force. It was natural that a struggle over priorities heated up between these two programs.

By the end of the war, the new secret weapon had been given the highest priority among all the orders to industry and transportation. Dornberger, von Braun, and the infantry forces leadership that supported them pushed the Ju-88 program into the background. This substantially reduced the combat capabilities of German bomber aviation. At that time, British and American aviation was destroying one German city after another, and Germany did not have the strength to deliver counterstrikes. There were not enough high-speed, high-altitude bombers with the necessary range. They pinned all their hopes for such a strike on the V-2 vengeance weapon—the A-4. Such a turn in favor of the Peenemünde programs during a period when Germany was at the brink of military catastrophe on the Eastern Front and losing the Battle of Britain can only be explained by the blind faith that Hitler and his immediate entourage had in the miraculous force of the new missile weaponry as a means of mass destruction and a new means of air defense. It was precisely faith and by no means a certainty. This faith not only accelerated Hitler's defeat, but to a certain degree helped eliminate the terrible threat of the Germans creating an atomic bomb before the end of the war. The scale of work on the A-4 and the need for large quantities of particularly scarce materials for its mass production indirectly prevented the Germans from creating a nuclear weapon.

After the war, the renowned German physicist and Nobel Prize laureate Wernher Heisenberg recalled, "In September 1941, we saw the path that had opened up before us—and it led to the atomic bomb."

In his book *The Virus House,* British researcher and journalist David Irving writes:

> In June 1940, when battle had ceased in France and occupation reigned for four years, Germany's positions in the nuclear race were very impressive and

> even frightening: Germany did not have large stores of heavy water, but to make up for this she seized the only heavy water factory in the world; she became the holder of thousands of tons of very pure uranium compounds and established control over an almost completed cyclotron; she had at her disposal cadres of physicists, chemists, and engineers not yet robbed of their vitality by all-out war; and her chemical industry was the most powerful in the world."[5]

If the Germans had managed to create an atomic bomb before the Americans and then put two or three bombs into two or three of the many hundreds of A-4 rockets launched at Britain, the world today might look completely different.

It is surprising that the primary reasons for the slow pace of work on the German atomic project were not technical. The lack of progress resulted instead from conflicts among high-level scientists and the regime's arrogant and condescending attitude toward a discipline that lacked rocket science's active promoters. From the first days of the war, the German economy was consumed by the immediate needs of one *blitzkrieg* after another. The Germans' early military successes in Europe and the Soviet Union led the Germans to believe in the complete superiority of their military technology. And if that was the case, then why spend funds and divert efforts to new labor-intensive developments and scientific research projects aimed at creating an even more perfect weapon?

But that was not the single cause of the German physicists' failure. On this point I concur with the very competent research of David Irving, who writes in *The Virus House*, "In late 1940, German physicists had not foreseen any serious difficulties on the way to the military use of atomic energy. . . . Having rejected graphite in January 1941, German scientists committed a fatal mistake. Now it is well known." This error worked to the advantage of missile specialists because there clearly was not enough graphite in Germany for both fields of endeavor. We and the Americans also used graphite control surfaces to control missiles up until the mid-1950s. Now it is well known that it is better to use other methods instead of control vanes of any material. But more than ten years of persistent work by specialists from the USSR and United States were required to switch to this method.

Irving writes,

> Who knows how the situation would have turned out if the mistake had been corrected in a timely manner. This mistake, which was fatal for the German atomic project, proved to be fortunate for humankind. It became the main obstacle and hindered the Germans from creating a critical reac-

---

5. David Irving, *The Virus House* (London: Kimber, 1967). The American edition was David Irving, *The German Atomic Bomb* (New York: Simon and Schuster, 1967).

tor using graphite and uranium, in other words, the same type of reactor as the first operating reactor in the world, which the Americans created two years later . . .

As far as one can tell from the published research, neither Russian nor American postwar researchers have fully appreciated how the Peenemünde rocketeers' invention of graphite control surfaces saved humankind. The Germans were forced to use up their extremely limited stores of pure graphite.

IN AUGUST 1945, when we were in Thuringia, we heard on the radio about the dropping of atomic bombs on Hiroshima and Nagasaki. We first of all tried to understand what they were talking about.

There were no Soviet specialists among us at that time who had the slightest involvement in atomic research. Nevertheless, our knowledge of physics helped us, in a group discussion, to assume that the Americans had succeeded in creating a bomb by converting part of the mass of a substance into energy, in accordance with Einstein's famous formula: $E = mc^2$. There and then, we started to question Helmut Gröttrup about what had been known in Peenemünde regarding German work on the creation of an atomic bomb. To what extent were the German directors of the long-range missile program—in particular Dornberger, von Braun, or their closest assistants—familiar with the possibilities of creating an atomic bomb? Long conversations with Gröttrup enabled us to understand that work on some sort of super-powerful explosive had been conducted in Germany. Gröttrup was well acquainted with the names Heisenberg and von Ardenne, which I mentioned as possible scientists who could have been working on an atomic bomb. Moreover, he said that in the summer of 1943 the Peenemünde directors had, under great secrecy, talked about some new powerful explosive. For the specialists at Peenemünde this was very important. They understood that the ordinary TNT used in A-4 warheads—in quantities of 700–800 kilograms per warhead—would produce an effect no greater than a conventional 1,000 kilogram bomb dropped from an airplane.

British and American aviation had already dropped countless such bombs on German cities. Nevertheless, Germany had continued to fight and had even expanded its development of new weapons. Gröttrup recalled that he had heard about the new explosive when von Braun had been sent to Berlin to consult with the infantry command about the prospects of increasing the power of missile warheads.

Upon his return, von Braun did not say with whom he had met in Berlin. Gröttrup, smiling, recalled that it had been nice to hear from his boss that the theoretical physicists, despite the very interesting problem they were working on, had absolutely no engineering experience—in contrast to the missile specialists, they could not imagine how they needed to organize their work in order to transition from naked theory to "living" objects.

In addition, and Gröttrup said this frankly, German physics and science as a whole had been severely weakened as early as 1937. More than forty percent of the professors had been removed from their university positions for disloyalty to the regime or for non-Aryan bloodlines. In this regard, German science was not well served, the persecution of scientists cost Germany dearly. But everything having to do with rocket technology was virtually unaffected because, as surprising as it may seem, Gröttrup could not recall any persecution of the specialists over the entire period of work at Peenemünde, except for one incident with the Gestapo.

The dismissive attitude of the military and political leaders of Germany toward the atomic project contrasted sharply with their close attention to the work conducted in Peenemünde. This is despite the fact that Wernher Heisenberg, in a June 1942 conference with Albert Speer, the senior Reich official who to a great extent determined Germany's economy, spoke plainly and directly about the military use of atomic energy and explained how the atomic bomb could be fabricated.[6] For military leaders, evidently, the authority of world-famous physicist Heisenberg was not sufficient for the atomic project to be given the same priority as guided missiles and flying bombs.

Heisenberg's report did not make the proper impression on Field Marshal Erhard Milch, who was responsible for aircraft armaments. Milch shortly thereafter authorized series production of the new "vengeance weapon," the V-1 flying bomb, while not giving Heisenberg the support he needed. The well-known "General's skepticism" emerged with regard to the new incomprehensible sciences. The V-1 was the brainchild of aviation and the V-2 was the weapon of the infantry troops. This was clear, intelligible, and obvious—you could look at it, put your hands on it. But converting mass into energy—this was abstraction. Better to wait!

But serious technical problems continued to occur even in work that had been conducted at Peenemünde on a broad scale. I have already mentioned that the first launch that could be described as a success took place on 3 October 1942. In order of construction, this was the fourth A-4 rocket. Oberth himself, who was in Peenemünde at that time, congratulated von Braun and the other rocket developers. The engine and the control system worked comparatively normally for the first time. On the occasion of the long-awaited success, a banquet was held at the Schwabes Hotel. A huge boulder with the inscription "On 3 October 1942, this stone fell from my heart—Wernher von Braun" was placed near the launch pad. (We had heard this story, but none of us paid any attention to the stone when we were visiting Peenemünde). A series of failures followed afterward, however. There were explosions during launch, explosions in the air, and failures of control surface actuators, gyroscopes, propellant and oxidant line valves, and the onboard electrical power circuits.

---

6. Speer was the Nazi Minister of Armaments and War Production in 1942–45.

In late 1941, the armaments ministry expressed interest in the large-scale series production of the A-4. In this regard, a large number of mutually exclusive missile variants were proposed, the majority of which were rejected on paper. Such a large number of specialists were drawn into this work that the optimization of the A-4 was greatly slowed down. Nevertheless, experimental launches of the A-4 continued, and the number of launches reached in 1943. During the launches, engineers identified and corrected the primary shortcomings of the engine, fuel supply system, and instrumentation. They began serious work on achieving the required firing precision.

In early 1943, the Germans discovered a great lag in the development of ground equipment and ground services for monitoring and supporting the flight tests. Simultaneously, troop units began experimental launches in which they not only had to master the new weapon, but also work out sighting methods to ensure firing accuracy. This research on ballistics and firing accuracy required the development of special onboard and ground-based radio equipment. In addition, optical devices were stationed on the flight path to monitor the trajectory. The launches exposed many shortcomings in the electrical circuits and control system equipment design. Successful launches at a nominal range of up to 287 kilometers alternated with explosions, fires in the tail section, and control system failures.

The missile was extremely unreliable—it had not been brought up to standard, and required substantial improvements. But as the situation on the fronts deteriorated, with the adventurism characteristic of Hitler, the political and military leadership of the Reich pinned their hopes all the more on this new weapon.

After the Battle of Stalingrad and the Germans' defeat in the Battle of Kursk, the Eastern Front was so unstable that a missile such as the A-4 could not be used to stop the Red Army's advance. Britain was another story. In the absence of a second front, the Germans could count on using the coastline of the North Sea or English Channel to create stationary launching areas to fire against Britain. There was a glimmer of hope that the British, having concentrated their attention on their own territory, would not dare participate in assault landing operations, while the Americans would not undertake anything alone. The Führer gave outlandish instructions to begin launching 1,000 flying bombs and missiles against Britain per day. After that the number of launches would gradually increase to 5,000 per day!

In May 1943, the question of priority between the V-1 (Fi-103) flying bomb and the V-2 (A-4) missile was to have been resolved. By that time there had been more than twenty-five A-4 launches, the last of which had been successful. In terms of striking accuracy and range, the flying bomb and the A-4 missile were approximately identical. Both weapons were suitable for striking targets the size of a large city. In this regard, London was difficult to miss. But British air defense units had learned to fight very effectively against the slow and low-flying V-1 flying bombs (cruise missiles in today's terminology). Anti-aircraft artillery and air defense fighters shot them down, and they ran into barrage balloons. New British radar facilities made it possible to detect V-1s long before they reached London. The *Wehrmacht* leaders understood that they could hardly break Britain by launching around one thousand V-1s per month, each having only an 800 kilogram warhead, of which barely forty percent reached the target. In total during the war, around 12,000 V-1s were fired against Britain.

The V-2 was another matter. The British air defense were powerless against this missile. Its flight speed and altitude precluded any possibility of warning or announcement of an air-raid alert.

The Germans needed to set up the large-scale series production of the A-4. When Peenemünde was first being established, it was assumed that A-4 missiles

NASA History Office.

V-1 rocket.

would be produced there on the island, or at least assembled after production and tested there. To this end, a high-capacity production facility was built and richly equipped with various production equipment and testing instruments. It soon became clear, however, that Peenemünde could not accommodate the required scale of production and production line processes. The factory that had been built in Peenemünde was therefore converted into an experimental *Versuchswerk* (VW—Research Factory). A total of approximately one hundred rockets were assembled there.

In July 1943, Hitler once again personally received the Peenemünde directors. He announced that the rocket program had become the top priority for the entire *Wehrmacht* and for German industry. Such a task required the development of technology and the organization of mass production. In Thuringia, in the vicinity of Nordhausen, construction began on an enormous subterranean factory with a projected production capacity of thirty A-4 missiles every twenty-four hours. This factory was called *Mittelwerk*. By mid-1944 it was already producing up to 600 A-4 missiles per month.

The construction and production of A-4 missiles at *Mittelwerk* near Nordhausen was perhaps one of the darkest and most tragic pages in the history of German rocket technology. Foreign workers, prisoners of war, and concentration camp prisoners were used to build and produce the missiles under the supervision of German specialists and Gestapo overseers.

Before work began underground, the workers were brought to the Dora concentration camp, which had been set up especially for this purpose next to a picturesque wooded mountain. Inside the mountain factory, the most rigid regime was established—the slightest violation of order and discipline was punished by death. Smoke billowed from the chimney of the crematorium at the Dora camp around the clock. Camp workers died from beatings, torture, diseases, exhaustion, and execution for the slightest suspicion of sabotage. Very few of the Dora prisoners who worked on the top-secret vengeance weapon would get out alive. Nevertheless, an underground center of the anti-Nazi resistance was active at the camp.

Nine thousand skilled German workers were sent to *Mittelwerk* as conscripted workers by the companies AEG, Siemens, Rheinmetall Borsig, Dynamit AG, Krupp, and Thiessen-Hitton. The Gestapo sent more than 30,000 prisoners from various concentration camps. The camp underground committee, which consisted of Russians, Czechs, French, and communist Germans, organized sabotage at the factory under the motto The Slower You Work, The Closer to Peace! The prisoners found ways to make the most delicate rocket assemblies useless.

The Gestapo managed to pick up the scent of the underground anti-Nazi committee, which was led by German communist Albert Kuntz. Among those arrested and thrown into the Gestapo torture chamber for interrogation were French officers, Polish partisans, Czech scientists, German communists, and Soviet prisoners of war. The names of these heroes of the rocket underground remain unknown to us. But the sabotage continued in spite of the reprisals and executions. There were also anti-Nazis among the German workers in the subterranean factory. One of them, the skilled metal worker Joseph Zilinskiy, who had worked at Peenemünde before being sent to *Mittelwerk*, managed to establish contact with Soviet prisoners of war. He was seized by the Gestapo and thrown into a cell in the Nordhausen barracks. He was to be hanged, but during a British and American aviation attack the barracks was bombed. He managed to escape and hid until the end of the war. People such as Zilinskiy, who by some miracle survived, have enabled us to learn the terrible details of the Nazi's subterranean missile production.

In October 1992, I visited a memorial museum created on the territory of the Dora camp. The young people working in this museum had gathered very interesting material about the camp's history, the construction of *Mittelwerk*, and the heroes of the resistance. They were searching for the names of the heroes of sabotage at *Mittelwerk* and Dora victims.

The first meetings at Nordhausen and then in Bleicherode with German specialists enabled us to find out the primary specifications (listed as follows) of the series production A-4 missiles that were produced at the subterranean factory and then delivered directly to troop units:

Launch weight	12.9–13 metric tons
Maximum engine thrust on the ground	26 metric tons
Maximum engine thrust at high altitude	30 metric tons
Total length	13.9 meters
Diameter	1.6 meters
Warhead mass	900–1000 kg
Maximum flight range	250–300 km
Propellant mass (80 percent ethyl alcohol)	3.6 metric tons
Oxidizer mass (liquid oxygen)	5 metric tons
Engine operation time	64–65 seconds
Maximum speed	1500 meters/second
Maximum altitude above the ground	95 km
Speed at contact with target	800 meters/second

In spite of the free labor coerced from prisoners, the cost of the A-4 was over 300,000 Reichsmarks! Per missile! And this only included cost of the propellant and oxidizer and the pay for troop units; it did not include the cost of the ground equipment.

In September 1944, the Germans began firing A-4 missiles on London, as well as liberated Paris and Antwerp. The V-2 attacks caused great terror among the Brits. The missiles closed in without any warning noise and were like bolts from the blue. The approaching missile with its fiery exhaust could only be observed by accident, several seconds before it exploded. Immediately after the combat use of the V-2 had started, the Brits conducted reconnaissance and then organized air attacks against the rocket launchers, which were difficult to camouflage. There was no other way to fight against V-2 missiles. Launch positions proved to be the weakest link of the missile strategy.

In the 1970s, the United States devoted intense attention to developing mobile railroad launchers for the Midgetman missile and, before that, the Minuteman. The USSR also developed intercontinental missile launch variants using railroad rolling stock. But mobile railroad launchers as protection against air attacks had been developed by the Germans as early as 1944 in Peenemünde. The A-4 missile was to have been launched from a simple rack mounted on a railroad flatcar. The mobile launcher consisted of alcohol and liquid oxygen tanks, launch equipment, and the equipment needed to perform pre-launch checks. However, the Germans did not succeed in bringing the mobile launchers to the point of combat-readiness. All of the actual launches were conducted from launch "tables" at fixed positions. The missiles were transported there and set up using a special erector, the so-called *Meillerwagen*.

Dr. Matthias Uhl.

Three V-2 rockets being prepared for launch.

The 485th Front Artillery Division was created for the combat use of the A-4. According to information provided by Dr.-Ing. Hans Kammler, the SS Lieutenant General responsible for all V-weapon control points, they had managed to launch up to fifteen missiles per day in September 1944 when he was commanding an "artillery corps." As they gained mastery of the missile technology, they shortened the pre-launch preparation cycle. On 30 October, twenty-nine missiles were launched. On 26 November and again on 26 December, the Germans launched a record thirty-three missiles!

According to information that is evidently close to authentic, from 5 September through 31 December 1944, a total of 1,561 A-4 missiles were launched. Of these, 447 were launched against London and forty-three were launched against Noordwijk and Ipswijk, the garrison and assembly bases in the Netherlands that supported the Allied deployment on the long-awaited second front.

Beginning in early 1945, bombing raids and the advance of British and American troops deprived the Germans of the most advantageous positions for firing on London. The majority of the missiles were then directed against targets in continental Europe. The Germans launched 924 rockets against supply bases in Antwerp; 27 against Liège; 24 against Lille; 19 against Paris and the Meuse valley; and 2 against Diest. In their memoirs, various World War II Allied commanders make no mention of substantial Allied military losses caused by missile attacks. As the rockets often missed their targets by tens of kilometers, they ended up damaging civilian morale much more than the Allied economy or armed forces.

According to various sources, the 2,000 missiles launched against London over a seven-month period killed over 2,700 people. There is no other reliable information from other locations concerning the casualties of A-4 rocket fire. But if you extrapolate based on the London statistics for the average number of deaths per launch, then one can assume that approximately 7,500 deaths resulted from the A-4 missiles. Those who were burned in the crematorium of the Dora camp or annihilated during the construction of Peenemünde and the rocket firing range in Poland should be added to these direct victims of missile technology combat operations, along with the anti-Nazis who suffered in the torture chambers and were executed. Their number far exceeds those who died as a direct result of the A-4's use as a weapon. Such are the tragic pages of the history of the first operational long-range, guided ballistic missile.

What were the Allied and Soviet intelligence services doing during these years? One can forgive our intelligence service for their ignorance of the scope of operations in Germany on missile armaments. The service had been crushed twice—first under Yezhov, and then again under Beriya. But the celebrated Western intelligence services had also overlooked a secret project that engaged tens of thousands of German civilians and military specialists. Orders from Peenemünde and *Mittelwerk* were fulfilled by dozens of companies scattered throughout the entire country. Experimental launches of missiles into the Baltic Sea had been conducted since 1940 and at the firing range in Poland since 1943.

It seems improbable, but nevertheless, before May 1943 neither reports provided by agents, nor information from prisoners of war, nor air reconnaissance, nor any other type of intelligence gathering had provided reliable information about the true scale of operations to develop the new secret weapon.

There is a plausible legend that in May 1943, in London, a meticulous aerial photograph interpreter, Royal Air Force officer Constance "Babs" Babington-Smith, discovered a small aircraft without a pilot's cockpit in one of the photos of Usedom Island. This was the Fiesler-103 flying bomb, later called the V-1. Other aerial photos taken afterward revealed the A-4 "little cigars." Only then did the British General Staff begin to analyze related information provided by agents in France, Poland, Norway, and Sweden. From these agents, the British learned that in December 1943 they could expect an attack on Britain by new weapons—flying bombs, and some sort of enormous rocket. Aerial photographic reconnaissance had already detected 138 possible launch pads on the northern coast of France and the Netherlands. The British obtained photographs of launch pads and other information from the French about the presence of special troop units for the maintenance of a special purpose weapon. These materials, which British intelligence did not assimilate until three-and-a-half years after the previously mentioned letter from anti-fascist engineer Kumerov, finally compelled the British to act. The skeptics who had been convinced that the intelligence reports were nothing more than rocket hysteria finally came around.

Comparing all of this information on the as yet unknown rocket weapon with information on the Germans' work on the "uranium project," the British wondered if there was a link between these two operations. Churchill had been fully informed about the Americans' work on the atomic bomb. Moreover, he believed that British scientists should be sent to the United States to work on this problem in order pre-empt the Germans at all costs. But what if these flying bombs or these "cigars" detected in the photographs were linked with the Germans' work on an atomic bomb? It was dangerous to investigate further, and Churchill agreed to a bombing strike against Peenemünde. The Royal Air Force developed a disorienting tactic—for many weeks before the strike, British and American pilots flew over Usedom on their return from bombing runs on Berlin. The island's air defense was given strict orders not to open fire and not to send fighters into the air so as not to attract the enemy's attention to the top-secret island. This continued until 17 August 1943.

On the eve of the Peenemünde attack, Marshal of the Royal Air Force Sir Arthur Travers Harris summoned the officers responsible for the upcoming operation and warned them about the particular responsibility of the crews and the extraordinary importance of destroying this target. "If the attack is not successful, it will be repeated the following night. In that case, however, it will not be possible to avoid great losses."

The first waves of bombers flew over Usedom in the late evening of 17 August 1943 without dropping a single bomb. The Germans below did not even sound the air-raid alarm. Suddenly flares lit up over the northern end of the island. This was the beginning of the first and most powerful bombing strike of the entire history of Peenemünde. Five hundred and ninety-seven four-engined bombers rained down thousands of high-explosive and incendiary bombs on the prohibited area and nearby settlement. One wave of bombers followed another, carpet-bombing the production buildings, test-rig facilities, and laboratory buildings. A total of 1.5 million kilograms of high-explosive and incendiary bombs were dropped. The local air defense proved powerless but night fighter aircraft urgently called in from Berlin shot down 47 American B-25 Flying Fortresses.

Seven hundred and thirty-five Peenemünde residents were killed—among them were many leading specialists, including chief engine designer Dr. Walter Thiel. After hearing of the scale of the attack, *Luftwaffe* Deputy Commander Colonel General Jeschonnek, who was directly responsible for the air defense system of that area, committed suicide. But Dornberger and von Braun did not lose heart. They assured the chief of Himmler's security service, SS *Obergruppenführer* Ernst Kaltenbrunner, that the Peenemünde survivors would be able to overcome the aftermath of the catastrophe. Operations were slowed down but not halted. The air war against Peenemünde confirmed again that it was quite impossible to stop experimental weapons development operations using conventional aviation bombers, even such powerful ones.

This example of the Peenemünde team's tenacity was one more piece of evidence against Douhet's celebrated doctrine which counted on the use of conventional means of air attack.

As a result of the Peenemünde bombing in August 1943, the *Wehrmacht* decided to create a backup research test range in Poland to continue the optimization of the A-4 and to bring it to the point of reliability for combat.

At the same time, the military was tasked with intensifying the training of troop formations to service combat launchers. To accomplish this, Himmler proposed using the SS Heidelager test range in Poland, which was located in the Debica area between the Vistula, Wisloka, and San Rivers. The test range's line of fire went north-northeast from the small town of Blizna to the bend of the Bug River in the Siedlce-Sarnaki area east of Warsaw. The range and all of its facilities were thoroughly camouflaged. Construction was carried out by the prisoners of the Pustkow concentration camp (approximately 2,000 people), who were later all killed.

The 444th Test Battery, the "Blizna Artillery Range," was located in the villages of Blizna and Pustkow. The 444th Test Battery conducted the first experimental launch in Blizna under field conditions on 5 November 1943, and the first combat use of the A-4 began just one year later.

During the test firings in Poland there was one failure after another. Some rockets failed to take off; immediately after ignition the circuit would reset itself. Some took off and immediately fell on their tails, destroying the launch pad. Others exploded at an altitude of several kilometers due to fires in the tail section, crashed due to control system failures, or broke up in the air due to aerodynamic heating of the oxidizer tanks. Only 10–12 percent of the launched missiles reached their targets. Series production at *Mittelwerk* was already running at full speed, and the Peenemünde specialists were desperately attempting to determine the causes of the mid-air breakups by performing series after series of new test launches. Such a method now seems anachronistic to us because, as a rule, we are tasked with ensuring the successful launch of a new missile on the first attempt. At that time there was no other way to gain experience. We went part way down that difficult path at Kapustin Yar from 1947 to 1950.[7]

The lack of multi-channel telemetry systems also played a role. Messina-1, the first radio telemetry system, had only six channels. But even its use was limited at the Polish test range due to radio silence. The Red Army offensive was rolling—the eighteenth and last test launch near the village of Blizna took place on 30 August 1944. The Blizna test battery was restationed in an area south of Liège, and the first combat firing was conducted from there. It was aimed at Paris. Three days later they began routine firing against London using long-range ballistic missiles.

Thanks to the actions of the Polish partisans and underground, the British intelligence service had received valuable information concerning the test range in Poland. They had even managed to send an airplane to pick up missile parts gathered by the partisans from missile crash sites. The Brits had also obtained the remains of rockets that crashed in Sweden.

There was no time to waste, and Churchill appealed for help directly to Stalin.

FROM CORRESPONDENCE BETWEEN CHURCHILL AND STALIN

Personal Top Secret Message from Mr. Churchill to Marshal Stalin:

1. There is reliable information to the effect that for a substantial period of time the Germans have been conducting missile tests from an exper-

7. Kapustin Yar was the location in central Asia where the Soviets test-fired their V-2 and V-2-derived missiles in the late 1940s and early 1950s.

imental station in Debica, Poland. According to our information this projectile has an explosive charge weighing approximately 12,000 pounds, and the effectiveness of our countermeasures depends to a significant degree on how much we can find out about this weapon before it can be used against us. Debica is located on the route of your victoriously advancing troops and it is completely possible that you will seize this site in the next few weeks.

2. Although the Germans almost certainly will destroy or haul off as much of the equipment located at Debica as possible, you will probably be able to obtain a great deal of information when this area is in Russian hands. In particular, we hope to find out how the missile is launched because this will enable us to determine the missile launch sites.
3. Therefore, I would be most grateful, Marshal Stalin, if you could give the appropriate instructions concerning the preservation of the equipment and facilities in Debica, which your troops might capture after seizing this area, and if you would then provide us with the opportunity for our specialists to study this experimental station.

13 July 1944.[8]

Churchill and Stalin exchanged six telegrams in 1944 on the participation of British specialists in an expedition to the German test station in Debica. Stalin gave instructions to allow the Brits to inspect the test range, though not as quickly as Churchill would have liked. Due to the particular secrecy of the correspondence between Churchill and Stalin, the texts of the letters were not accessible until long after the deaths of both leaders.

In July 1944, we Soviet missile specialists who had been working in NII-1, formerly RNII, knew nothing about the test range in Poland and still had virtually no idea about the A-4 missile. As is evident from Churchill's letter, the Brits only had vague notions about the missile. All of the instructions that Churchill mentions in his letter were given directly to the General Staff. Accordingly, our army intelligence services received orders to be particularly vigilant in gathering intelligence on the Debica area, which in August 1944 was still 50 kilometers from the front line.

At the same time, People's Commissar of the Aircraft Industry Shakurin received instructions from Stalin to prepare a group of Soviet specialists who could study everything that could be found on that test range before the British specialists showed up there.

The first expedition, comprised of military intelligence under General I. A. Serov, was sent to the liberated area of the alleged test range hot on the heels of

8. *Perepiska Predsedatelya Soveta Ministrov SSSR s prezidentami SShA i premier-ministrami Velikobritanii vo vremya Velikoy Otechestvennoy voyny, 1941-1945 gg., 2-ye izd.* (Correspondence of the USSR Council of Ministers Chairman with the U.S. President and British Prime Ministers During the Great Patriotic War, 1941-1945, Vol. 1, 2nd ed.) (Moscow: Politizdat, 1986).

battle. Included in this group from our institute were Yu. A. Pobedonostsev, M. K. Tikhonravov, and several of their immediate technical assistants. They dug around Poland for a rather long time under heavy guard. After our group had been working in Poland for about a week, the British specialists arrived, including a representative of British intelligence who had a detailed map of the area showing the coordinates of the launch site and numerous sites where the missiles had fallen. Upon his return, Tikhonravov told us that our military intelligence officers had driven all over the test range and had confirmed that the Britsh map was right on the money. Their intelligence service had provided accurate information.

From the author's archives.

Mikhail Klavdiyevich Tikhonravov, 1945.

In many respects for our future activities, Churchill's appeals to Stalin were truly decisive. If not for his letters, our victorious army would have moved right past these Polish marshlands and forests without investigating what the Germans had been doing there. With the help of the Brits, we were able to recover A-4 missile parts for the first time. (We of course did not know the designation "A-4" at that time.)

Within days after the captured missiles were delivered from Poland to NII-1 in Moscow, some wise person commanded that they be kept secret from Soviet rocket specialists. It was sometimes impossible to understand the logic of our intelligence services.

All of the missile parts were placed in a large assembly hall at the institute. Only the chief of the institute (General Fedorov), his science unit deputy (our "patron" General Bolkhovitinov), and an information security officer were granted access. Even Pobedonostsev and Tikhonravov, who had seen everything in Poland, loaded everything into the airplane, and brought it with them, were initially barred from entering. But gradually common sense began to prevail. Isayev, and then I, Pilyugin, Mishin, and several other specialists were allowed to inspect the German secret weapon. Entering the hall, I immediately saw a dirty, black, funnel-shaped opening from which Isayev's lower torso protruded. He had crawled head first through the nozzle into the combustion chamber and, with the aid of a flashlight, was examining the details. A gloomy Bolkhovitinov sat nearby.

I asked, " What is this, Viktor Fedorovich?"

"This is something that can't exist!" he answered.

We had simply never imagined a liquid-propellant rocket engine of such proportions at that time!

According to Tikhonravov, who had delivered this engine from the Polish swamp, its location had also been indicated on the British intelligence map. The Brit who brought them to the swamp said that a local resident had passed along the site coordinates. He, in turn, had received the coordinates from Polish partisans. Not far away they had found blown up aluminum tanks, pieces of the exterior steel casing, and white shreds of prickly fiberglass. They didn't manage to get everything out of the swamp. The explosion of the propellant components had scattered missile parts all over the area.

The Brits were very interested in the remains of radio equipment and control system instruments that had remained intact. They had gathered several large cases of all sorts of parts to be sent immediately to Britain via Moscow. Upon the arrival of the British cases in Moscow, we were given the opportunity to inspect the contents the night before they were transferred to the British Mission. Pilyugin, two other engineers, and I did just that at the Khoroshevskiy barracks.

A group headed by Bolkhovitinov—consisting of Isayev, Mishin, Pilyugin, Voskresenskiy, and I—received the assignment to reconstruct the general form of the missile, its methods of control, and primary specifications based on the fragments that had been recovered. A year later, already working in Germany, I determined that for the most part, we had correctly reconstructed the missile, and this greatly facilitated our subsequent activity.

Early in 1945, we received information from Poland about new interesting parts that had been discovered in the area of the same test range. This time institute chief General Fedorov decided to head a search expedition himself. He took along the leading specialist on radio systems, my colleague Roman Popov. His group also included the leading specialist on solid-propellant missiles, Colonel Leonid Emilyevich Shvarts. They flew out of Moscow on 7 February 1945 in a Douglas. Near Kiev the aircraft became cloud-bound, evidently lost its bearings, and then crashed. All twelve passengers and crew perished. For me the loss of the remarkable radio engineer Roman Popov was particularly painful. His death effectively ended our work on a radio-guided target location system for rocket-powered interceptor aircraft.

During the first months of 1945, we put together an approximate representation of the A-4 missile, but we still had no concept of the true production scale or the combat-effectiveness of this "vengeance weapon." Specialists from the Air Force who were working with us at NII-1 were particularly interested in the reliability of the missile—a drone aircraft that was fully automatically controlled. We did not resolve these questions until after we arrived in Germany.

In December 1944, General Kammler conducted a review of the A-4 missile's reliability. Over the review period, 625 missiles were delivered to troop units. Of these, 87, or 12.3 percent, were immediately returned to the factory due to defects in the control system. Of the 538 remaining missiles, 495 were launched. Of this number, 44 launches were recorded as failures. Here, 41 percent of the failures were attributed to the control system, 13 percent to the propulsion system, 13 percent to fires in the tail section, and 2.9 percent to explosions upon launch. Thus, out of the 625 missiles, 131 were clearly unfit for launch. The Germans did not have data concerning missile crashes and breakups during the descending atmospheric segment. According to our subsequent experience launching A-4 missiles at Kapustin Yar in 1947, crashes during the latter segment of the trajectory occurred at least 15–20 percent of the time. Consequently, one should consider that no more than 400 of the 495 missiles launched reached their target, or 64 percent of the missiles delivered from the factory.

In spite of the V-weapon's low degree of reliability and effectiveness, the *Wehrmacht* and Reich leaders did not advise Hitler to reduce resources for the production of missile weaponry and increase expenditures for aviation and infantry armaments. On the contrary, work at Peenemünde to perfect the A-4 and on new projects that were capturing the imagination were developed with new vigor, in spite of the approaching inescapable defeat.

According to the testimony of German Minister of Armaments and Munitions Speer before the International Military Tribunal in Nuremberg, in technical production and economic terms, the war had been lost as early as early summer 1944. Total production was already insufficient to meet the demands of the war. Speer commented, "The Germans cannot recall without pain the astounding achievements of their researchers, engineers, and specialists during the war and how these achievements proved to be in vain, especially since their enemies could not oppose those new types of weapons with anything remotely comparable to them in the least."[9]

There is no point in feeling sorry for the German scientists and engineers because they ran out of time and their "achievements were in vain." Whether they wanted it or not, they accelerated Germany's defeat by diverting large amounts of resources for prospective developments from Germany's depleted military-economic production potential. And they had even more interesting projects than the A-4.

During his first visit to Peenemünde in May 1945, Aleksey Isayev and a group of colleagues from NII-1 went through all kinds of trash trying to find any remnants of missile documentation. All of their searches had been unsuccessful. But then, according to Isayev, one of our colleagues who had stepped away from the group to relieve himself behind a woodpile let out a howl and returned with a thin booklet—a report. A diagonal red stripe ran across the slightly damp cover with the frightening inscription *Streng Geheim* (Top Secret). The collective panel of experts that was set up there on the spot determined that this document was the design for a rocket-powered bomber.

Isayev told me about this rare find in Berlin after returning from Peenemünde. He was an engineer with an original way of thinking, who was captivated by

---

9. E. Schneider, *Itogi vtoroy mirovoy voyny* (Results of the Second World War II) (Moscow: Izdatelstvo inostrannoy literatury, 1957).

extraordinary, new ideas regardless of who proposed them. In a half-whisper so that no one could overhear, he confided, "Blow my brains out! What have they invented here?! It's an airplane, but not our pitiful BI, with a 1.5 metric ton bottle. This one had 100 metric tons of sheer fire! That damned engine hurls the airplane to a frightful altitude—300 or 400 kilometers! It comes down at supersonic speed, but doesn't break up in the atmosphere—it glances off it, like when you throw a flat stone across the water at an acute angle. It strikes, skips, and flies farther! And it does this two or three times! Ricocheting! Remember how we used to compete at Serdolikovaya Bay in Koktebel? The one who got the most skips won. That's how these aircraft skip along the atmosphere; they dive down only after they have flown across the ocean in order to slice their way into New York! What an impressive idea!"

The newly discovered report was immediately "classified" a second time. In the presence of witnesses, Isayev slipped the report under the shirt of his most reliable collaborator, then instructed this person to return to Moscow in their B-25 Boston without reporting to General Sokolov.

As far as I was able to understand later, this was not the design of the A-9/A-10 missile, which was designed for a range of 800 kilometers. The report discussed the ranges required to strike New York. From today's standpoint, we can say that the layout of the vehicle described in the report—found in the woodpile in Peenemünde in May 1945—anticipated the structure of the American Space Shuttle and our Energiya-Buran system.

Let's interrupt the narrative about Peenemünde and take a closer look at what they had found in the "woodpile." After the report arrived in Moscow via the special flight of the Boston bomber, it was personally delivered to our patron, General Bolkhovitinov. Together with engineer Gollender, who had a good command of German, Bolkhovitinov studied the sensational contents.

The report had been issued in Germany in 1944. Its authors were the Austrian rocket engine researcher E. Sänger, who was already well known before the war, and I. Bredt, who was unknown to us and was later identified as Irene Bredt, a gas aerodynamics specialist.

Eugen Sänger was known for his book *Raketen-flugtechnik* (The Technology of Rockets and Aviation), which he published in 1933. It had been translated and published in the Soviet Union. Back when he was a 25-year-old engineer, Sänger was captivated by the problems of rocket technology. He was one of the first serious researchers of gas dynamic and thermodynamic processes in rocket engines.

You can imagine how Bolkhovitinov and other NII-1 specialists felt as they leafed through the top-secret report, one of 100 printed copies. Judging by the distribution list, it had been sent to the leaders of the *Wehrmacht* main command, the ministry of aviation, to all institutes and organizations working in military aviation, and to all German specialists and leaders who were involved in rocket technology, including General Dornberger in the army department of armaments, who also served as chief of the Peenemünde center.

The title of the report was "*Über einen Raketenantrieb für Fernbomber*" (On a Rocket Engine for a Long-Range Bomber). This paper analyzed in great detail the technical capabilities for creating a manned winged rocket weighing many tons. The authors convincingly showed by constructing nomograms and graphics that with the proposed liquid-propellant rocket engine with a thrust of 100 metric tons it was possible to fly at altitudes of 50–300 kilometers at speeds of 20,000–30,000 kilometers/hour, with a flight range of 20,000–40,000 kilometers. The physical and chemical processes of high-pressure and high-temperature propellant combustion were studied in great detail, along with the energetic properties of propellants, including emulsions of light metals in hydrocarbons. The work proposed a closed, direct-flow, steam power plant both as a cooling system for the combustion chamber and as a means to activate the turbopump assembly.

The problems of aerodynamics for an aircraft with a speed ten to twenty times greater than the speed of sound were new for our aerodynamics specialists. The report went on to describe the launch, takeoff, and landing dynamics. In an apparent attempt to interest the military, the report included a highly detailed examination of bombing issues, considering the enormous speed of a bomb dropped from such an aircraft before it approached the target.

It is interesting that Sanger had already shown by the early 1940s that launching space aircraft without auxiliary means was unacceptable. He proposed launching space aircraft using a catapult with a horizontal track that would enable the aircraft to reach a speed greater than the speed of sound. Commenting on the calculation and visual graphics of flight, Sänger and Bredt wrote:

> Takeoff is conducted using a powerful rocket complex fixed to the ground and operating for approximately 11 seconds. Having accelerated to a speed of 500 meters/second, the aircraft lifts off from the ground; with the engine at full power, it climbs to an altitude of 50–150 kilometers on a trajectory that is initially at an angle of 30 degrees to the horizon, and then becomes lower and lower. . . . The ascent lasts from 4–8 minutes. During this time, as a rule, the entire fuel supply is consumed. . . . At the end of the upward phase of the trajectory, the rocket engine shuts down; using its stored kinetic energy and potential energy, the aircraft continues its flight in its characteristic glide along a wave-like trajectory with attenuating amplitude. . . . At a previously calculated moment, the bombs are dropped from the aircraft. Tracing a large arc, the aircraft returns to its airfield or to another landing pad and the bombs, which are flying in the original direction, come down on the target. . . . This tactic makes the attack completely independent of the time of day or the weather over the target, and deprives the enemy of any capability to counteract the attack. . . . The problem that we posed, which until now no one anywhere had solved, entails firing on and bombing targets located 1,000–20,000 kilometers away. A formation of 100 rocket-propelled bombers . . . within the course

> of several days would be capable of completely destroying areas approaching the size of world capitals, including their suburbs, located anywhere on the face of the earth.

The total takeoff weight of the bomber was 100 metric tons, of which 10 metric tons was the weight of the bombs. The landing weight was assumed to be 10 metric tons. If the flight range were reduced, the weight of the bombs could be increased to 30 metric tons. They proposed that the subsequent work to implement the design of the rocket-propelled bomber be divided into twelve stages, in which the bulk of the time would be devoted to firing rig optimization of the engine, rig testing of the interaction of the engine and aircraft, launcher testing, and finally, all phases of flight tests.[10]

In 1945, the work of Sänger and Bredt was translated, and in 1946 it was published under the title "Survey of Captured Technology" by the Military Publishing House of the USSR Armed Forces Ministry, under the editorship of Major General of the Aviation Engineer Service V. F. Bolkhovitinov; a large number of copies were printed.

Being in Germany at the time, Isayev and I had no idea that this report had caused quite a stir after its delivery to NII-1 in May 1945. We could only imagine the feelings experienced by our patron, who was considered a dreamer in higher aviation circles, but who was respected for his enthusiasm in the face of extremely bold proposals, a quality that was very unusual for a chief designer. Together with the engine specialists from RNII, we had only obtained a reliable liquid-propellant rocket engine with a thrust of 1.5 metric tons in 1943. Isayev dreamed of bringing the engine up to a thrust of 2–3 metric tons in a year or two. But then, in 1944, a V-2 engine with a thrust of almost 30 metric tons was recovered in Poland. Added to this now was Sänger's report, which outlined the design for an aircraft with engine thrust of 100 metric tons!

When Bolkhovitinov's deputy, MAI professor Genrikh Naumovich Abramovich, flew into Berlin from Moscow in June, he was already familiar with Sänger's work. Being a very erudite theoretician, he said that such an abundance of gas-kinetic, aerodynamic, and gas-plasma problems required a profound scientific analysis. He believed it would take ten years—God willing—before it came down to the business of designers. "It's easier to make rockets than that airplane."

Yes, that proposal was at least twenty-five years ahead of its time. The first space aircraft in the form of the Space Shuttle flew in 1981. But it launched vertically as the second stage of a rocket. To this day there is no authentic aerospace vehicle with a horizontal launch.

---

10. For an inside look at the Sänger-Bredt bomber, see Irene Sänger-Bredt, "The Silver Bird Story: A Memoir," in R. Cargill Hall, ed. *Essays on the History of Rocketry and Astronautics: Proceedings of the Third through the Sixth History Symposia of the International Academy of Astronautics, Vol. I* (Washington, D.C.: NASA, 1977), pp. 195-228.

In modern-day Germany, they have designed an aerospace system that is called "Sänger" in honor of the pioneer of this idea. The largest German aircraft firms participated in the work on this program. The spacecraft was designed on the basis of forward-looking, but realizable technology and was intended to transport various cargoes into space while lowering costs and ensuring safety, reliability, and all-purpose use. It differs fundamentally from the 1940s design in that horizontal acceleration is not achieved by a catapult, but by a special booster aircraft that carries the actual spacecraft, which will be capable of inserting 10 metric tons of payload into near-Earth orbit—the same amount specified in Sänger's original design—at an altitude of up to 300 kilometers. Working in 1944, Eugen Sänger certainly could not have imagined the materials, engines, and navigation and control methods that German scientists with access to advanced space technologies are working on now.

In 1947, in conversations with Gröttrup, we were trying to determine Peenemünde's attitude toward Sänger's design during the war. The gist of his response was something like the following: First, the consensus was that work on Sänger's design might hinder the A-4 program and the other programs at Peenemünde that were purely rocket-oriented. Second, they believed that such a design would require at least four to five years of intense work before the first flight; and third, it was an aircraft—the design interested the *Luftwaffe*, but rocket technology was under the management of the infantry command. Even here institutional partiality was at work!

It is interesting to compare the different assessments of the development cycle for the Sänger aircraft. Peenemünde estimated up to five years, while G. N. Abramovich's subsequent assessment was up to ten years. The present Germany began work on the "Sänger" in 1986 and scheduled the first demonstration flight for 1999—a thirteen-year development cycle! And this was more than fifty years after Isayev's group extracted the top-secret report from the woodpile. Currently, work on the project has been practically halted due to the European Space Agency's lack of funding.

Though Eugen Sänger would never see an aircraft bearing his name, he nevertheless received international recognition during his lifetime. In 1950, he was elected the first president of the International Academy of Astronautics, and in 1962 the USSR Academy of Sciences awarded him the Yuriy Gagarin Medal.

In Peenemünde, there was serious work underway on another large cruise missile. By December 1944, the territory of Germany had been invaded by the Red Army from the east and the Allies from the west. The defeat of the Nazis was inevitable. Nevertheless, the stubborn specialists in Peenemünde launched an A-9 cruise missile under the designation A-4b on 27 December. The launch was unsuccessful. From our vantage point today, the failure can be easily explained—it was simply unavoidable, the knowledge and experience to realize this design did not exist. They started this work with the particular courage that comes with ignorance. The time for the realization of such designs had not yet come, especially since it was already too late to be working on them in Peenemünde. One had only to glance at a map of the military situation.

Nevertheless, in 1944, Dornberger decided to consolidate in Peenemünde the projects for the development of automatically guided anti-aircraft missiles that were scattered among different agencies and companies. The combat use of air defense missiles had been scheduled to begin in 1942–43. The existing types of radar at that time—Burund, Hansa, Brabant, Percival, and Lohengrin—were used to support launches, guidance, and monitoring. The launch area design was code-named Vesuvius.

Each *Wasserfall* anti-aircraft guided missile battery consisted of one radar and four launchers. The utopian plan for the protection of Germany called for 870 *Wasserfall* batteries and 1,300 *Schmetterling* batteries. It is astounding that this pipe dream was considered possible!

By mid-1945, the unfeasible production plans also called for no fewer than 2 million *Taifuns* to be produced per month. The *Taifun* was developed during the last year of the war to battle large Allied bomber formations. It was the smallest liquid-propellant rocket the Germans developed, with a length of just 1.9 meters and a diameter of 10 centimeters. The engine generated 500 kilograms of thrust, imparting to the projectile—which weighed just 9 kilograms—a speed four times greater than the speed of sound! The *Taifuns* were supposed to be launched in salvos from a launcher with forty-six launching rails. Here the influence of our *Katyusha* was clearly evident. But the *Taifun* engine had not been optimized, and the Germans had not yet mastered the technology of solid-propellant rockets.

In contrast to the Allies and us, the Germans understood that a guided missile capable of generating supersonic speed is the most effective means of combating aircraft. The *Wasserfall* could have been produced earlier, but little attention had been devoted to it—the doctrine of vengeance had dominated. The *Wasserfall* required very large expenditures. They believed that a battery of eight launchers using thirty-five rockets could repel an entire bomber squadron.

The *Schmetterling* was an air defense missile produced by the company Henschel, with tests conducted in Peenemünde. The *Rheintochter* was still in development. It was a two-stage, solid-propellant rocket produced by the company Rheinmetall Borsig, but it possessed only subsonic speed. The anti-aircraft guided missile *Enzian* was developed on the basis of the Me 163 rocket-propelled fighter at the research center in Oberammergau. Thirty-eight of these missiles were fired. It used a Walter engine with a thrust of 1.5 metric tons and *Rheinmetall Borsig* solid-propellant launch boosters. The *Enzian* is one more example of the transfer of scientific-technical ideas through super-secret barriers, even during wartime.

While we were developing the ideas of radio guidance for the BI-1 aircraft in early 1944, Roman Popov, Abo Kadyshevich and I arrived at the idea of making this aircraft automatically guided. Popov and Kadyshevich worked on using the newest American radar for this purpose, and I attempted to create a small work force to develop an autopilot. The work proved to be considerably more labor-intensive than we had imagined at that beginning stage when an interesting idea entices inventors into the meat grinder of problems. The discontinuation of work on the BI-1, the virtual elimination of the danger of German bombings, and then the tragic death of Popov halted further work.

The Germans discontinued similar work because the ideas of the *Wasserfall* were more reasonable. The development of the *Wasserfall* anti-aircraft guided missile went farther than the rest. Greater efforts by control systems specialists were diverted for this work because the problem of hitting an airplane proved to be much more complex than firing A-4 missiles "against areas." To optimize the *Wasserfall* control system, its equipment was installed on an A-4, and an experimental launch was conducted in March 1944. The launch was conducted vertically from the island of Greifswalder-Oie. Due to a failure of the control system, the missile turned to the north and fell in the south of Sweden. Fragments of the missile were delivered to Britain and provided the Brits with the first more or less accurate notions about the A-4 missile. At that time, no one in Britain knew that this launch was a control system test for an air defense missile.

By December 1944, only the *Wasserfall* and *Schmetterling* remained on the list of air defense missiles that had been kept for production and testing in Peenemünde. Counting on dragging out the war, the Germans developed plans for their series production during 1945 and 1946. But it was impossible to oppose the Allies' powerful bomber strikes from the west and the complete air superiority of the new models of Soviet aircraft. Nevertheless, over the course of a year the Germans managed to conduct around one hundred experimental launches of the *Wasserfall*. According to Gröttrup, the *Wasserfall* and *Schmetterling* documentation was completely destroyed during the evacuation of Peenemünde. In Germany, we were certain that the Brits were most interested in the *Wasserfall*.

Work on the A-9, the winged version of a long-range missile, continued in spite of the catastrophic situation on the Eastern and Western Fronts. On 24 January 1945, a successful launch of the A-4b finally took place. This was the first launch of an experimental long-range missile with wings.

In December 1944, Hitler awarded the Knight's Cross, one of the highest Nazi decorations, to five Peenemünde scientists, including von Braun, for exceptional service in the design, manufacture, and use of V-2 missiles.[11]

On 14 February 1945, the last A-4 missile was launched from Peenemünde. The Eastern Front of Hitler's Reich was collapsing. After their decorations were conferred, the Peenemünde directors received no more orders and began to prepare for evacuation on their own initiative. All of the equipment and documentation was packed into cases marked "EW." The accompanying documents noted that this was the property of an *elektrotechnisches werk* (electrical factory). Convoys of automobiles and special trains carrying specialists, archives, and equip-

11. During the Third Reich, the Iron Cross had eight grades. The lowest two grades were Iron Cross First and Second Class, and the highest grade was the Grand Cross. The five grades in between were the Knight's Cross and its variations.

ment, headed by Dornberger and von Braun, left Usedom Island on 17 February 1945. They evacuated to the areas of Nordhausen, Bleicherode, Sonderhausen, Lehesten, Witzenhausen, Worbis, and Bad Sachsa. The primary archives with the results of thirteen years of research and work were hidden in the tunnels of *Mittelwerk* and nearby potassium mines. The main group of Peenemünde directors was sent to the Bavarian Alps. On 4 May, the troops of the Second Byelorussian Army Group entered the area of Peenemünde. On 2 May 1945, the Peenemünde directors went out toward the Americans and surrendered willingly. On the blindingly sunny day of 2 May 1945, when my comrades and I were jubilantly signing our names on the still-smoldering Reichstag walls, the Americans captured some of the most valuable spoils of the war: more than four hundred of the main scientific-technical employees of Peenemünde; documentation and reports; more than one hundred missiles ready to be shipped to the front that had been stored at *Mittelwerk* and on spur tracks; and combat launchers, along with the military personnel who were trained to operate the missiles!

The next stage in the history of rocket technology had begun. It could rightly be called the Soviet-American stage. German specialists participated in the work of this stage in the USSR and in the United States.

Some old hands from Peenemünde who were still alive in 1992, along with a few admirers of the Hitler era in modern Germany, decided to commemorate the fiftieth anniversary of the first successful launch of the A-4 on 3 October 1942. For this occasion they planned a large festival in the Peenemünde area with the participation of foreign guests. The celebration was advertised as the "50th Anniversary of the Space Age." The British public strongly protested the commemoration, and Chancellor Kohl had to intervene. Mass demonstrations were prohibited, and the regional minister who had encouraged this festival was forced to retire. The British acted graciously. The event coincidentally took place twenty-four hours before the real anniversary of the Space Age—the thirty-fifth anniversary of the October 4 launch of the first artificial satellite! Meanwhile, Russia's new reform-minded authorities ignored the 35th anniversary of the launch of the first artificial satellite, and moreover did not interfere in the British-German brouhaha.

Chapter 18

To Thuringia

The Allied armies had occupied Germany. But they were faced with a subsequent regrouping in accordance with the decision of the Yalta Conference. We were supposed to vacate the western areas of Berlin and, by the same token, the Americans were vacating Thuringia, the same Thuringia where Nordhausen, already known to us from German accounts, was located.[1]

All of the personnel from Peenemünde, all of the documentation and unique equipment were evacuated to Thuringia. Our authorities were in no hurry to bring the troops out of the western part of Berlin because they needed time to disassemble and transfer machine tools and valuable equipment from the factories in the western part of the city to our zone. Two motor rifle divisions had been sent for the sole purpose of disassembling *Siemensstadt* (Siemens City). Now dust was stirred up not by battles, but by the hundreds of Studebakers and other vehicles hauling captured equipment along streets that had not yet been cleared.

Meanwhile, the Americans were also in no hurry to take their troops out of Thuringia. They needed to search for and bring out as many German specialists as possible, particularly missile and nuclear experts. They needed to retrieve rockets and various rocket equipment from the subterranean factories in Nordhausen and transport this material outside of the Red Army zone. Everyone working on both sides of the still undesignated boundaries was in a hurry. They asked their commanders not to rush to remove the checkpoints and the guards on these boundaries. Nevertheless, officers and soldiers interacted spontaneously and with friendship, and as a rule, exchanged watches, cigarettes, belts, and military insignia from garrison caps.

On 9 May all the armies jubilantly celebrated victory. The war had been won. Now we were faced with winning the peace. The central streets of Berlin were decked out with the flags of the four Allied powers. On 4 June, a meeting was set up between the commanders of the occupying troops, who were supposed come to an agreement about practical measures for the management of Germany after its unconditional capitulation.

---

1. At the Yalta Conference in February 1945, the major Allied powers made plans for dividing up Germany into four zones of occupation (American, Soviet, British, and French).

Tempelhof airfield was preparing for a meeting of distinguished guests—the commanders of three powers. Our commandant's staff had restricted the passage of Soviet officers through the western portion of Berlin. But after some wrangling, we obtained the necessary passes and June became a very hectic time, especially for me.

During the "special regime" period when the allied commanders were meeting, I went to visit Isayev in Basdorf. This quiet hamlet housed an experimental factory, laboratories, and test rigs produced by the Walter company. These facilities produced the liquid-propellant rocket engine for the German Me 163 fighters.

Early in the war, Messerschmitt was developing a fighter-interceptor that was very similar in specifications to our BI. The Germans had also, before this, tested the Heinkel He-176 with a *Walter* engine in Peenemünde—they had even executed a takeoff! Several dozen of these aircraft were manufactured, but they saw virtually no action in air battles.

The *Walter* engine was very reminiscent of the engines that Dushkin and Shtokolov had begun to develop at RNII, that Isayev further developed, and that Glushko, working independently in his *sharashka* in Kazan, was making more reliable than any of them had been able to do. Korolev was testing engines developed together with Glushko on Pe-2 airplanes in Kazan.

In this field, our developments and those of the Germans were running in parallel. It was therefore not surprising that Isayev, after getting a look at deserted Peenemünde, had settled down for a long time in Basdorf with an entire team of Muscovite and Khimki engine specialists.

Isayev had clearly gotten bored in Basdorf and his mood was somber. Isayev's compatriots—the expansive Ivan Raykov; the imperturbable Anatoliy Tolstov; and Arvid Pallo, who had joined them—enthusiastically told me about their interactions with the German *Walter* employees.

"Before our troops showed up they had managed to bury the most technologically complex and original parts of their engines, chambers, and turbopump assemblies. The ones who buried them have taken off for the West. We had to track everything down using persuasion, bribery, and sometimes threats. After they had determined that we were engine specialists, the Germans restored their test rigs and conducted several firing tests with us. We achieved a thrust of up to 1,500 kgf and there is still some in reserve. Everything they've got is really high quality."

"So what's the point?" asked Isayev, interrupting his compatriots, who were captivated by the detective story. "I know now that we and Sasha Bereznyak, our dear patron in Bilimbay, the brilliant Glushko in Kazan, Kostikov and Tikhonravov in Likhobory, and completely independent of us, the renowned Messerschmitt and Walter have gotten off-track! Rocket-aircraft such as our BI and the German Me 163 aren't needed! In nine out of ten cases the pilot doesn't have time to attack the enemy. Our engine will burn up all its fuel and die at the most critical moment. Bakhchi, may he rest in peace, was a hero and an enthusiast. But for hundreds of young pilots, one single thought will pulsate in their minds from the moment they take off at night: How and where will I land in order to save the aircraft and my life? Our developers need to find another application. It seems the Germans have found it, and we simply have not matured. We don't need a rocket-powered fighter with a pilot, but an unmanned rocket. I heard about the *Wasserfall* in Peenemünde and a little bit here in Basdorf. This rocket *is* an aircraft fighter. That is precisely what our engines will be needed for. But we, the engine specialists, won't be the most important contributors—it will be you, the guidance specialists. The engines—they're a given. The design of a rocket, the body, that's much simpler than for an airplane. But how to find an airplane and hit it—that's the main thing!"

Isayev was morose. The thoughts that tormented him had taken shape here in quiet Basdorf.

"Blow our brains out! And Messerschmitt's too! We have squandered four years developing airplanes that nobody needs. The Germans managed to go farther, but the result is the same! We were discussing unmanned rocket-fighters with the late Roman Popov back in Khimki, after returning from Bilimbay. He and Abo Kadyshevich showed me these multi-ton radar systems and told me that if a rocket were to lift them into the air, they would be worth more than the airplane they shot down."

Our evening discussions about the future, which we held during those quiet evenings in Basdorf over steins of flat local beer, led us all to conclude that aircraft with liquid-propellant rocket engines were not necessary. Isayev was already full of new ideas. He asked me to get in touch with Moscow upon my return to Berlin and arrange for him to be recalled to the NII-1 institute—otherwise we would form an expedition and make our way west to Nordhausen without waiting for orders.

After returning to Berlin, I discovered that our headquarters in Adlershof had been greatly expanded. I received orders to fly to Moscow for a brief report and to receive new instructions.

Our NII-1, which was enjoying friendly relations with the Air Force command (the NII chiefs were aviation generals), had received an American Boston B-25 two-engine, high-speed bomber as a gift. It was much more pleasant to fly in than the Douglas transport planes. The view from the navigator's cockpit was magnificent, and it was faster—the flight from Berlin to Moscow took just a little over five hours.

I was in Moscow for just two days. I managed to meet with Bolkhovitinov, who sounded depressed. There was an "opinion" in the People's Commissariat that the NII-1 leadership should be replaced and the NII should be reoriented to pure science—to areas such as gas dynamics and air-breathing jet engines. Viktor Fedorovich asked me to immediately familiarize myself with Nordhausen and return to Moscow.

I was trying to understand who in Moscow would be conducting work on subject matter that was purely rocket-oriented; in other words, who needed the V-2 and all the German secrets immediately. His opinion was, "Nobody needs the V-2. We need jet aviation, and as fast as possible. Rockets are the future, but at the People's Commissariat they don't think that's the business of aviation."

I met with Pilyugin and Voskresenskiy. Neither objected to joining me in Germany, but Bolkhovitinov would not let them go for the time being.

Pilyugin grumbled, "Why are we rummaging in packages and papers and racking our brains over what's what? We need to be there, on site, to understand it."

I spent one night at home in Sokolniki on Korolenko Street. I saw our newborn child, my second son, for the first time—he was only two months old. Katya had cares that I had already managed to forget: firewood, kerosene, redeeming ration cards and new quota books.[2] Bathing the baby was quite an event. You had to heat the water in the kitchen on a kerosene stove and haul it to the tub in the other room.

But the mood of my resilient wife and all Muscovites was joyful. Victory! Now everything would be different.

On 14 June, I was joined on the flight to Berlin by NII-1 chief engineer N.V. Volkov and G. N. Abramovich, Bolkhovitinov's deputy who was a professor at the Moscow Aviation Institute (MAI) and already a well-known scientist in the fields of gas dynamics, thermal processes, and air-breathing jet engines.[3] But as Abramovich explained to me, his interests in Berlin would be broader. "I need to take a look at how their scientists are working in general."

Back in Berlin, I was once again involved with packing up and shipping cargo to Moscow. By 28 June I had "polished off" all the Adlershof "leftovers". I spent two entire days loading the Douglas with cases filled with measuring equipment. At that time, I still did not know that this shipment was being sent to an institute to which I would not return. Nevertheless, ten years later I was pleased to learn that Rauschenbach's group, which worked at NII-1 with M.V. Keldysh, had put this rich arsenal of measuring technology to good use.

While I was involved with loading and shipping and acting as tour guide for the higher-ups who had flown in, Abramovich caused quite a stir in Berlin. Having landed in Germany with the rank of an engineer colonel, he had managed, with the assistance of his aviation connections, to secure his own personal vehicle with a military administration license plate. It was a light gray Mercedes in excellent condition.

The Mercedes came with a German driver, whom everyone called Alfred. As a soldier Alfred had made it as far as Smolensk before being demobilized due to illness. Before serving in the army, he had been a circus performer—his part was to ride a motorcycle around a vertical wall. He drove the Mercedes magnificently. After much wrangling, Abramovich managed to obtain a document certifying that citizen Alfred Hessler was serving as a driver in the Soviet Military Administration (SVA).[4] The now-official driver wore his old circus costume—leather jacket,

2. Quota books were used for non-food consumables such as kerosene and firewood.

3. MAI—*Moskovskiy Aviatsionnyy Institut.*

4. The provincial Soviet military organizations in occupied Germany were known as *Sovetskaya Voyennaya Administratsiya* (SVA) or Soviet Military Administration. SVA organs reported to the top-level Soviet governing authority in postwar Germany, the *Sovyetskaya Voyennaya Administratsiya Germanii* (SVAG), the Soviet Military Administration in Germany.

leather breeches, and gleaming gaiters—and looked very impressive in the Mercedes. Alfred was always correcting us when we were talking about the car. "It's not pronounced 'Mersedes', but 'Mertsedes'. That's how real automobile connoisseurs say it." It turns out that that was the name of the Benz automobile firm's founder's daughter.[5]

In late June, we learned that the orders had finally been given to the American troop units to clear out. Consequently, our troops would now occupy Thuringia.

Professor Abramovich was a very sophisticated man. He was captivated not only by his professional activity in Germany, which involved studying problems that were dear to him, but also by the country, its old culture, its people, and their postwar psychology. Without much trouble he talked me into traveling to Nordhausen by car. We would mix business with pleasure by making a two-day trip from Berlin to Nordhausen through the cities of Dresden, Annaberg, Aue, Zwickau, Gera, Jena, and Weimar.

Before our departure, we arranged with Isayev for him to travel to Nordhausen with his main group of engine specialists over the shortest route, through Magdeburg. We arranged to meet in Nordhausen on 14 June.

Our journey really did prove to be interesting. I strolled through the already cleared streets of Dresden. It was astonishing how rapidly peaceful life had been restored in the city. Streetcars were beginning to run between heaps of majestic ruins. In places where by some miracle the first floors had been preserved, work was starting up again in stores, cafes, and pharmacies. There were many signs proclaiming, Checked—No Mines. The air in the suburbs was fragrant with roses. The air army command was still based here, and we stopped to fill up with aircraft gasoline. To Alfred's great joy we filled the tank and three canisters. Until then he had filled the tank with methyl alcohol and had been upset by the obvious drop in engine power.

After Dresden, Alfred drove the Mercedes confidently while we monitored the route using an excellent tourist guidebook. We were surprised that it contained so many unclassified details. We had the maps of Germany that our General Staff had issued to troop units, but the readily available German maps were much more informative.

Just over a month had passed since the war's end, and we were cruising along through villages and towns without seeing any destruction. If it weren't for the passing columns of our shifting troop units, the swing-beam barriers on certain roads, and patrol guards checking documents, one would have wondered, "Was there really a war?"

At the military commandant's office in Annaberg, we stopped to have dinner and find a place to spend the night. We were warned that up ahead on our route

---

5. The Mercedes-Benz company claims that 'Mercedes' was the daughter of the company's first client, an Austrian businessman. The company was originally known as Daimler, becoming Daimler-Benz in 1926.

there was a blown-up bridge and a village that had been wiped off the face of the earth. We would need to bypass both. What had happened there? We learned the answer to that question there and then.

After a good dinner with Rhine wine, we heard about the episode in the village from an elderly man who turned out to be a Russian "displaced person" who had worked in SVA as a translator. To him, this characterized the American method of combat and the basic tenet of American military operations: Above all, save the lives of your own soldiers.

An American mechanized column was moving deep into Thuringia, meeting virtually no resistance. After entering this unfortunate hamlet—I don't remember what it was called—the advance guard was suddenly fired on by automatic weapons and hunting rifles. Later it was determined that a small detachment of *Hitlerjugend* (Hitler Youth) had settled in that village. They had responded to Goebbels' appeals and decided to become "werewolf" guerrillas. Their fire did no harm to the Americans.

If a Red Army unit had been here, these "werewolves" would have been annihilated on the spot or taken prisoner. But the Americans didn't want to risk the life of a single one of their guys. Without firing a shot, the powerful mechanized formation withdrew several kilometers. The "werewolves" decided that their village had been saved from the occupying forces, but they were sorely mistaken. The commander of the American unit reported the situation in such a way that a bomber formation was sent to his aid. It turned the ill-fated village and all of its inhabitants into formless heaps of smoking ruins. Only after this treatment from the air did the Americans continue their "victorious" advance.

We made a small detour to get a look at this demolished "fortress" and discovered an intense reconstruction project at the site of the former village.

Chapter 19

Nordhausen—City of Missiles and Death

We arrived in Nordhausen the evening of 14 July. The Seventy-seventh Guard Division, part of the Eighth Guard Army, had just taken over the city from the Americans and had already been billeted in the city and surrounding areas. The offices of the commandant and the *Bürgermeister* were already in operation. With some difficulty, we found Isayev's team of engine specialists. They had arrived a day earlier and had taken up residence in a remote and devastated villa close to the site we were most interested in—Kohnstein Mountain, where the subterranean *Mittelwerk* factory was hidden.

Isayev had already managed to establish contact with division intelligence and Smersh. The division command had already placed guards at all the obvious entrances to the subterranean factory and also at the Dora death camp. The *Bürgermeister* promised to find and gather any Germans who had worked at the factory and have them meet with us.

Meanwhile, we wandered around the city and discovered that American military Jeeps were still tearing around at breakneck speed carrying obviously tipsy soldiers with heavy pistols dangling from their broad belts. During their two-month stay in Nordhausen, the American soldiers had acquired a lot of girlfriends. In spite of orders about the demarcation of the occupation zones, it wasn't easy to back out of the next date, and our patrol received these strict instructions: "No conflicts with Allied servicemen until border guards have been set up."

Half the night we talked with Isayev about our impressions and adventures. In spite of our fatigue, here in this dark, ravaged villa hidden in the middle of an overgrown, mysterious garden, we felt very uncomfortable.

The city of Nordhausen had been largely destroyed by Allied bombers. The British and Americans, with assistance from our aviation, had tried to prevent missile production. While they succeeded in destroying the city, the brutal bombing attacks had not inflicted any damage on the subterranean factory and the neighboring Dora concentration camp.

In the morning, we found that a whole line of people wishing to offer their services had assembled at our villa after being summoned by the local authorities. We began with a Soviet officer who introduced himself as "Shmargun, former prisoner of war, liberated from the camp by the Americans." According to his statement, he was a first lieutenant and a political commissar who had been taken prisoner in 1944. After being moved to various places, he had been sent through Buchenwald to Dora. Outfitted in the uniform of an American soldier, he did not look at all like the walking skeletons usually seen at death camps. We asked the expected question: "Why did you survive?"

"Before the Americans arrived there was a lot of work—we were ordered to gather up and burn more than 200 corpses that were delivered from the factory to the camp. They needed us to be alive for that work. But we didn't manage to burn them all. About 100 bodies were still lying around when the Americans forced their way in. The Germans fled. The Americans fed us and gave us clothes. Some of the skeletons and I refused to leave with the Americans and decided to wait for our own troops.

"Now I can show you around the camp. I know some Germans who worked at the factory and didn't leave. They have agreed to help in investigating everything that was going on there. I can be in contact with 'that side.' There were a lot of good guys among the American officers. There are also a lot of Russian girls in town. They were domestic workers and worked on farms. They know the language well, and can serve as interpreters until they are repatriated. I know places where the SS hid the most secret V-2 equipment that the Americans didn't find. We prisoners knew a lot."

This helper immediately won us over, but our entire previous upbringing made us vigilant: "Might this be an American agent?" Isayev and I decided that if our Smersh rep hadn't bothered him, then in the interests of the work at hand—to hell with vigilance!— let him work and help us. After all, we had come here for secrets; we didn't bring the secrets with us.

We began by inspecting the horrible Dora death camp. The Americans had already put things in order here. All of the dead had been buried, and those who had survived had been treated and fed—the living dead were back on their feet. Special troop units were preparing the camp for Russians who were former prisoners or had been driven off into Germany. They would be sorted out and then repatriated.

Shmargun led us to a distant wooden barracks hut, where in a dark corner, after throwing aside a pile of rags, he jubilantly revealed a large spherical object wrapped in blankets. We dragged it out, placed it on a nearby cot, and unwrapped the many layers of blankets. I was stupefied—it was a gyro-stabilized platform of the type that I had seen for the first time in Berlin at the *Kreiselgeräte* factory. At that time, "civilian" Colonel Viktor Kuznetsov, who was also seeing the instrument the first time, explained its layout to me. How had a gyro-stabilized platform, which still hadn't become a standard V-2 instrument, ended up in this death camp prisoners' hut? Shmargun could not give me a clear explanation. According to what he had heard from others, when the camp guards fled, some Germans who were neither guards nor *Mittelwerk* personnel brought a beautiful case to the barracks, covered it with rags, and quickly fled. By the time that the Americans arrived, the surviving prisoners had discovered the case and opened it; one of them said that it was very secret. They decided to put it away until the Russians arrived. They used the case to pack up various things that they had begun to acquire after liberation, and when they found out that Shmargun was staying to wait for the Russians, they revealed the secret to him and packed everything up in dirty blankets so that the Americans would be less suspicious.

As we could see, the operation went brilliantly. Now Isayev and I were responsible for this priceless windfall. We wrapped it back up in the blankets, since no other container was available, transported it to the division headquarters, and asked them to store it there until we could take it to Moscow. Approximately six months later there was a struggle over the possession of this gyro platform that led to the first rift in the relationship between Viktor Kuznetsov and Nikolay Pilyugin, the friends that I made shortly after the instrument's discovery. But we'll get to that later.

After a brief inspection of the horrible Dora camp, we hurried off to inspect *Mittelwerk* itself. I must honestly confess that we hurried to leave the camp not because we had completely run out of time. The horrors that Shmargun began to tell us about and the live witnesses who had arrived from somewhere were so out of synch with the radiance of the hot July day and our frame of mind as impassioned hunters who had finally seized real spoils. We could not help wanting to cast off this hallucination. They showed us the area where the bodies had been placed before being fed into the crematorium and where they had raked out the ashes. Now there were no traces of ashes anywhere. When the Americans had been here a commission had been at work documenting atrocities and war crimes. Before our eyes the camp was now being converted into a dormitory for displaced persons. But the ashes we could not see were beginning to pound both in our hearts and in our temples.

A group of Germans was waiting for us in front of the entrance to *Mittelwerk*. They had turned up as a result of the *Bürgermeister*'s efforts. A young, thin German with delicate facial features separated himself from the group. He boldly approached and introduced himself, "Engineer Rosenplänter from Peenemünde." He explained that everyone had been evacuated from Peenemünde to Nordhausen, and they had settled not far from here in Bleicherode. At first von Braun and Dornberger, whom he knew personally, had lived there. They had left Bleicherode and moved farther west.

Before the arrival of the Russians, the Americans had sent almost all of the specialists to the towns of Worbis and Witzenhausen. Rosenplänter and several dozen other specialists had refused to move, and the American officers, having checked with their lists, didn't make them go. Certain others had been taken despite their unwillingness.

Rosenplänter said all of this rapidly; he was very agitated. Shmargun could not keep up with the interpretation, so someone drove up to the camp and brought a Russian girl who could interpret more rapidly than the Germans spoke. This interpreter charmed everyone. Her name was Lyalya. From that day on we declared her our official interpreter-secretary, and we formalized her status with the military authorities. Rosenplänter offered to familiarize us with V-2 technology, but he did not know *Mittelwerk* and recommended another specialist from Peenemünde who had often visited *Mittelwerk* to perform monitoring tasks. Rosenplänter denied that they had anything to do with the atrocities that had taken place there.

Our first inspection of the legendary subterranean *Mittelwerk* missile factory took almost two days. Literally translated, *Mittelwerk* means "middle factory" or "factory located in the middle" (it was located in the middle of Germany). The construction of the factory began in 1942 under the codename *Mittelbau* (middle construction). This was before the successful launches of the V-2 (A-4) missile. They didn't need to go extremely deep into the ground. The construction workers successfully used the natural terrain. The wooded hill that locals proudly called Kohnstein Mountain rose up almost 150 meters above the surrounding terrain four kilometers from Nordhausen. The limestone rock forming the interior of the mountain yielded easily to mining work. Four opened galleries had been cut in the mountain along the diameter of the base, each was a bit longer than three kilometers. Forty-four transverse drifts connected the four galleries. Each gallery was a separate assembly factory.

The two galleries on the left side of *Mittelwerk* were BMW-003 and JUMO-004 aircraft turbojet engine factories. These engines had already been made fit for series production in 1942. And here the Germans had surpassed the Brits, the Americans, and us. But (luckily for us, of course) as a result of somebody's foolishness, they did not use this advantage and did not release into large-scale production the twin-engine jet Me 262 Messerschmitts, which were equipped with these engines. It wasn't until the end of the war that these aircraft appeared on the fronts in small numbers. In postwar memoirs, German generals noted that Hitler had been personally opposed to using these airplanes for a long time. This demonstrates how a dictator's stubbornness yields invaluable benefit to his mortal enemies.

The third gallery at *Mittelwerk* was used for the production of the V-1 "winged bombs," which in modern terms would be called cruise missiles. The mass production of V-1s began in 1943. Only the fourth gallery was dedicated to the assembly and testing of A-4 missiles. Rolling stock bringing in materials could roll directly into each gallery from the surface. Rail cars loaded with the finished product exited at the other end.

The gallery for A-4 missile assembly was more than 15 meters wide, and its height in individual bays reached 25 meters. In these bays they conducted the so-called vertical *Generaldurchhalteversuchsprüfung*. We later translated this as "general vertical tests" and used it as the official term for this type of test for a long time.[1]

---

1. The literal meaning of the term *Generaldurchhalteversuchprufüng* is "general endurance test."

Horizontal tests were usually carried out before vertical ones, although these did not have the prefix "general." The transverse drifts were where the assemblies and subassemblies were fabricated, integrated, inspected, and tested before they were mounted on the main assembly.

Our inspection of the galleries and drifts was hampered by the fact that the lighting had been partially damaged, from what we were told, by order of the Americans. Only "duty" lighting was on. We had to be very careful while walking around the factory to not fall into some processing pit or hurt ourselves on the remains of missile parts that hadn't been cleared away. We noted the large number of missile components scattered around in disarray. It was easy to count dozens of "tails," side panels, middle sections, and tanks.

A German who was introduced as an assembly engineer-tester said that the factory had worked at full power until May. During the "best" months, its productivity was as high as thirty-five missiles per day! At the factory the Americans had seized only fully assembled missiles, taking more than 100 which had been piled up at the factory. They had even set up electrical horizontal tests. The assembled missiles had been loaded into special railroad cars before the arrival of the Russians and hauled to the west—to their zone. "But it's still possible to gather assemblies for ten and maybe even twenty missiles here."

The Germans told us that the special equipment used purely for missile testing had been hauled away. But the ordinary machine tools and standard, general-purpose equipment in the shops had remained untouched. Even the most state-of-the-art metal-cutting machine tools had been passed over by the wealthy missile-secret hunters from across the Atlantic.

In the gallery, Shmargun directed our attention to the overhead traveling crane that covered the entire width of the bay. It was used for the vertical tests and for the subsequent loading of the missiles. Two beams were suspended from the crane over the width of the bay. They were lowered when necessary to a human being's height. Nooses were secured to the beams and placed around the necks of prisoners who were guilty or suspected of sabotage. The crane operator, who also played the role of executioner, pressed the raise button and immediately an execution of up to sixty people was carried out via mechanized hanging. Before their very eyes, the *polosatiki*, as the prisoners were called, were given a lesson in obedience and fear under the bright electrical lights 70 meters below ground.[2]

Hearing this horrible story, Isayev nudged me and pointed at the Germans. Earlier they had crowded closely around us, but now they huddled in a bunch and retreated into the darkness. Here Rosenplänter intervened and told us that they had been warned about an underground organization operating at *Mittelwerk*. Prisoners who had worked on assembly had learned how to slip in a defect so that it would not immediately be detected and would not become evident until after the

2. *Polosatiki* literally means "people in stripes," referring to their prison uniforms.

missile had been shipped, often during its tests, before launch, or in flight. Someone had taught them to solder the electrical connections unreliably. This was very difficult to check. The German control personnel were not capable of checking tens of thousands of soldering points per day. The Gestapo asked the Peenemünde engineers to devise a way to automate control.

As far as Rosenplänter knew, they had not come up with anything. Up to twenty percent of the missiles were rejected during the final tests at *Mittelwerk*. All of the rejected missiles were sent to a small "rehabilitation" factory, *Werk Drei* (Factory Three), to determine the cause of the failure and make repairs. *Werk Drei* was located near the village of Kleinbodungen, not far from Bleicherode. There would still be electrical equipment there for horizontal tests if the Americans had not hauled it away.

In a sort of justification, Rosenplänter said, "Only Germans worked at *Werk Drei*. There weren't any prisoners there. If the Russian command is interested in the reconstruction of an A-4 missile, it would be best to use this small factory for that purpose." Later we did just that. Especially after dozens of our process engineers and dismantlers subsequently descended on *Mittelwerk*. Their primary objective was to disassemble and haul away processing equipment that was of any value.

Much later—in early 1946 as I recall—a German artist came from Erfurt to speak to General Gaydukov, chief of the Institute Nordhausen. He brought a large collection of watercolors and pencil drawings depicting subterranean production activity at *Mittelwerk*. According to the artist, any photography or filming of the factory and the surrounding areas was forbidden on the threat of death. But the leaders of the A-4 program believed that a creation as great as *Mittelwerk* should somehow be immortalized.

They had sought out the professional artist and caricaturist, and with the help of the Gestapo, had brought him to the factory to draw the entire primary missile assembly process and do as many sketches as possible in color. He labored diligently, but at times he got so carried away that drawings appeared of prisoners being beaten, executions, and visits to the factory by high-ranking guests such as Ernst Kaltenbrunner himself.[3] We looked intently at these drawings filled with doomed individuals in striped uniforms, among whom were certainly dozens of heroes whose names humankind will never know.

How did he manage to keep these pictures? "Very simply," he explained. "A special officer of the Gestapo took some pictures from me. But he wasn't interested in a lot of them. I was supposed to hand them over to the factory management, but I wasn't able to—and now I would like to offer them as a gift to the Russian command." General Gaydukov accepted this unexpected gift with gratitude. Eventually the album of these pictures was sent to Moscow. But as to its

3. From 1942–45, Kaltenbrunner was head of the *Sicherheitsdienst* (SD), overseeing both the Gestapo and the system of Nazi concentration camps throughout Europe.

whereabouts now, I do not know. Perhaps it is in some archives and someone will manage to find them.

While we were investigating *Mittelwerk*, a new group of specialists arrived in Nordhausen from our NII-1, including Professors Knorre and Gukhman; Dushkin, the chief designer of the first liquid-propellant rocket engine for the BI aircraft; and Chernyshev, a chemist and rocket propellant specialist. Late that evening, when we had made our way back to town, tired and dusty, dreaming of rest, this information-starved team pounced on us and demanded to be let in on all our secrets. While Abramovich conversed with them, Isayev, who dearly loved practical jokes, pulled me aside and said, "We've got to get rid of them or else our hands will be tied. With this professoriate we'll have nothing but trouble."

"But how? We have no right to just drive them out of Nordhausen."

"I have an idea. We'll frighten them with British-American intelligence—I'm sure they're hunting for Soviet specialists, documents, and great State secrets."

Isayev pulled off the performance in his best tradition. In the middle of the night, the entire host of scientists was invited to our dark mysterious villa. There Isayev announced to them that in 20–30 minutes, an agent, whom we had recruited from British intelligence, would arrive, having crossed the border to discuss secret documentation for the *Wasserfall* missile and provide information about where these missiles had been hidden. Moreover, he knew where von Braun himself was. It would be very good if our comrades just arriving from Moscow would get involved in the operation to abduct von Braun.

During these explanations, a prearranged knock was heard at the window overlooking the dark garden. Isayev grabbed his pistol and commanded, "Quick, everybody into that room and don't make any noise. Chertok will conduct the meeting."

I received the "agent," whose role was brilliantly played by Shmargun dressed in a quasi-American uniform. At first we talked a bit in German. Then I began to yell in Russian that we could not promise so many dollars, this was out-and-out robbery, and so forth. Shmargun, our agent, threatened that his bosses already knew about the arrival in Nordhausen of high-ranking Soviet rocket engine specialists. As a sign of his good relationship with us, he asked me to warn them that it would be better for their safety to leave here for the time being. I thanked him for his valuable information and said that this service would be paid for. Our "agent" quietly withdrew. Isayev released everyone from the adjacent room and asked triumphantly, "Did you hear that?"

But we didn't stop there. We took the frightened company under our guard to the apartment where the commandant's office had settled them—they immediately discovered that their suitcases had been opened and searched! Isayev, feigning anger, demanded answers from the landlady. She explained that some officers had come and demanded that she show where her tenants were staying. The landlady had been coached in advance on how to answer. Needless to say, the entire group of specialists wished us success and set off in the direction of Berlin the next morning. Abramovich shared a good laugh with us and then followed them to Berlin in an escort vehicle. He left the Mercedes and Alfred at my disposal. In our joy, we decided that such a successful performance called for a celebration that evening at the local café-cabaret that was still operating after the Americans' departure. There we wanted to develop our plan for future operations.

The café, which was located in a cozy bomb shelter, proved to be a noisy establishment that served beer and bootleg schnapps instead of appetizers and coffee. The joint was already filled with smoke, American officers, and black soldiers, and on an improvised stage, a brunette past her prime, dressed like a gypsy, sang something unintelligible in a husky voice. Evidently we were the first Soviet officers to come here. As soon as we sat down at the only free table, one of the American officers jumped up and yelled something in the direction of the bar. A fellow in white quickly flitted from behind the bar and deftly placed foaming mugs in front of us. The singer came running over to us, gauged our ranks by our shoulder boards, and without asking permission planted a smacking kiss on Isayev's cheek. "At last, the Russians have arrived! What shall I sing for you?"

The American officer said something to her in the tone of an order.

"He knows that I am Russian and wants me to interpret," she explained. "He welcomes the Russian officers to the land that they, the Americans, liberated from our common enemy. Horrible crimes occurred here. He hopes that we will be friends. To victory and comrades in arms!"

We reached for our mugs, but he managed to add something to his beer and ours from a bottle that he was holding in his outstretched hand. One of the American officers talked a lot, demanding the whole time that the singer interpret for us. This is what he managed to tell us.

The Americans, who had advanced from the west as early as 12 April, that being three months before us, had taken the opportunity to familiarize themselves with *Mittelwerk*. The subterranean factory had been shut down only twenty-four hours before their arrival. They were staggered by the sight of it. There were hundreds of missiles under the earth on special railroad flatcars; the factory and spur tracks were completely preserved. The German guards had fled, and the prisoners had not been fed for at least two days before the arrival of the American troops. Those who were capable of walking moved slowly. They approached the Americans to get food and did not hurry. It was as if they were doing everything in their sleep. The singer interpreted further, "They told us that more than 120,000 prisoners had passed through the camp. First they built—gnawing away at the mountain. Those who survived and newly arrived prisoners worked in the subterranean factory. We found prisoners in the camp who had fortuitously survived. There were many corpses in the subterranean tunnels. Our soldiers were horrified when they saw all of this. We put many Germans to work cleaning up and bringing order. It will be easy for you to work there now. To our victory, to our friendship!"

We did not notice that another Soviet officer had appeared at our table. He was clearly not "civilian," because his chest was covered with decorations and medals. He put his arm around my shoulder and quietly said, "I'm from the Smersh division. In the morning, you and the lieutenant colonel should stop in at headquarters."

I had to wake up Isayev a little early the next morning. We managed to conduct some blitz-strategy and work out a plan of action: "Do not under any circumstances make excuses—demand and attack!" This was our frame of mind when we arrived at headquarters. But they weren't even thinking about dealing with us for our "improper" behavior the previous night. The deputy division commander's political commissar, the executive officer, and the Smersh officer we had spoken with the night before very kindly explained, "The headquarters of the Eighth Guards Army is located in Weimar. It is under General V. I. Chuykov, who for the time being has been assigned to head the Soviet military administration in Thuringia. You are obliged to coordinate your future actions on the use of German specialists, and especially contacts with Americans, with Soviet Military Administration in Germany (SVAG) representatives. Smersh has reported through its channels that the American intelligence services are conducting broad-ranging operations to seize German specialists. According to reliable information, some of your drinking companions last night were not combat officers, but men who have been assigned to "wrap up" the seizure of German specialists, look for remaining missile paraphernalia, and monitor the actions of Russians hunting for German secrets. Subsequently, we found out that the Americans' operation to seize German missile secrets was being conducted by a group under the codename 'Paperclip'."

We laid out our plans: "A group headed by Major Pallo is leaving today for the town of Saalfeld. According to German accounts from *Mittelwerk*, there is a V-2 engine firing test station near the village of Lehesten. Engines for assembly came from there after the firing tests."

We requested assistance with transport and asked that instructions be given to the commandant of Saalfeld to provide us with living quarters in the town of Bleicherode, where we planned to assemble a group of German specialists. We requested that they provide workrooms for the specialists, establish storage areas for valuable equipment, set up guards and make arrangements for food and communications. Then we intended to call for assistance from Moscow. But for the time being, it was advisable not to allow anyone to enter *Mittelwerk* so that what the Americans had left behind would not be dragged off every which way.

"And one more thing," I added. "A certain Shmargun, a former prisoner, is helping us."

"That's our concern," the Smersh officer interrupted me. "You can trust him. According to our information, the Americans didn't have time to remove the equipment hidden by the SS in the potassium mines. That's somewhere here in the surrounding area. A German told us that about fifteen kilometers from here, almost at the border, there is a lot of secret equipment hidden in a forester's cabin. The forester, an ardent Nazi, has fled, but supposedly there are local forestry lookouts guarding this cabin. We do not advise going there alone. If you come up with something, we will help. But be careful—the other side is also conducting a hunt.

After our "dressing-down" at headquarters, Isayev and I seated Rosenplänter in our Mercedes and gave Alfred the command: "Onward, to Bleicherode!"

That was the morning of 18 July 1945.

Chapter 20

Birth of the Institute RABE

The route to Bleicherode went first through forests on a narrow road, then along the steep and dangerous curving lanes of the villages of Pustleben, Mitteldorf, and Oberdorf. The very picturesque road climbed up into the forested mountains and, leaving the forest, crossed a railroad and entered a pristine little garden hamlet. This town, Bleicherode, became my workplace for more than a year-and-a-half.

Right on the central square the red flag showed us the location of our commandant's office. He came out to meet us, introducing himself as Captain Solodyankin. He was a typical frontline veteran with decorations and medals, around forty-five years old, with a face that showed great fatigue, though one could just barely see a kind smile.

He had already been in charge of the town here for two days, and was overwhelmed. "I don't know the language, and there's a long line of Germans coming to me. I got rid of the *Bürgermeister.* They said he was a Nazi. Some have shown up and introduced themselves as communists and social democrats. But who the hell knows? I appointed one of them the new *Bürgermeister.* He is selecting his police department and various other services. I have already received the instructions about helping you, but I don't have a lot of people. I advise you to go to the division headquarters. General Goryshnyy's Seventy-fifth Guards Army is billeted nearby."

We took his advice. It turned out that the general had also been notified of our arrival. He gave us a very warm reception.

"The commandant is arranging for your accommodations and you will be provided with meals according to all the norms for officers in the division. As far as gasoline, your Germans, and guards, we'll also take care of that tomorrow.

We returned to the commandant's office. There Rosenplänter, who had come with us, having seen the respect with which they received us, was conversing in a rather harsh tone with the new *Bürgermeister.* He explained to us, "The best villa in town is on Lindenstrasse—the Villa Franka. Wernher von Braun lived there for some time after Peenemünde. Now a German pilot lives in part of the villa. He is ill, which is why neither the Americans nor the Russians took him. I told the *Bürgermeister* that they should resettle him, but he doesn't know where."

"Let's go and have a look."

After a five-minute drive on a cobblestone road going up a hill, we got out of the car on a small square by the main entrance to a three-story villa. The massive doors—plate glass behind ornamental wrought-iron bars—would not give. The *Bürgermeister* ran off somewhere and brought back an elderly German woman—Frau Storch. "She was a maid here. She knows everything and is prepared to help you." Frau Storch had the keys. We entered. But where was the German pilot? Suddenly we were almost run over by a little kid about five years old on a tiny bicycle. He turned out to be the pilot's son. We learned that the villa had another half with a second entrance. Isayev was outraged that part of the residence was still occupied. Rosenplänter rapidly muttered something. I announced that the house suited me and let Alfred unload our meager luggage.

The villa was magnificent. The first floor had a large drawing room. It was a library with bookcases made of dark wood. There were deep armchairs in front of an elaborate fireplace and a separate smoking room with ashtrays of varying sizes. Passing from the drawing room through heavy doors, we entered a fragrant garden. There were magnolias, roses, and a pool with a fountain, which for the time being was not functioning.

"Aleksey," I said, "the fountain—that's your bailiwick. Let's get some rest and then you can fix the jet." Isayev promised he would.

A marble staircase led from the vestibule to the second floor. Here there were four bedrooms, two bathrooms, and two half-baths equipped with various lavatory sanitary facilities—four toilets in all! The floors were covered with large carpets, and the walls were decorated with ornamental rugs and paintings of local landscapes and scenes of nature. Heavy red velvet curtains hung at the wide windows. We entered the largest bedroom. The bed was mahogany and designed to accommodate, as we determined, four. It had snow-white featherbeds instead of blankets. And the ceiling! The ceiling was a mirror. Lying blissfully in bed you could admire yourself in the mirror. Isayev could not contain himself. He threw back the featherbed and plopped onto the bed, sinking just as he was, in dusty boots and full uniform, into the froth of feathery-white bedding. With a leisurely attitude, he pulled out a soft pack of his favorite *Byelomorye* cigarettes, which were in short supply here, and lit up.

"You know, Boris, it's really not all that bad in this 'fascist beast's den of iniquity.'"

At that moment, alarmed by our long absence, Rosenplänter appeared, accompanied by Frau Storch and the *Bürgermeister*. Finding Isayev in the bed, they were completely dismayed. "Is Herr Officer very ill? Should we bring a doctor?" We calmed them down and announced that we would take the villa. We asked only that the other half be made available. "We are going to have a lot of guests!"

"*Jawohl!*" was the response, to which we had already become accustomed. The third floor turned out to be a mansard; Frau Storch said that sometimes the maid or guests spent the night there.

We asked, "Why isn't there any hot water?"

From the author's archives.

The former Villa Franka, where Wernher von Braun stayed after evacuating Peenemünde, and home to Major B. Ye. Chertok for eighteen months.

"Oh, for that you need to go down into the basement and heat up the boiler." Isayev could not restrain himself. "Come on, let's go light it now."

In the basement there was a large bunker filled with coal. We heated up the boiler and each of us luxuriated in a separate bathroom. Next, wrapped up in fluffy bathrobes (where did they get this stuff?!), we went down into the library and celebrated our new digs over an improvised lunch.

For almost a year, Villa Franka in Bleicherode was converted into an officer's club and headquarters where we summed things up, developed plans of action, broke bread together, and celebrated holidays. It is difficult to remember who came up with the idea of calling our breakaway group—which for the time being consisted of twelve Germans under the command of Lieutenant Colonel Isayev and Major Chertok—an institute. The Germans were delighted with this idea and announced that they could quickly put together specialists and an entire staff. But what should we call this new invention? After a brief "Soviet-German" discussion, we came up with a name: the Institute RABE. The precise translation of the German word *Rabe* is "raven." Our acronym stood for ***Ra****keten****b****au und* ***E****ntwicklung* Bleicherode (missile construction and development in Bleicherode). Our "cover" had emerged—we became a place where German specialists scattered by the war could take refuge.

This was clearly a guerrilla operation on our part that could lead to diplomatic complications with the Allies, especially since the border was only six kilometers away, and immediately beyond the border was a town where, according to our intelligence, the American command had assembled several hundred German specialists. Isayev and I visited division commander General Goryshnyy in order to get his approval for our guerrilla operations. The general honestly confessed that he was not in a position to advise us in this business and that we needed to go to Weimar, to the SVAG for Thuringia being set up there. He then asked us to give a lecture about missiles to his division officers. "Then my people will be more than willing to help you."

Soon thereafter, at his command, around 100 officers from his division and the attached artillery brigade assembled to hear us speak. I told them as much as I could about the V-2 missile and Isayev explained the operating principles of the liquid-propellant rocket engine. We were pleasantly surprised at the genuine interest with which these combat officers—who had gone through the entire war from Stalingrad to Thuringia—listened to us. They bombarded us with questions, thanked us for our reports, and asked to be called upon to render whatever help they could. We could understand their situation. The round-the-clock battles with their intense physical and emotional stress were over. The constant horrible, oppressive feeling of risking their own lives and the lives of their soldiers had lifted. The heavy burden of responsibilities and cares had disappeared. Now there was simply nothing to do. Such a quiet, peaceful life in a strange land had disrupted their normal frame of mind. The Germans had transformed from enemies into something like allies. Young German women had no objection at all to paying attention to Soviet combat officers, who had both cigarettes and butter, and who on the whole had been brought up a lot better than the Americans.

Having enlisted the support of the "entire division," as Isayev said, we could now demand attention from the high command. So we set off for Weimar. We knew this pristine town, which had not suffered at all from the war, mainly as a place associated with the name Goethe.[1]

Colonel General V. I. Chuykov was in command at Weimar, and Ivan Sazonovich Kolesnichenko had just been assigned as chief of the SVAG Directorate for the Federal Land of Thuringia. He cordially received us and listened to us attentively. Then, in spite of the long line of visitors waiting in the reception room, he summoned several officers and explained to them who we were and began to consult with them. "Here we must set up a peaceful life based on new democratic principles in an alliance with all the anti-Nazi forces. We must uproot vestiges of Nazism in the ethos of the German people and redirect the entire economy onto a peacetime footing. And what do you propose? To restore Hitler's military technology! And where? Here, in Thuringia! And what will we tell the Allies as soon as they find out about the creation of a missile institute?"

---

1. Johann Wolfgang von Goethe (1749-1832) is widely recognized as the greatest writer in the German literary tradition. He lived in Wiemar for the majority of his working life, and the city became a prominent cultural center on account of his influence.

We were totally crestfallen. But one of the officers subordinate to Kolesnichenko was on our side. He expressed an idea that for some reason had not occurred to me. "The institute needs to be registered as a new scientific institution. Under the military administration's supervision, scientists will be gathered whom we do not want to be left unemployed. In addition, they will help us reveal the mysteries of Hitler's secret weapon so that we will have evidence of war crimes. We not only should not object, but we should support such an initiative at all costs!"

These arguments and Isayev's and my eloquence overcame Kolesnichenko's wavering. He took on the responsibility and gave us permission to set up an institution according to all the rules—with an official stamp, letterhead stationery, telephones, and a German staff. The Military Administration of Thuringia took on the responsibility for the provisions and monetary support for the Germans in the initial stages.

Having established communications in Weimar, we wandered around the town musing about everything that we had undertaken. All of a sudden, after such an administrative coup, Isayev said that now he wanted to leave for Lehesten to have a look at the engine technology there and then return to Moscow. "It's time to start a revolution in our patron's way of thinking at NII-1. To this day there in Moscow they do not want to understand the scale of German work on rocket technology. We admit that during the war we were not up to that. But the Americans were farther behind the Germans than we were! We need to take advantage of this. I must reeducate our patron. If not he, then who else in Moscow is capable of open-minded thinking?"

I OBJECTED, BUT SINCE I KNEW HIS NATURE, I understood that he had already made his decision. We agreed simply that he would not leave until help flew in from Moscow.

Upon our return to Bleicherode, we assembled our German team, announced the SVAG's decision, and gave the assignment to develop the institute's structure. Isayev announced that Major Chertok was being appointed chief of the institute. Having thus taken over command, I immediately announced that engineer Rosenplänter would be the director of the new institute. His deputy for general matters would be engineer Müller, whose duty it would also be to establish conditions suitable for scientific research work. The priority task for all personnel would be to restore, first on paper, the technology that had been developed at Peenemünde. To do this we needed to search diligently for everything that had been hidden or left behind after the American's withdrawal, and recruit rocketry specialists who had worked in Peenemünde or other places. Everyone set about their work with great enthusiasm.

Within twenty-four hours, the former employees of the company that managed the regional electrical power grids vacated their three-story building. We began our hectic preparations to move in. Each Institute RABE employee received a photo-identification pass in accordance with all the rules. Trucks rolled up to the building and the rooms were outfitted with workbenches and stands for the laboratory equipment that had not yet been acquired.

My office was luxurious, even by modern standards. The desk had deep leather chairs. There were watercolor landscapes on the walls, and a bouquet of fresh roses on the conference table. Telephones were provided for internal and local communications and communications with Berlin. A couple of days later, a "field telephone" appeared providing us with direct communications to the military administration and the commandant's office. A typist-stenographer appeared in the reception room, which led into the office of the German director. She introduced herself simply as "Fräulein Ursula."

Several days later, I tracked down the interpreter Lyalya, who had found refuge in the division political department. She had helped us during the first days of our stay in Nordhausen and I placed her in the reception room as a backup and supervisor for Ursula. But Lyalya showed character—she proclaimed herself chief secretary of the office of Major Chertok and Director Rosenplänter, and no one dared enter our offices without her consent. We soon restored peace among the female staff and the Russian-German secretariat began to operate efficiently.

By agreement with the division, we were allocated a guard service, and no one could enter the building without presenting his or her pass to a soldier wielding an automatic weapon.

From Nordhausen they transported the priceless windfall that had been stored there—the gyro-stabilized platform. But our staff still did not have anyone who would dare start studying it. Therefore, they placed it in the future gyroscopic instruments laboratory, which they locked and sealed.

From the author's archives.

Main building of the Institute RABE in Bleicherode, 1992.

Every day, first thing in the morning, Rosenplänter would report to me and present newly hired leading specialists. Though many of them were good engineers, judging by their documents, they had a weak understanding of what they were supposed to be doing at the institute. Nevertheless, I agreed to accept them on a "wait and see" basis while operations were beginning. It wasn't until the end of July, after the hectic days of establishing the institute, that we returned to the search for "secret treasure." In Nordhausen, they were still talking about the potassium mine and the forester's mysterious mountain cabin. We consulted with the town commandant and concluded that for such operations it would be best to obtain a car with a driver-gunner from the division or commandant's office. The potassium mine was well known in the area and was located quite near Bleicherode. Our driver arrived and Isayev and I headed over to the mine in the division Jeep, whose driver was armed with an automatic weapon and proud of it. In the yard we came upon some sort of meeting. The miners were wearing their helmets with lights mounted on them, but they weren't the least bit dirty. After all, this was a white potassium mine, not a coal mine. They were all quite surprised at the arrival of Soviet officers. They clustered around us until the man who had introduced himself to us as the director gave a shout and the miners dispersed. We explained the objective of our trip. The director called out some names, and about ten men assembled by our car. We found out that a few cases had been placed in some remote dead-end drifts at a depth of 500 meters. An SS detachment had brought them just before the arrival of the Americans. They said that the equipment in the cases was rigged with explosives.

After briefly wavering, we decided to risk descending into the mine. We gave the driver instructions to wait for ninety minutes—if we hadn't returned by then he was to "high-tail it to the commandant's office." The director apologized that he had pressing business, and entrusted the foreman to accompany us. The ten miners surrounded us, each of them with something similar to a pickaxe. We then entered a spacious cage. A chime sounded and we descended at a surprising speed. Isayev had been in mines as a youth, but they were coal mines. I had only read about them and seen them in movies. This was nothing like what we had expected to see.

Underground it was well illuminated, and the walls sparkled with the white crystals of potassium salt. We walked without bending over, standing fully erect, surrounded by miners. After walking briskly for fifteen minutes, we were standing in front of a stack of green cases. Isayev and I began to carefully examine them. We grew bolder and grabbed the top case by the convenient side handles. The catch clips were wrapped and unusual seals hung from them. I asked a miner if I could borrow his pickaxe. He shook his head and gave it to me. We broke off the seals and opened the case. There were no explosives, and the contents were clearly radio equipment. What was this? I remembered the broken instruments that had been shipped to Moscow from the Polish firing range in 1944. To calm the miners, Isayev and I took down another case, lifted and shook it, and placed it one top of the other. We checked the time—only thirty minutes remained before the prearranged deadline. We needed to get back! We asked the miners to carry the two cases to the elevator and leave the other six behind. Our Jeep would not hold them all in any event. We decided we would pick them up on the next run. When we reached the surface our driver was smiling. "Five more minutes and I would have raced over to the commandant's office."

We explained to the director that this material was now the property of the Red Army. We left a receipt for the two cases, asked that the others be stored, and promised to pick them up the next day. As a precaution, we said that there might be explosives there, and therefore other specialists would be coming for them. The next day, per my request, two officers from the division, including a sapper, brought back the remaining cases. An examination at the institute showed that we had received sets of Viktoria-Honnef radio-control equipment for lateral radio correction and range control. It was the first contribution to the equipment of our radio laboratory.

IN SPITE OF THESE ADVENTURES, Isayev was dying to investigate engines, and soon thereafter he left Bleicherode and went to Lehesten.

Abramovich, whom we had last seen in Nordhausen, kept his promises. At the time of his departure, after a cursory familiarization with *Mittelwerk*, he had promised that when he arrived in Moscow he would persuade Bolkhovitinov and everyone else that he could to have reinforcements sent to us. I did not even have time to miss Isayev before a vigorous team arrived, headed by Nikolay Alekseyevich Pilyugin, the future two-time recipient of the Hero of Socialist Labor award and future academician, director, and general designer for one of the most powerful Soviet space electronic instrument firms, *NPO Avtomatiki i Priborostroyeniya* (Scientific-Production Association of Automation and Instrument Building), the company that developed the control systems for many combat missiles and space launch vehicles.

Pilyugin flew in with the rank of a colonel, despite the fact that his military service record stated that he was a non-combatant. He was accompanied by Leonid Aleksandrovich Voskresenskiy, who had risen from the rank of private to lieutenant colonel. Voskresenskiy, who had the astounding ability to sense and foresee the behavior of a missile system during the most diverse off-nominal situations and failures, subsequently became Korolev's legendary deputy for testing.

Semyon Gavrilovich Chizhikov flew in with the rank of first lieutenant. Designer Chizhikov was my long-time comrade from the Bolkhovitinov OKB, Factory No. 22, Factory No. 293, and also the factory in the distant Ural village of Bilimbay where the BI interceptor was built during the difficult war years.

Among the new arrivals there was only one real service officer—Engineer First Lieutenant Vasiliy Ivanovich Kharchev. He was the youngest member of our group, having graduated from the N. Ye. Zhukovskiy Air Force Academy in 1944. I had been his adviser when he was doing his diploma project, and at that time I had been convinced of his exceptional abilities and impressed with his penchant for

From the author's archives.

Boris Ye. Chertok (left) and Nikolay A. Pilyugin in occupied Germany—Bleicherode, winter of 1946.

coming up with new technical ideas. While the ideas weren't always feasible, they were very interesting and original.

Two weeks later, Vasiliy Pavlovich Mishin, Aleksandr Yakovlevich Bereznyak, and Yevgeniy Mitrofanovich Kurilo appeared. Vasiliy Mishin, who was inclined toward integrated design and theoretical work, immediately set about obtaining materials from the Germans concerning the theory of rocket flight. He later became Korolev's first deputy, and after Korolev's death in 1966, he became the chief designer of Korolev's firm and was selected as a full member of the USSR Academy of Sciences. Sasha Bereznyak was the initiator of BI development, an enthusiast only for the winged versions of missiles, and the future chief designer of cruise missiles. After surveying our work in Bleicherode, he announced that he was leaving because there was nothing for him to do, but said he would join us again after Isayev returned from Lehesten.

I set about distributing duties among the new arrivals. In spite of his high rank, I appointed Pilyugin as the first deputy chief of the Institute RABE and chief engineer.[2] After receiving his office, he immediately asked the Germans for a set of precision mechanics tools and ordered that all electromechanical instruments, no matter where they came from, pass through his office workshop. I visited Semyon Chizhikov at Villa Franka. Taking into account his exceptional organiza-

2. The title 'first deputy' was common in Soviet institutions to denote someone who was 'first among the deputies.'

tional skills, I appointed him as general assistant, which put him in charge of all problems concerning transportation, housing, food for the officer staff, and interaction with the commandant's office and the local German authorities. He had quite enough to worry about, because the so-called domestic problems were becoming more complicated with every passing day. I also had several long conversations with Vasiliy Kharchev. We agreed that we would set up our own independent intelligence service. The primary mission of this group would be to search for authentic missile specialists and entice them, or even abduct them, from the American zone.

BY THE END OF AUGUST, our institute was already becoming, by provincial standards, a robust, large-scale institution. We created laboratories for gyroscopic instruments, control-surface actuators, electrical circuits, ground-based control consoles, and radio instruments. We also set up a design bureau. There was an excellent Photostat machine in the semi-basement, and an exemplary, while yet empty, technical documentation archive nearby. The first report on the institute's activity soon appeared on our official letterhead and in the institute's files.

An officers' mess hall had opened up at the Villa Franka. In exchange for meals and a small salary, a language teacher from the Baltic States was conducting daily lessons in conversational German with our officers. While Semyon Chizhikov proved to be the most linguistically challenged, he was somehow understood better than the others when it came to dealing with the Germans concerning routine business problems. He was already well-known in the area as a wholesale buyer of provisions and schnapps and as a specialist in the repair of automobiles.

I asked Kurilo to inspect the factory in Kleinbodungen and begin restoring the production of missiles. He arranged the transport of the the assemblies that had been left behind by the Americans at *Mittelwerk*, found several skilled workmen who knew the assembly process, and began to develop real missile production. When we had made a list and tallied up the riches, it turned out that out of all the tail and middle sections, tanks, instrument compartments, and nose sections we could assemble at least fifteen, and maybe even twenty missile bodies. But the situation was a lot worse with the innards. We still did not have a single control system instrument that we could use. There were also no engines and no turbopump assemblies that we could clear for installation.

BACK IN MOSCOW, hectic decisions were being made related to our work. The Main Artillery Directorate had tasked the Guards Mortar Units command with finding captured German missile technology. As mentioned previously, this whole thing started with General Sokolov's commission, which was the first to arrive in Peenemünde. In August, General N. N. Kuznetsov appeared in Nordhausen with a large retinue. In accordance with a command from Berlin, he occupied the sole ancient palace in the town. There they founded, as we used to joke, the missile headquarters of the Guards Mortar Units or 'GAU Representation,' which had been ordered to direct the study and expropriation of German missile technology.

After determining "who was who," Kuznetsov announced that the Institute RABE and all of us were subordinate to the GAU military command, as per the decision of the Central Committee, which had instructed the military to head this activity until the industry could sort out which one of the Peoples' Commissars would be in charge. Having discussed these problems in our own aviation circle, we decided that it wasn't worth it to put up a fight. After all, in Germany at that time, the military were our masters. The aircraft industry had really abandoned us or forgotten us, and no one else had picked us up yet. Soon thereafter we made the acquaintance of General Lev Mikhaylovich Gaydukov, who had arrived for an inspection. He was a member of the Guards Mortar Units military council and also a Central Committee department manager. He impressed us as an energetic man who was full of initiative, and we liked that he made no secret of the fact that he would support in every way the expansion of our operations in Germany, to the extent that he would even issue the appropriate Central Committee and governmental resolution.

In August 1945, the Institute RABE became fully established and began to expand its activity. The following month, commissions and all sorts of plenipotentiaries from Moscow—ranging from the serious to the idly curious—began making regular visits to determine who we were and what we were doing.

After returning to Moscow, Gaydukov really got down to business. The first result of his efforts was the arrival of a group that consisted of future chief designers Mikhail Sergeyevich Ryazanskiy, Viktor Ivanovich Kuznetsov, and Yuriy Aleksandrovich Pobedonostsev, as well as Yevgeniy Yakovlevich Boguslavskiy and Zinoviy Moiseyevich Tsetsior. We had now been brought up to full strength with radio and gyroscopic specialists. This was an interagency group organized at Gaydukov's initiative by decision of the Central Committee.

When Viktor Kuznetsov saw the gyroscopic platform that we had in our laboratory, he announced that it should be sent immediately to his institute in Moscow. But it didn't work out that way. Pilyugin categorically objected. This was the first serious conflict between the two future chief designers. Later, in the early 1960s, disagreements as to who should make gyroscopic instruments and what they should be like led Pilyugin, who had obtained a powerful production base, to begin successfully developing and producing gyroscopic instruments and highly sensitive elements for inertial navigation systems.

Yevgeniy Boguslavskiy, who remained with us at the institute to the very end, immediately immersed himself in the mysteries of the lateral radio correction and range radio control systems.

TO CONSOLIDATE ITS POLICIES, the GAU sent cadre officers from its staff and troop units. And so, the following people turned up first in Berlin, and then at our place: Georgiy Aleksandrovich Tyulin, who would later became director of the head central institute of the USSR Ministry of General Machine Building (*Minobshchemash* or MOM) and then MOM's first deputy minister; Yuriy Aleksandrovich Mozzhorin, who also became director of the head institute after Tyulin; Colonel Pavel Yefimovich Trubachev, a future first regional engineer and chief of military acceptance at the head institute NII-88; and Captain Kerim Alievich Kerimov, a future chairman of State Commissions.

A special regiment of the Guards Mortar Units was billeted in the town of Sondershausen. The regimental commander introduced himself to me as the future commander of a rocket troops subunit. The military leadership was finally wide awake. But the situation with the German specialists was clearly not satisfactory. We needed to take urgent action to entice them from the Western zone.

Chapter 21

Operation "Ost"

There were more and more complaints from the higher-ups. This time it was coming from General Kuznetsov, GAU's authorized representative, to the effect that his officers were not receiving materials and satisfactory technical information from the Germans. One of the high-ranking artillery officers who had visited Bleicherode told me bluntly, "We have the impression that you Russian aircraft specialists already have a better understanding of this technology than the Germans here. But the mortar and artillery specialists aren't receiving the materials to study."

They also accused me of setting a very high pay scale and food rations for the Germans. We obtained rations and issued them with the consent of the SVAG, who were striving to show that German scientists were being provided with both moral and material support, despite the burdens of the postwar period. Germans with engineering and doctorate degrees at the Institute RABE received the following food rations every fourteen days: sixty eggs, five pounds of butter, twelve pounds of meat, and a fully sufficient amount of bread, sugar, vegetable oil, potatoes, cigarettes, and alcoholic beverages. Their monthly salary of 1,400 Marks was issued without any deductions. Soviet scientists back home in Moscow during those years would not have dreamed of such conditions. Despite our incentives, however, we still did not have "the right Germans" working for us. We needed to stir up a second wave and get some real specialists.

The institute had now become a well-organized enterprise, and we were not ashamed to invite people to undertake good and interesting work. We undertook efforts in two areas. The first, which Pilyugin took on, was to offer scientists and highly qualified engineers in specific specialties the opportunity to work for us through the SVAG Thuringia directorate, where there were already many good administrators. Prior direct involvement in rocket matters was not a prerequisite. This personnel program quickly led to the discovery of prominent specialists who made valuable contributions, even though they had not previously worked at Peenemünde.

This is how Dr. Kurt Magnus joined us. Dr. Magnus was a first-class theoretician and engineer in the field of gyroscopes and theoretical mechanics. He quickly acquired an understanding of the gyroscopic platform technology and announced that he would take on all gyroscopic problems. He summoned Dr. Hans Hoch, his friend at Hettingen University, who was a theoretician and brilliant experimenter in automatic control. Magnus convinced Hoch to stay and work with us in Bleicherode. After briefly wavering, we assigned Hoch general stabilization theory and the "*Mischgerät*" laboratory.[1]

This pair, Kurt Magnus and Hans Hoch, assisted us greatly. Unfortunately, Dr. Hoch died in Moscow in 1950 from suppurative appendicitis. Dr. Magnus became a renowned scientist and gyroscope expert, whose works were translated into many languages. His fundamental monograph, *Kreisel*, is still a first-class textbook for those studying gyroscopic technology.[2] In October we acquired Dr. Manfred Blasing, a former employee of Askania who appealed to the departments of that firm in the various zones until he managed to get on board with us. We entrusted him with the directorship of the control-surface actuators laboratory. The ballistic calculations associated with controlling missile flight were entrusted to Professor Waldemar Wolff, who had been the chief ballistics expert for Krupp. Soon thereafter, an aerodynamics specialist from Dresden, Dr. Werner Albring, joined him. He took on the aerodynamic problems of the atmospheric segment.

But we still needed authentic Peenemünde missile specialists. For this I set up a "secret" second program, which I entrusted to Vasiliy Kharchev. His task was to establish a network of agents, and if necessary, personally penetrate into the American zone to intercept specialists before they were sent to the United States. At Kharchev's suggestion I assigned this program the code name "Operation *Ost*" (East).

Semyon Chizhikov was instructed to supply Kharchev with cognac, butter, and various delicacies "on account" for Operation *Ost*. The division chief of staff agreed to open and close the border between our zone and the American zone at Kharchev's request. Pilyugin undertook a special mission—I no longer remember what town he went to—and brought back many dozens of wristwatches to be used as souvenirs and "bribes" for the American border guards. Vasiliy Kharchev could barely sleep because of his intensive German and English studies.

The first success of Operation *Ost* was to win over and bring onto the RABE staff an authentic specialist on the combat firing of V-2 missiles, Fritz Viebach.

The Americans unexpectedly gave our Operation Ost a boost. Early one morning, I was awakened by a telephone call from the town commandant. He reported that his patrol had stopped two Jeeps with Americans who had apparently burst into town and were trying to abduct German women. The latter had raised such a ruckus that our patrol had arrived. The arrested Americans were raising Cain over at the commandant's office. They explained that these women were the wives of German specialists who were supposed to be sent to America. I asked the comman-

1. A *Mischgerät* (mixing unit) was an amplifier-converter on the circuit connecting the command gyroscopes and the control-surface actuators.

2. Kurt Magnus, *Kreisel: Theorie und Anwendungen* (Gyroscope: Theory and Application) (Springer-Verlag: Berlin, 1971).

dant to serve the Americans tea and offer them some Kazbek cigarettes and promised to be there soon.

I woke up Chizhikov and Kharchev and ordered them to find some cognac, some good snacks, and to set the table at once. When I appeared at the commandant's office, the din was terrible. The four American officers, each trying to out-yell the others, were communicating with the commandant through two interpreters—a German interpreted from English to German and a Russian lieutenant from German to Russian, and vice versa.

Introducing myself as the Soviet representative for German missile specialists, I asked our American friends to calm themselves and take a break from their tiring work by joining us for refreshments at the Villa Franka. They responded with an "Okay," and the cortège set off for our villa.

Chizhikov and Kharchev had not let me down. When the Americans looked at the table their eyes lit up. All four of the young Yankees broke into smiles and exclamations of approval followed. After numerous toasts, the Americans were quite tipsy. Excitedly slapping each other on the shoulders, we declared our friendship and found out that in September and October all of the German specialists that the Americans had named as war criminals would be sent from Witzenhausen via France to the United States. But several of their wives or mistresses had remained in the Soviet zone, in particular in Bleicherode, and the Germans categorically refused to go without them. On behalf of the command, the Americans requested that the Soviets help them return these women to them.

Finally we came to an agreement. The American representative would present the commandant of the town with instructions as to which women he wanted transported from Bleicherode and the surrounding areas; this list would include the names of the German specialists associated with the women. We said that we would permit the transfers, but only under the condition that an officer of ours be present to hear each woman voluntarily agree to leave—especially if she had children. Then and there we composed a protocol in Russian and English and signed it. It is now difficult to say to what extent the two texts were identical.

First Lieutenant Vasiliy Kharchev was introduced to the American major who was taking part in the transfer. The operation to send off five women and three children took place peacefully, without any yelling. Kharchev explained to the alarmed women that they were free to act as they wished. He said the Soviet command was prepared to petition for their return to Bleicherode if they did not like it in the American zone. What could these women do? After all, they were not indigenous, but had been evacuated from Peenemünde, and their landlords were glad to see them go because it was considerably more profitable to house Russian officers.

Kharchev quickly befriended the American officers guarding the Germans in the border towns of Worbis and Witzenhausen, and established a rapport with the women.

Just a week later, we received a report through our new female network of "agents" that Frau Gröttrup, the wife of a German specialist, wanted to meet with us. The meeting was set up near the border. Irmgard Gröttrup, a tall blonde in a light-colored sport suit, appeared with her 8-year-old son. "If we run into any trouble I'll explain that we were out walking and got lost." She immediately made it clear that she would make the decision, not her husband. She supposedly hated fascism. She had even been arrested on a number of occasions, and her husband, Helmut, had been as well. They wanted to know what the Russians would promise them.

She said that Helmut Gröttrup was von Braun's deputy for missile radio-control and for electrical systems as a whole. He was prepared to come over to us under the condition of complete freedom. I said that I needed to receive the consent of a general in Berlin before we could give her an answer, and we would first like to meet with Herr Gröttrup. Frau Gröttrup said that we should hurry because they might be sent to America in a week or two. Three days later, we pulled off the transfer of the entire family: papa, mama, and the two Gröttrup children—without the consent of Berlin of course.

The Gröttrups settled in a separate villa and were offered a very high salary and food rations compared with those of the other Germans. But there was one condition: to keep order and to take part in the creative work of restoring missile technology, our specialist Colonel Kuteynikov, who had a good knowledge of German, would live with them at the villa. The primary interaction with the Soviet management should go through him. The institute's German directorate and staff were not happy with this arrangement. They were unhappy that the institute that they had created was being transferred to an associate of von Braun who would drive out everyone who disagreed with him.

Gröttrup was clearly better informed than the others about the operations at Peenemünde. He had been close to von Braun, and he spoke very skeptically about the German contingent at our Institute RABE, except for Kurt Magnus and Hans Hoch. The others he simply did not know. To avoid stirring up passions, we agreed that we would create a special "Gröttrup Bureau" at the institute. Its first task was to compile a detailed report on the development of A-4 missiles and other projects at Peenemünde.

In retrospect, I would say that we were right on the money with Gröttrup. It is true, however, that Frau Gröttrup, who had grown bolder, proved to be not nearly as modest as she had seemed during our first meeting. Soon thereafter she acquired two cows, "for the children and to improve the nutrition of the institute's Russian management." She managed to obtain orders for products in extremely short supply, which Semyon Chizhikov grudgingly had to pay for and deliver. But an unexpected report from Colonel Kuteynikov sent us into a state of shock.

Attached to the villa where we had settled the Gröttrup family was a stable. The Frau was impatient to put it to use as originally intended. And so one night two rather decent horses appeared there. Colonel Kuteynikov, a man already advanced in years, evaluating the situation from all sides, reported that the Frau wanted to go horseback riding, not with her husband, but escorted by himself or another Soviet officer. After all, they might otherwise be detained at the nearest checkpoint.

When Colonel Ryazanskiy heard about this, he said sardonically, "Colonel Kuteynikov's mother is a ballerina and his father is a dancer. He will certainly stay put in the saddle. Let him go first." I was also losing patience. Ryazanskiy chided, "And what will your wife say if she gets a photo and a letter saying how pleasantly her little husband is passing the time, prancing around on horseback with a German Amazon?" Kuteynikov had worked as a military inspector in radio factories during the entire war. He was a pretty good radio engineer, and had risen to the rank of colonel in the engineer technical service. He had been attached to the Gröttrup family against his will, and was seriously offended. "My wife barely survived the siege of Leningrad. Now she is seriously ill and I have to worry about German mares. You can all go to hell!"

Kharchev couldn't stand for it either. "My sister is graduating from Moscow State University. She goes to the university wearing men's boots and is selling her last dress in order to buy food for our sick mother. My young wife Tamara had to quit her studies at theater school because she can't make it without my help—and here we are getting saddle horses!"

Chizhikov was tasked with developing an operation to exchange the horses for service automobiles for the institute. But Frau Gröttrup nevertheless secured one of the automobiles, by rights a service vehicle, for herself. She drove herself around from place to place. At one stop a ton of apples was procured for us; at another, a pig was slaughtered and sold to her; at a third, a "representational" sports car rented for the institute; and at a fourth stop, milk with enhanced fat content was ordered for the officer's mess hall. She would burst into our officer's mess, inspect our food, and demand that we immediately fire the cooks, whom she had caught stealing. In spite of Herr Gröttrup and Kuteynikov's protests, she sacked and replaced the stenographer/typists. Throughout all of this she was raising her children, learning Russian, and riding over to Villa Franka on a motorcycle to play Liszt, Beethoven, and Tchaikovsky on the grand piano.

But Operation *Ost* was not without its mishaps. Immediately after the operation to bring Gröttrup across the border, Kharchev announced that it was now time to bring Wernher von Braun himself across the border. Before giving the go-ahead, we discussed with Pilyugin, Voskresenskiy, and the local military intelligence whether such an attempt would be permitted. The nominal intelligence service immediately distanced itself, fearing a scandal with the allied forces. They were sure that "heads would roll" at the military administration if the operation was found out. "Therefore, you can act at your own peril and risk. Your shoulder boards aren't legitimate anyway. If they take them away, you won't be losing anything."

We decided to take the risk. As always, we sent Kharchev on the operation. Driving up to the border checkpoint early in the morning, Kharchev greeted the Americans and announced that he needed to cross into Witzenhausen to meet with the American officers who had visited our villa. They all exchanged wrist-

From the author's archives.

Irmgardt and Helmut Gröttrup with their children.

watches and Kharchev parted with the only bottle of genuine Moscow vodka in our entire garrison. Touched by such generous gifts, the Americans invited him into their Jeep and took it upon themselves to deliver him into town. This was the first hitch in the operation. What else could he do? Instead of taking Kharchev to see his acquaintances, they drove him to the American commandant of the town. The duty officer announced Kharchev's arrival and received instructions to send the Russian first lieutenant to the commandant's personal quarters. We asked Kharchev to repeat many times the story of what happened next, especially in the evening after dinner, expecting new exciting details on each retelling.

"So they take me into a big bedroom. The commandant himself is lying there in the big wide bed, just like the one here at the Villa Franka on the second floor. And on the other half of the bed is this beautiful woman. And between them is a German shepherd. It looks like they're having breakfast: there are bottles and all sorts of food on the bedside table. He throws off the featherbed, shoos the dog away, and tells me to climb into bed. 'For a Russian officer, my border neighbor, I begrudge nothing!'"

At this point in the story, we usually asked questions: "And did the beautiful woman throw off her blanket too?" Kharchev, red with indignation, would usually get all muddled and lose his train of thought. But he always firmly insisted that he did not accept the invitation, and that in broken English had confirmed that he was there on business.

Kharchev told us that, finally, after throwing on a robe, the commandant went out with him into the adjacent study. There they drank whiskey or something else for a long time; Kharchev argued that they needed to share the German specialists because they were war spoils. The commandant explained that a special assignment was in charge of guarding them. Kharchev then tried to turn the conversation to von Braun. The commandant said that he was the most important war criminal and was being very heavily guarded. After the conversation, they put Kharchev back into that same Jeep and as quick as the wind they drove him back to the checkpoint, where he hopped into his own vehicle and an hour later reported this adventure to us.

WHEN WE LET HIM GO REST AND SOBER UP, Pilyugin reproachfully said to me, "He's your ward. What would have happened to us all if he had given in to temptation? The commandant or somebody else there would have taken pictures of him in bed and then presented the photos to 'the right people'!"

Several months later, when Korolev was working in our company, he just about split his sides laughing over that story. But in contrast to Pilyugin he said, "What a fool Kharchev was not to accept the American's offer." As regards von Braun, Korolev was pleased that the operation had failed. He did not conceal this. Having seen the conditions in which Gröttrup lived and worked among us, he could imagine what would have happened if the most important German missile specialist had also showed up. Korolev thought that Von Braun clearly would have gone against the ideas that he was developing here, and changed the plans he had made through all his suffering during the last years in Kolyma and in the Kazan *sharashka*.

Having heard somehow about our attempt to make contact with von Braun, Gröttrup was quite amused. He said something to the effect that it was out of the question to think that von Braun would voluntarily come over to us. Von Braun was, of course, a very good engineer, a talented designer, and an effective organizer of ideas. But he had also been a baron, a member of the Nazi party, and even a *Sturmbannführer*.[3] He and his mentor, General Dr. Dornberger, met several times with Hitler and received high state honors. Reich leaders Goebbels and Kaltenbrunner kept close tabs on his work. In his youth, Wernher von Braun had dreamed of space travels under the influence of Hermann Oberth's works, but life forced von Braun to use his talent for purely military objectives. We asked Gröttrup to discuss the beginning of von Braun's career. He agreed, but with great reluctance. It ended with his wife inviting a small, select circle to coffee with whipped cream. She promised that Helmut would tell us a bit about the very beginning of their work.

Von Braun's story is no longer novel, thanks to numerous publications by American, German, and Soviet researchers, and even television documentaries. For that reason I will not cram my memoirs with yet another version of von Braun's biography. Gröttrup said that one of von Braun's very good traits was his striving to

3. *Sturmbannführer* was equivalent to the rank of major in the SS.

attract the most talented people. In doing so, he did not take age into account, and he was not afraid of competition.

Von Braun had been named technical director of Peenemünde-*Ost* at the age of twenty-five! For Germans this was quite unusual. But this shows how highly they valued his talent, initiative, and rare intuition. According to Gröttrup, von Braun was very attentive to senior, experienced specialists. He made fundamental decisions after having first gathered and listened to diverse opinions. There was no voting; von Braun always had the last word, but he managed not to offend the other staff. In spite of his youth, his authority was not called into question.

Fifty years after this assessment of von Braun's management style, I met with an American engineer named Jerry Clubb, who was a participant in the Apollo-Saturn lunar program, which had been directed by the no longer twenty-five but fifty-five-year-old, world-renowned creator of the first long-range ballistic missiles. The American talked about von Braun's working style in the United States in the same way that I had heard Gröttrup describe it in 1945 over a cup of coffee with whipped cream.

We did not manage to pick up Baron Wernher von Braun through Operation *Ost*, and I think that this was good both for us and for him. Despite all of his capabilities, what he achieved in the United States would have been impossible for him to achieve in the Soviet Union. It is true, another prominent scientist, professor, doctor, and also baron, Manfred von Ardenne, who had worked in the Soviet Union from 1945 to 1955 at the Electrophysics Institute in Sukhumi, was awarded the title Hero of Socialist Labor for his participation in the creation of the Soviet atomic bomb. It should be noted, however, that von Ardenne had not been a member of the Nazi Party, and he had not created weapons of mass destruction under the Nazis.

Von Braun had been valued and trusted by Dornberger and the higher military leadership of the infantry forces who had financed the construction of Peenemünde. He did not have to fear intrigue against himself, and he was able to work confidently as the technical director. Dornberger, who had become a general at Peenemünde, had always shielded von Braun. They were a powerful duo.

In spite of various nuances in political views, the main leadership staff had worked rather harmoniously and very selflessly. Everyone understood that it was too dangerous to express one's innermost thoughts. Any conversations concerning the possibility of using rockets for space travel were also dangerous because the Gestapo had ears everywhere. Such conversations were viewed as sabotage, the diversion of efforts from the Führer's most important assignment.

While we were talking with Gröttrup, the conversation somehow turned to the forced labor at *Mittelwerk* and the atrocities that had been carried out there. "What was your attitude toward the production of missiles using people condemned to death?" we asked. Immediately Frau Gröttrup interjected. No, she and Helmut had had no idea that such horrors that were taking place there. She said that this was also the case with the majority of the specialists. But von Braun and the

Peenemünde production personnel had been in Nordhausen more than once, and they of course had seen everything. This was yet another reason why he would avoid contact with the Russians. After all, the majority of the Dora camp prisoners were Red Army prisoners of war. Gröttrup recalled that Von Braun, during the hurried evacuation of Peenemünde, had been clearly afraid and had commented, "We fired missiles against England, but it's the Russians who will take revenge."

Gröttrup said that he had been arrested by the Gestapo because of indiscreet dinner conversations, but von Braun and Dornberger's intervention had saved him.

A report Gröttrup wrote toward the middle of 1946 was the most complete and objective account of the history of Peenemünde and the technical problems that were solved during the development of the first long-range ballistic missile, the A-4.[4]

The prehistory of German rocket technology, the interrelationship between "rocket and state", and the problems that were solved outside the scope of the A-4 program are discussed in greater detail and more reliably in the historical works and memoirs that emerged significantly later. In this regard, I would refer the reader to the memoirs of Walter Dornberger: *V2, der Schuss ins Weltall* (V2, A Shot into the Universe), which was published in the United States in 1954, and the richly illustrated book, *Wernher von Braun, Crusader for Space: A Biographical Memoir*, by Ernst Stuhlinger and Frederick I. Ordway, von Braun's coworkers in Germany and the United States.[5]

Michael J. Neufeld, a professional space historian working in the United States, has most completely and objectively researched the history of Peenemünde. In 1996, I was in Washington giving a lecture related to *Rakety i lyuidi* (Rockets and People) at the Smithsonian Institution's National Air and Space Museum. On this occasion, Neufeld, who at the time was the curator for World War II history at that museum, presented me with a copy of his book, *The Rocket and the Reich: Peenemünde and the Coming of the Ballistic Missile Era*, published in 1995.[6] This is truly a very serious scientific historical study.

---

4. Author's note: In the chapter "What is Peenemünde" (chapter 17) I used material from Gröttrup's report.

5. Walter Dornberger, *V2, der Schuss ins Weltall; Geschichte einer grossen Erfindung* (V2, A Shot Into the Universe: The History of a Big Invention) (Esslingen: Bechtle Verlag, 1952). An English-language version was published in 1954. See also, Ernst Stuhlinger and Frederick I. Ordway III, *Wernher von Braun, Crusader for Space: An Illustrated Memoir* (Malabar, FL: Krieger, 1994).

6. Michael Neufeld, *The Rocket and the Reich: Peenemünde and the Coming of the Ballistic Missile Era* (New York: The Free Press, 1995).

Chapter 22

Special Incidents

The energetic but uneventful activities that characterized the formation of the Institute RABE were punctuated by a few "strength tests." There were many conflicts among the town's indigenous residents, the German specialists enlisted for work, and Soviet military and civilian specialists arriving with "special assignments" from Moscow. In most cases, these conflicts had to do with the forced reduction of living space. More than a thousand new residents had converged on the small, quiet town, and there had been no new construction. The housing conflicts were for the most part successfully resolved through the personal efforts of the military commandant's office, the administration of the town Bürgermeister, and the Institute RABE administration.

Transport was more difficult. Sometimes high-ranking envoys from Moscow appeared who hampered work and who for some reason wanted to join in and get some benefits. They demanded personal automobiles, private apartments, and guides to show them local points of interest. Isayev and I, and later Pilyugin, had to take on the task of throwing them out of Bleicherode as quickly as possible.

But there were incidents that we categorized as "critical." Approximately two weeks after the official founding of the Institute RABE, a Smersh lieutenant colonel from the division that had been billeted in Nordhausen appeared in my comfortable office. He reminded me of our meetings during our first days in Nordhausen and was interested in what we had turned up using the information we had received from Smersh about possible hiding places of secret equipment that the Americans had not discovered. It turns out that we had forgotten about the "forester's cabin."

"According to information that we received from the Germans in the first days after we entered Nordhausen," he said as he unfolded a large-scale map of Germany, "somewhere in the area of the forest preserve, which used to be strictly guarded under the pretext of protecting the wild animals, there is a forestry base that locals called the 'forester's cabin.' We haven't been able to find anyone from among the locals who has been at that base. The new German authorities have found forestry service employees who themselves did not have access to that base. They said that in actual fact there wasn't a 'forester's cabin' there, but a comfortable house with all the conveniences that was used for recreation and hunting by high-ranking officials of the Reich. Eyewitnesses, who are not always trustworthy, told us that they suspected that a bigwig was visiting the cabin when the normal police patrol on the roads was reinforced with a special SS motorcycle patrol. But the big shots didn't feel like hunting during the last year of the war. There were increasing numbers of wild boar, wild goats, and roe deer in the forests, but the law-abiding Germans never did any poaching for meat. Someone reportedly saw a column of trucks escorted by an SS motorcycle patrol on the road to the cabin just before the American troops occupied Thuringia.

The Smersh officer continued, "I think that the Americans visited these areas, but did they find everything? I won't be able to help you. We're being transferred to the east. They're about to begin the demobilization and reduction of Smersh. According to our information, the last vestiges of Vlasov's men are hiding somewhere in these forests.[1] They are considerably more dangerous than the German Nazis. If you go looking for the "forester's cabin," I'd advise you to arrange an armed escort with the commandant's office or the Seventy-fifth division. Just in case."

"Are you sure," I asked, pointing to the map, "that the alleged secret cabin is in our zone and not the American zone?"

"I can't say for sure. They still haven't laid a precise border over the hilly, forested terrain. Our border guards and the Americans are posted on the roads, but there are no checkpoints, barriers, or soldiers in the forest."

That very same day at the Villa Franka I convened an improvised "military council" to develop a plan to uncover the secret of the "forester's cabin." Isayev was certain that there would be nothing there to do with engine technology—everything that the Germans had in that field he had already studied at the *Walter* company in Basdorf, and now it was time to settle down in Lehesten. Therefore, he had no intention of wasting a day on some mythical "forester's cabin" with the added risk of ending up in the American zone. But the others supported me, and we formed a two-car expedition. The division command allocated a Jeep and driver, an "unemployed" combat intelligence officer, and a soldier with an automatic weapon—just in case. Accompanying me in my Mercedes was my German guide, whom the military commandant and *Bürgermeister* had recommended, and engine specialist Ivan Raykov, who joined us despite Isayev's objection.

"One automatic weapon and three pistols is quite sufficient to defend ourselves against an attack of wild boars," I commented.

We left early in the morning. Soon thereafter, on the gravel road that dodged around the forested hills, I was no longer able to orient myself using the map and had to completely trust the guide. After about an hour of driving that was difficult

---

1. Andrey Andreyevich Vlasov (1900-1946) was a Soviet military commander with anti-Stalinist sentiments, who after being captured with his army by the Germans in July 1942, formed the Committee for the Liberation of the Peoples of Russia and the Russian Liberation Army aimed at overthrowing Stalin's regime. The Germans allowed Vlasov's 50,000 troops to go into battle against the advancing Red Army at the end of the war. He was handed over to the Soviets in May 1945, tried, and hanged.

for the Mercedes, Alfred hinted that the motor was overheating and said we might not have enough gas for the return trip in these road conditions. As if in response to his apprehension, having ascended to the top of the next hill, we came out of the dark forest into a sunny clearing. The road ended at the open gates of a small country estate, in the middle of which stood a small two-story house with a double-gabled, red tile roof. Along the edges of the clearing stood haystacks and mangers for the forest animals. Everything appeared to be well-kept. But the most unexpected sight was the Jeep parked in front of the house.

The division intelligence officer yelled out, "Turn the cars around! Drivers, stay in the cars! All others stay by the vehicles." He turned to the soldier with the automatic weapon, "Petro! Cover me in case something happens. I'll go alone first to investigate."

Evidently having seen or heard our procession, an elderly woman came out of the house. She was waving her arms very aggressively, making it clear to us that we were to beat it. Summoning every shred of my knowledge of German, I emerged from where I had taken cover and tried to communicate with her.

"This is a nature preserve," she yelled. "You are scaring the animals. The American soldiers came here and went off into the forest to hunt. They took away my husband's car. He walked to Nordhausen two days ago to ask for help. Now you've come to hunt too, and we are going hungry—but we never shoot the wild animals!"

The Jeep had a military license plate and the stars and stripes marking. How did it get here? we thought, through our zone?

As if in response to our musings, a burst of automatic fire resounded in the forest followed by several single shots. We lit up cigarettes and decided to wait for the hunters.

Soon thereafter, four Americans emerged from the forest. We couldn't figure out what their military ranks were. They were carrying a roe deer by the legs and dragging the carcass of a wild boar along the ground. When they saw us, the Americans were not embarrassed, but greeted us gregariously. In Russian-German-English, but primarily pointing with our finger at the map, we showed them that this was our zone and they didn't have the right to kill animals here. In response the Americans guffawed, slapped us on the back, and offered us the pick of the roe deer or the wild boar. After loading their booty, the Americans departed rolling down the road that went over the other side of the hill. We understood then that we were located somewhere on the border.

I attempted for some time to explain the objective of our visit to the distraught proprietress. After we had handed over our entire supply of cigarettes "for her husband," she led us to the barn, moved aside a pile of some sort of brooms, and pointed to a trapdoor in the wooden floor. In the dry cellar we discovered what was clearly well-packed electrical equipment. We couldn't make it out in the darkness. The next day, a team that included German specialists—without an automatic weapon-wielding soldier but with a wealth of rations—brought the precious cargo from the "forester's cabin." It turned out to be two sets of relay boxes and control panels intended for pre-launch tests and the launching of V-2 missiles. I managed to find the Smersh officer who hadn't yet left Nordhausen and thanked him for steering us to the cabin. He said that we had been marvelously lucky. Subsequent clarification had shown that the "forester's cabin" was considered part of the American zone.

A MONTH AFTER THE INSTITUTE RABE HAD BEGUN TO FUNCTION, the weakness of the organizational and financial management of the institute's German side started to become evident. Director Rosenplänter was a very energetic engineer, but he had a poor understanding of supply problems, and he evidently carried no prestige among the local businessmen and merchants who were setting up business ties with the institute. His deputy for general affairs devoted all of his attention to procuring rations for the Germans and improving their living conditions. I was therefore very glad when, on the recommendation of the SVAG, Herr Schmidt, a former director of the Peenemünde supply and cooperation service, came to us from Weimar. Obviously no spring chicken, stout and nattily attired, his sincere manner and pleasant smile instilled respect for the institution he represented.

We appointed Schmidt to the post of commercial director, and put him in charge of transportation, supply, equipment for the buildings, and what later came to be called "social welfare." He energetically set about his business and immediately showed his competence by outfitting the laboratories with first-class instruments and excellent standard benches. A broad range of electrical power sources

From the author's archives.

View of the former Institute RABE building.

appeared and new top-notch metal-working machine tools began to arrive at the factory in Kleinbodungen and in the institute's shops.

In early September 1945, Pilyugin and I went to Dresden for three days to place orders for gyroscopic instruments. When we returned, we found Isayev in a gloomy mood. He was chain-smoking his favorite *Byelomor* cigarettes, and finally he told us about what had happened in our absence.

In the middle of the workday, Schmidt, who was usually very polite and respectful, literally burst into Villa Franka where Isayev was resting after returning from Lehesten. He yelled at Isayev, "Your soldiers are raping German women!" After calming down a bit, Schmidt explained that the woman he was living with (his wife had remained in another town) was strolling with a female acquaintance in the forest. A Russian soldier had attacked them, raped his friend, and disappeared. There was nothing for Isayev to do but to report Schmidt's complaint to the military commandant. Captain Solodyankin, remembering Marshal Zhukov's threatening orders, reported the incident to the commander of the division billeted in the area of Bleicherode. There they quickly determined that they should look for the perpetrator in the artillery brigade attached to the division, which was getting ready to return to the Soviet Union. To wipe away the division's shame, the brigade commander ordered all of his men to stand in formation in the town square. He led the victim past them to identify the perpetrator. She pointed at a Kazakh soldier, who was arrested on the spot. The accused confessed during interrogation that his goal was to get revenge against the Germans for his two brothers who died at the front. He himself had never gotten the opportunity to kill a German in combat. Having learned of their impending evacuation from Germany, he decided to avenge the death of his brothers in that way. The next day a military prosecutor was to arrive from Erfurt, and a military field trial would be held. Iron discipline was now being restored in the army, and it was not out of the question that they would sentence him to be shot. "And it's all my fault," Isayev said, "I don't know what possessed me to hurry over to the commandant with that report. I needed to calm Schmidt down. Maybe we could have agreed not to take the matter to the military tribunal."

The next day there really was a trial and they sentenced the soldier to be shot. I reported this to Schmidt over the telephone. He immediately rushed over to Villa Franka along with his girlfriend, the victim. She was not young, and was tall and well built. She seemed to be a woman who could stand up for herself.

Schmidt and his girlfriend shouted over each other that they were shocked by such a harsh sentence. They were prepared to forgive the soldier and asked me to intervene to save his life. I immediately called the division commander and asked if something could be done to save the soldier since the victim was stunned by the severity of the sentence, wanted to petition for a pardon, and had forgiven him. The general was silent for a moment and then said that he would call me back in a half hour. Thirty minutes later, he reported that he had talked with Berlin. They had explained to him that Moscow had given instructions to establish rigid order among the occupation troops. Our critical incident had already been reported to the commander, and the sentence would be carried out for the edification of others, regardless of the German woman's petition. Schmidt and his girlfriend were stunned by this message.

That evening, after a hard day, we gathered by the fireplace in the spacious hall at Villa Franka. It had come to pass that several officers of the Seventy-fifth Guards Division billeted in Bleicherode who were hungry for cultured society had joined our informal officers' club. Our circle in Germany, which consisted of trade union or "civilian" officers, somehow attracted real officers who were decorated with combat ribbons and medals, but who had grown tired of the war.

The combat officers explained their interest in our activity, "You here are involved in interesting and important work, but we're a bunch of layabouts. You know very well what you will be doing when you return home. But we have been fighting for more than four years, and we happened to survive—now we have nothing to do. The combat guards division is enjoying comfortable conditions without the war, but it's doomed to disintegration."

On those evenings when we gathered by the fireplace, the souls of our society were the "hussars," Isayev's name for Voskresenskiy, Boguslavskiy, and Rudnitskiy, who lived in the Villa Margaret on the highest hill in Bleicherode.[2] Somehow, it turned out that Guards Captain Oleg Bedarev, a combat officer, became the latest member of our "hussars' squadron." He had marched to Thuringia all the way from Stalingrad with his combat girlfriend Mira, a Guards captain in the medical service. Pilyugin called military physician Mira and Oleg Bedarev our "Guards lovebirds," and we included them among the welcome members of our evening meetings and fireside chats, where we would share fragrant kirsch and precious packs of Kazbek cigarettes purchased at the central Berlin military exchange.

We, including Oleg, Mira, and their combat friends, traded stories about our everyday life in the war for hours, constantly interrupting each other. Sometimes Oleg would stop himself and say, "Let's not talk about that." To distract us from any bad memories, he would pick up his guitar and perform songs that he had composed himself. Oleg was also a poet and the editor of the division newspaper. His poetry was very unsettling to the political department, which made it difficult for him to get promoted.

Mira explained, "The political workers consider his poems to be demoralizing and decadent. Not once does he mention the Party or Stalin in them."

The tribunal's sentence stunned Oleg and Mira as much as it did us.

"I was ordered to examine the condemned soldier and to certify that he was in the proper state of health to be shot. It turns out that some regulation or another

---

2. The phrase "hussar" was originally used to describe ornately attired officers of the Hungarian light cavalry from the fifteenth century. Chertok uses the word for his elegantly dressed associates.

calls for that. He was very calm and said that he was no longer disgraced before his two dead brothers. As a doctor, I doubt that he was capable of overcoming a powerful woman if she resisted. He was nineteen years old and had been in the army for a year and a half. All the way from Stalingrad to Berlin, I have signed so many death certificates that I can't keep count. This is the first time I have had to sign a certificate of health before an execution."

Isayev had smoked an entire pack of *Byelomor* cigarettes that evening; after Mira's story, he announced that there was nothing left for him to do in Bleicherode. The Institute RABE would get along without him, and he had already seen and understood everything he needed to in Lehesten. It was time for him to return to Moscow.

The soldier's death was especially hard on Mira. She took Isayev's arm, grabbed the bottle of kirsch, and they retreated to the garden.

Voskresenskiy turned to Oleg, "Be careful that we don't make off with your beautiful wife. Who knows, we might haul her off to Moscow."

"I'm not afraid. They didn't make off with her on the front lines, and I'll return with her to Moscow before you. I just got my orders—our division is being temporarily redeployed to the area of Wittenberg, and from there back home. Tomorrow we begin the evacuation."

Our party broke up late. When they said goodbye, Mira gave Isayev a long embrace and kissed him.

We had a proper celebration of Isayev's departure soon thereafter. On 10 September 1945, he went to Berlin and soon thereafter flew back to Moscow. He never returned to Germany. Two weeks later, I took a trip to Wittenberg with two of our "hussars," Voskresenskiy and Boguslavskiy. We sought out Mira and Oleg. They had not unpacked their suitcases and were preparing to load onto a special train that was heading back to the Motherland. I jotted down a possible address and phone number for Mira in Moscow.

It wasn't until the spring of 1947 that Katya and I were able to host a gathering of old friends in Moscow on Korolenko Street. Mira and Oleg and two "hussars"—Boguslavskiy with his wife Yelena and Voskresenskiy with his future wife, also Yelena—squeezed into our cramped communal apartment. Isayev did not come to the party. Mira arrived in a stylish wool suit. She appeared very different than how I remembered her in uniform as a medical service Guard's captain in 1945. Oleg, however, in his old, second-hand jacket with no decorations, was a sorry sight. He was drinking a lot without taking part in the toasts and hardly touched the appetizers that Katya painstakingly arranged for him on a plate.

"Why didn't Isayev come?" Mira asked me before leaving.

"I invited him, but he didn't want to come without Tatyana, and she is ill," I answered.

Mira knew that Tatyana Isayeva and Oleg Bedarev were suffering from the same disease—alcoholism. The medical profession was powerless when it came to treating acute alcoholism. The condition was exacerbated by the transition from the rigid, goal-oriented discipline of military service to life in a peacetime dictatorship. His wife's illness was one of the reasons for Isayev's hurried departure from Germany.

HERE IT IS PERTINENT TO MENTION that Isayev and Arvid Pallo, whose group we had sent from Nordhausen to Lehesten on 15 July, successfully orchestrated engine firing tests. Soviet engine specialists—engineers and mechanics—mastered the technology that they had seen in Germany to such an extent that they were able to conduct firing tests in different modes without the Germans' assistance. I managed to visit Lehesten in August and for the first time witness a stunning spectacle—the open plume of an engine firing with 25 metric tons of thrust. Our GAU chief, General Kuznetsov, who considered himself officially responsible for the Lehesten site in addition to the Institute RABE, had not yet seen the firing tests and demanded that director Rosenplänter and I accompany him on a trip to Lehesten. I postponed this trip several times under various pretexts.

In late September, Aleksandr Bereznyak visited us. He had become familiar with all of the German aircraft firms in the Soviet zone. Bereznyak's head was full of ideas, and he had hurried to Bleicherode to meet with Isayev. But Isayev was already in Moscow. Bereznyak talked me into going with him to Lehesten. The general was still making his persistent demands, so we decided to combine our trip. On Sunday, 30 September, we left Nordhausen in two cars.

General Kuznetsov's Opel-Kapitän was the lead car, with a military driver at the wheel. Kuznetsov seated Rosenplänter next to the driver to serve as his guide. The general sat in back and demanded that our interpreter be seated in his car so that he could converse with Rosenplänter. Lyalya had to obey the general. Bereznyak, Kharchev, and I were in the second car, our Mercedes, with Alfred at the wheel. We set out and were soon lagging way behind the lead car. Alfred repeatedly shook his head reproachfully and muttered something about the general's car exceeding permissible speeds on the narrow, winding roads. Suddenly Alfred let out a shout. He was the first to see that the Opel-Kapitän had slammed into a tree. The injured were taken to the hospital in Erfurt. All three men were severely injured, but the German doctors assured us that they would live. "But Fräulein Lyalya has been fatally injured. The base of her skull was fractured and she has multiple injuries to her spinal cord." I made a telephone call to Bleicherode and requested that Dr. Musatov, the chief of the division medical-sanitary battalion, come immediately. We had become friends with him. His front-line surgical operations were legendary in the division.

When Musatov arrived, the chief surgeon of the hospital, Professor Schwartz, came out to speak to us. The hospital was top-notch. Before the surrender it had been the military hospital for SS officers. We went into the operating room with the professor. Both of General Kuznetsov's legs were bandaged and already in casts, and he was still in a state of shock. Rosenplänter had multiple injuries to the head and face. He was lying there with his head completely bandaged. The driver was unconscious. Both his legs, an arm, and several ribs were broken. The nurses were

From the author's archives.

"Our beloved Lyalya."

bustling about them. Off to the side lay our beloved Lyalya, unattended, completely naked.

Kharchev was beside himself. He snatched his pistol from its holster. He fired into the ceiling and shouted that this was premeditated. "If you don't save her life I am going to shoot everybody here." But Professor Schwartz and the other doctors and nurses did not even flinch. Evidently their dealings with SS officers had hardened them. Our surgeon, Major Musatov, deftly disarmed Kharchev. He examined Lyalya, spoke with Schwartz, and told us, "They did everything they could. Now they have given her the opportunity to die peacefully. She doesn't have more than an hour to live." Kharchev began to sob. The doctor motioned to one of the nurses. She approached Kharchev, rolled up his sleeve, and gave him an injection.

Lyalya was given a funeral with honors. She had not only worked at the institute, but had helped on her own initiative in the division, and she had enjoyed great success with her artistic talent. A standard pyramid with a red star was erected at her gravesite in the garden next to the institute. On one of the facets was a portrait of Lyalya—a beautiful, talented young Russian woman from near Tula, whose life came to such a tragic end in Germany. I wrote to Isayev about Lyalya's death. He was shaken—all the more so because he was making arrangements to fulfill his dream of her working in our Moscow institute after repatriation.

Chapter 23

In Search of a Real Boss

We soon started to receive pessimistic letters from Isayev in Moscow: "Horrible, horrible, horrible! . . . We don't have anything sensible to work on! Our patron is no longer interested in our research. He wants to go back to work at the Academy." Using the darkest images to describe the work life and unsettled daily life in Moscow, Isayev reminisced about the Villa Franka and Bleicherode. "Our work in Bleicherode—that was just a beautiful dream. Here the primary "missile-related concerns" are firewood, early freezes, and agricultural work. . . . Your Katya is a real trooper! She wants you to come home. She met with the patron himself. He said that it was your own fault that you had stayed too long in Germany. Nobody was making you stay there. You were loyal to him during the most difficult time and now suddenly you have 'sold your own sword'; and to whom?—to the artillerymen."

This was November 1945. It was hard to read Katya's letters about the difficult life in postwar Moscow with two little boys. And what's more, the youngest was always getting sick. Katya had to travel all the way across Moscow to NII-1 for the rations to which she was still entitled, rush off to the children's clinic for milk for the sick baby, and repair the eternally burned-out electric hotplate. It was cold in the apartment—it was even difficult to dry underwear. The autumn wind blew in around the windows. Often the water didn't make it up to the fifth floor, so she had to run down to the water fountain on the street. After returning with the water, she had to repair the electric wiring. There was a short circuit somewhere and the fuses were always burning out.

After General Kuznetsov's accident, which had him confined to a hospital bed for a long time, it was understandable that receiving news like this prompted me to petition for leave. I reached an agreement with Pilyugin that he would take over all the business of managing the Institute RABE for about two weeks. But it turned out that Moscow had not forgotten us.

General Gaydukov ordered that none of us were to leave Germany. He was evidently one of the first among our military leaders to appreciate the prospects of the work we had undertaken and its scope. He understood who was propping it all up, and he lobbied in Moscow for energetic support of the Institute RABE.

We could sense this. First, there was an increased stream of specialists at our disposal who had been sent on temporary assignment from various agencies.

Second, leave was categorically forbidden. And finally, a Communist Party Central Committee plenipotentiary arrived in our midst. He announced that the staff of the Central Committee was monitoring our work and were pleased with us, but we needed to accelerate our work so that everything could be wrapped up in early 1946. The People's Commissars hadn't yet agreed as to who should be involved with long-range missile technology. For the time being, the Central Committee had entrusted all of the work in Thuringia to Gaydukov. As far as the calamitous situation with our families, the Central Committee plenipotentiary assured us that all the necessary commands would be given to our Moscow institutions. The commands were given, but how could they really help? All they could do was provide financial support and rations for the wives and children. In this situation, Gaydukov made the only correct decision, which had exceedingly important consequences for the future development of missile technology in the Soviet Union. In order for missile technology, which had been rejected by the aircraft industry, to acquire a real boss, it was necessary to report to Stalin and request his instructions. But regardless of Stalin's instructions, the specialists who had already immersed themselves in these problems needed to be retained. And to prevent them from being dragged off to various agencies, the best scenario was to gather them all into a single collective in Germany and let them work together there until the necessary decisions had been made in Moscow.

One could not report to Stalin empty-handed. He needed to be reminded of his correspondence with Churchill in 1944, and we needed to show that we had thoroughly investigated the secret V-weapon and that our specialists were already working in Germany. But that was not enough.

Sometimes the decisive actions of a single individual, especially if they have been covered for many long years by a Top Secret stamp, go unmentioned by historians or publicists. To prepare for his meeting with Stalin, Lev Mikhailovich Gaydukov studied the history of RNII and assessed the previous work of Korolev, Glushko, and other repressed specialists who were still alive. He determined their whereabouts and prepared a list of those whom he believed needed to be dragged out of the special prison where Beriya had driven them. The majority of these specialists had been officially freed in 1944, but they were still working in the same *sharaga*.

Wise people advised Gaydukov to get to Stalin in such a way that Beriya would not find out about it ahead of time under any circumstances. It is difficult to determine how Gaydukov managed to do this, but somehow he did. Gaydukov reported to Stalin about the work going on in Germany and the need to begin working on this program in the Soviet Union. Stalin did not make a concrete decision regarding whom he would entrust with this work, but he authorized Gaydukov to personally familiarize the People's Commissars with his proposal to take on the responsibility for the development of missile technology—whoever among them agreed could prepare the necessary draft resolution. In conclusion, Gaydukov requested that Stalin free the specialists on the list that Gaydukov had

From the author's archives.

Lev Mikhaylovich Gaydukov—member of the Supreme Command Headquarters Guards' Mortar Unit's military council, Institute Nordhausen chief, NII-4 deputy chief.

placed before him on the desk, so that the rocket efforts could be intensified. Stalin's decree, which contained the list, determined the fate of Korolev, Glushko, and many other people for many years.

Being in Germany in 1945, we of course had no way of knowing about this behind-the-scenes activity. Many years later, Yuriy Aleksandrovich Pobedonostsev's fragmentary hints led me to assume that he was one of those who had helped Gaydukov prepare the list and develop his tactics. One false step could have halted the development of missile technology in our country for many years.

It is also evident that Malenkov helped arrange Gaydukov's meeting with Stalin. Malenkov had been Gaydukov's mentor when the latter had been a Central Committee staff worker under his jurisdiction. At that time Malenkov was still the chairman of the State Committee on radar and air defense technology, which made it his duty to be interested in the anti-aircraft guided missiles that had also been developed in Germany in Peenemünde.

One way or another, Gaydukov had received instructions directly from Stalin to meet with the People's Commissars. There were three choices: Shakhurin, People's Commissar of the Aircraft Industry; Vannikov, People's Commissar of Ammunition and director of the First Main Directorate which was in charge of the atomic problem; and Ustinov, People's Commissar of Armaments.

All of the People's Commissariats were preparing for conversion into ministries. Gaydukov knew this could also mean changes among the directors. Stalin might recall some blunders in the development of wartime technology and not approve a particular People's Commissariat becoming a ministry. Such a danger threatened Shakhurin. During the war he had been considered Stalin's favorite. He had reported to Stalin more often than the others, and to a greater degree than the other People's Commissars had enjoyed Stalin's help in organizing the mass production of new aircraft technology, which by the end of the war had surpassed German technology both in quantity and in quality (with the exception of jet aircraft). Stalin had then received a report about our jet aircraft construction lagging behind, and this had affected his attitude toward the leadership of the

People's Commissariat of the Aircraft Industry. Turbojet technology became a priority task for Shakhurin. All of his chief designers were tasked with creating jet aircraft. Under these conditions it was impossible to also put the "missile noose" around the neck of the aircraft industry.

In response to Gaydukov's appeal, Shakhurin refused to participate in the missile weaponry development program, in spite of the fact that missiles, in their design, production process, instrumentation, and electrical equipment, were closer to aircraft technology than to other types of weaponry. This refusal was the beginning of a confrontation between aircraft and missile technology that would last many years. Aviation had the advantage until strategic missiles with nuclear warheads emerged.

But Shakhurin's refusal also had other direct consequences for those of us who had worked in Germany. In late 1945, the Deputy People's Commissar of the Aviation Industry, Dementyev, ordered all specialists from the aircraft industry, including Pilyugin and myself specifically, to stop our work in Nordhausen, Bleicherode, and Lehesten and return to Moscow. Gaydukov protested this order, and told us bluntly that no one would be released. Only Isayev was allowed to travel, and later his colleague Raykov. For the time being we had gone our separate way from aviation.

Gaydukov's appeal to People's Commissar Vannikov had been motivated by the fact that a missile is also ammunition. It's the same thing as a projectile, just big and capable of being guided. But Boris Lvovich was not born yesterday. His son Rafail, who had served in the first military rocket unit at Sondershausen, had enlightened his father to the fact that a guided missile was anything but a projectile. It was a large and complex system. Moreover, Vannikov explained to Gaydukov that he was now responsible for developing atomic weapons—now that's a real piece of ammunition—and they had better things to do than deal with missiles. Gaydukov needed to keep looking.

That left People's Commissar of Armaments Dmitriy Fedorovich Ustinov. Neither he himself nor his deputies had had anything to do with missile technology during the war. They had been responsible for artillery, "the god of war"; infantry small arms; and machine guns and large-caliber guns for aircraft, tanks, and ships. They hadn't even been involved with the Guards Mortar Units' *Katyusha* rocket launchers.

Nevertheless, Ustinov thought about it. What kind of prospects did this field have? He knew that being a perpetual supplier of armaments for aircraft, tanks, and ships meant always playing a supporting role. The aircraft, tank, and ship specialists would play the starring roles. Classic conventional artillery had practically reached its limit for all parameters—range, accuracy, and rate of fire. No qualitative leap was foreseen there.

And anti-aircraft artillery? the experience of the war had shown that thousands of rounds were needed to shoot down a single aircraft. Ustinov had received a report about German anti-aircraft missile programs, and he could foresee that sooner or later they would supplant anti-aircraft guns. In addition, the People's

Commissariat of Armament had been instructed to develop fire control radars (SON) in place of optical instruments to control anti-aircraft fire.[1] This was something similar to missile technology. What is more, even a hardcore artilleryman such as GAU chief Artillery Marshal Yakovlev had now become interested in missile technology. If GAU was to be a customer, as it had been during the war, then the Commissariat needed to preserve its strong military friendship in the new field. This was very important.

Ustinov gave his preliminary consent to take guided missile weaponry into his Commissariat, but asked that no final resolutions be prepared for Stalin's signature for the time being. The problem needed to be studied in greater depth. He gave this assignment to his first deputy Vasiliy Mikhaylovich Ryabikov.

In the initial period of Soviet missile technology development, we were fortunate to have devotees and individuals with initiative and daring. Despite the fact that the Germans' massive use of missiles against London had not yielded the anticipated effect, the Guard's Mortar officers were able to quickly appreciate the prospects for a completely different qualitative leap. The few NII-1 aircraft employees who had been in Nordhausen converted to the new "missile religion." They refused to follow the orders from their immediate and very highest superiors to curtail their activity in Germany. Finally, after the usual postwar delays, we had found a strong boss in the industry for the new technology.

After receiving Ustinov's instructions, Ryabikov flew to Berlin and made his way to Bleicherode. He was very calm, attentive, thoughtful, and evidently an intelligent and experienced leader. Pilyugin, Mishin, Ryazanskiy, and Voskresenskiy assessed Ryabikov in similar terms. He visited *Mittelwerk* and became familiar with the unusual history of the Institute RABE's establishment. He listened to our stories about the history of A-4 development, the history of Peenemünde, and our ideas about the future. We even spoke about designs to increase range and accuracy.

Toward the end of his visit, we arranged a farewell dinner during which Vasiliy Mikhaylovich said very frankly that his technical worldview had been significantly changed by everything that he had seen and heard. Now he saw that a completely new outlook had emerged in weapons technology. He promised to inform Ustinov about all of this. With regard to our work in creating a scientific-research institute on our former enemy's territory, he said, "You are simply champs to have devised and organized such a concept. I will support you everywhere."

Ryabikov's visit inspired us all. We were assured that in addition to having military guardians, we could also rely on a solid scientific-industrial and technological base and strong industry leadership.

---

1. SON—*Stantsiya Orudiynoy Navodki.*

Chapter 24

Korolev, Glushko, and Our First Encounters in Germany

We were always glad when new people arrived from the Soviet Union to expand the common front of our operations. When Yuriy Aleksandrovich Pobedonostsev called me in late September from Berlin and asked me to receive Sergey Pavlovich Korolev and tell him about our work, I did not associate Korolev with any previous events. I told Pobedonostsev that a lot of officers came to see me at the institute, and if they were specialists in the business, then we didn't begrudge anyone, everyone found work. In the daily commotion of work I forgot about this conversation.

Several days later, Lieutenant Colonel Georgiy Aleksandrovich Tyulin telephoned from Berlin. He was there as the GAU representative in charge of receiving and dispatching to various locations the military and civilian specialists sent to Germany to study technology. His mission, under the appellation "Tyulin's Domain," was located in Oberschöneweide and was well known by the military administration. The Institute RABE's example proved to be infectious. In September 1945, Tyulin began to assemble a group of specialists in Berlin to study anti-aircraft guided missile technology. These specialists would later become the core of the Institute Berlin, which was similar to our Institute RABE. When Tyulin reminded me about Pobedonostsev's phone call in connection with Lieutenant Colonel Korolev's upcoming visit (without prefacing this notification with any other comments), I concluded that Korolev must be one of those honored combat commanders who was being sent on special assignment from the Guard's Mortar Units, the artillery, or aviation. Lately those three branches of the armed forces had been the primary suppliers of seasoned specialists from the troops stationed in Germany.

Many years later when Korolev's name had gained worldwide recognition, recalling our first meeting, I asked Pobedonostsev and Tyulin why they had not told me who this Korolev really was who was traveling from Berlin to Bleicherode. They had not even indicated—as they usually did with senior officers—whether he was being dispatched on special assignment by industry or the army. The reply I received from both of them was essentially, "Why do you ask such a naïve question now? Back then we were forbidden to explain anything to you."

Pobedonostsev knew Korolev very well because he had worked with him at GIRD and RNII starting in 1930. Their families had lived in the same apartment building on Konyushkovskaya Street in Krasnaya Presnya. They had worked and socialized together almost everyday until Korolev's arrest on 28 June 1938. I first met Pobedonostsev in 1942 when he, A. G. Kostikov, and L. S. Dushkin arrived in Bilimbay to observe the firing tests of the liquid-propellant rocket engine developed at RNII. After the annihilation of the first RNII chiefs, Kleymenov and Langemak, Pobedonostsev was saddled with the heavy burden of organizing the program on solid-propellant rocket projectiles and launchers.

In 1944, Pobedonostsev and I often dealt with each other on projects at NII-1 under our common patron, Bolkhovitinov. I met with Pobedonostsev almost every day at the big dining table in the mess hall for the NII-1 managerial staff. This dining hall was a meeting place for the NII-1 employees who had known Korolev very well throughout all of his previous work. Among them was Mikhail Klavdievich Tikhonravov, Korolev's collaborator on the first rocket programs at the Moscow GIRD. But never during that time—not at the table, or anywhere else—did I hear the names of Korolev or Glushko mentioned.

In Germany I associated with Pobedonostsev very often, but here as well Korolev's name had never been mentioned, until the telephone call from Berlin. Some unwritten law had placed a taboo on the names of those who had been repressed. One could mention them and speak of them only during closed Party meetings and at various political activist meetings that came out immediately after their arrest. Here one was obliged to say, "We had failed to notice that we had been working side by side with enemies of the people." "Good form" at that time required that everyone stigmatize enemies of the people, and in doing so, in a fit of self-criticism, list all the shortcomings that one could possibly think of in the work of one's group, department, or entire institute. Then, having pledged our allegiance to the great Stalin, who warned us in time about the acute situation of the uncompromising class struggle, we were supposed to say that we would rally around "the great cause," we would correct the shortcomings that had been committed, we would strengthen, and we would "fulfill and over-fulfill" the plan ahead of schedule.

After a revelatory campaign and a series of similar public speeches, the names of the "enemies of the people" were to be erased from one's memory. If they were authors of books or magazine articles, then those books and magazines were removed from libraries. Usually they were hidden in a so-called "special archive" and issued only in extreme necessity with the permission of a governmental representative who was an employee of the state security apparatus.

That was the situation from 1937 until the beginning of the war. During the war some repressed military leaders and designers were freed, but nevertheless, the taboo remained in effect practically until the end of Stalin's life.

During those years, I committed a gross violation against the system by storing the scientific works of "enemies of the people." In 1935, I saw a book at a kiosk at

Factory No. 22. On the book jacket was an illustration of something akin to an aircraft bomb. At the time I was caught up with the problems of aircraft automatic bomb releases and other problems of aircraft armaments, so I shelled out one ruble and 50 kopecks to buy the book.

It turned out to be a work by Georgiy Erikhovich Langemak and Valentin Petrovich Glushko, *Rockets: Their Construction and Use,* written in 1935 and published by the Main Editorial Office of Aviation Literature in a run of only seven hundred copies. After a cursory perusal, I had found almost nothing in the book that interested me, but I understood that there were people and organizations who were working on flying machines that had nothing to do with airplanes. In 1937 there was talk of mounting a new type of weapon on our airplanes—rocket-propelled projectiles. As the director of an equipment and armament design team, I had to learn what this entailed. I remembered this book, found it in my bookcase, and got so carried away with it that I consider that time in 1937 to be the beginning of my association with the problems of rocket technology. But I was once again swamped with aircraft issues, and the book was once again hidden in the bookcase.

During our evacuation to the Urals in 1941, my meager library remained behind in the old wooden house where I had lived with my parents at the "quilt" factory. It wasn't until 1942 when I was in Bilimbay in the Urals that I remembered this book. Isayev and Bolkhovitinov had just returned from Kazan, where they had met with some liquid-propellant rocket engine developer named Valentin Glushko at the NKVD's special prison. I told Isayev about the publication, and he decided to look for the book in the Sverdlovsk libraries. It didn't turn up. As soon as I returned to Moscow, I checked my library and found to my great joy that it was still intact. Only one volume of my unique pre-Revolutionary eight-volume collection of Gogol's works had disappeared. Langemak and Glushko's book was still there!

In 1944, when our OKB-293 was combined with NII-3, I used that book to supplement my knowledge on the principles of rocket technology. Having allowed my vigilance to slip, I once brought the book to work at NII-1. One of my new employees, an old hand from RNII who had known Langemak and Glushko personally, saw their book on my desk and was seriously worried. He warned me that I should take the book home and never bring it back to the institute. "They shot Langemak—Glushko is alive, but convicted. This book is in the special archives in the institute library and is impossible to obtain. You are violating the rules; if any unpleasantness gets started, you risk losing a good book." Naturally I once again hid this book behind other books. But I can boast that, having retained this book, I am now the owner of a rarity.

Thus, we all knew about Glushko as early as 1942 after Isayev's trip with Bolkhovitinov to the special prison in Kazan, which the initiated called a *sharashka*. I learned about Langemak from reading his and Glushko's book, but I had no information about Korolev before the phone call from Berlin. Pobedonostsev knew all about Korolev, but in his dealings with the "uninitiated" he piously observed the rule of the taboo.

Tyulin later told me that before Korolev, Glushko, and the other *zeki* freed in accordance with Gaydukov's list arrived in Berlin, he was strictly warned by the "security apparatus" that a certain group would be flying into Berlin but that no one under any circumstances was to know that they were former prisoners.

Upon meeting Korolev for the first time in my office at the Institute RABE in late September or early October 1945, I knew nothing about him except for his full name. When he entered my office, I stood up to greet him, as is appropriate for a major receiving a lieutenant colonel. We exchanged greetings and introduced ourselves. More than a half-century has passed since our first meeting. We had countless meetings over that period of time. The majority, at least in terms of details, have faded from my memory, but this first meeting has stuck. Independent of our normal thinking process, there is a sort of subconscious "standby" memory system that involuntarily switches to "record." This recording cannot be erased and can be replayed many times.

The brand new officer's uniform fit the man who had entered very well. I would have thought that a cadre officer was standing before me, but the absence of medals on his clean tunic immediately gave him away as a "civilian" officer. The only unusual thing was the fine officers' chrome leather boots that he wore instead of our usual kersey boots. His dark eyes, which had a sort of merry sparkle, looked at me with curiosity and attentiveness. I immediately noticed Korolev's high forehead and large head on his short neck. There is an expression: "He sucked his head into his shoulders." But Korolev wasn't sucking anything in—nature had made him that way. There was something about him like a boxer during a fight. We sat down. He sank into the deep armchair and with obvious pleasure stretched out his legs. That's usually what one does after sitting behind the wheel of a car for a long time. "I would like to have a brief overview of the structure and operation of your institute." I always had a diagram of the institute's structure in a file on my desk. Of course, it was drawn by Germans and had German inscriptions.

Korolev started to examine the diagram without showing much attention or respect, or so it seemed to me, leading me to believe that he didn't like the fact that it was inscribed in German. His first question immediately found our weak spot. "So who here is responsible for mastering launch technology and for launch preparation?" I explained that Lieutenant Colonel Voskresenskiy was studying that matter with a large group of Germans, two or three of whom who had actually fired the systems. The military would soon be forming a special subunit to study firing technology in its entirety. For the time being we had concentrated all of our efforts on having something to fire. We needed to recreate the missiles themselves, and the chief problem was the control instruments. As far as the engines were concerned, we had found many of them in Lehesten, and firing tests were already being performed successfully there.

He looked at me merrily and decided to open up just a bit. "Yes, I have already been in Lehesten. The people there—including my old friends—are doing excellent work."

"A-ha," I thought, "So you're an engine specialist. But from where?"

Several trivial questions followed, evidently more out of politeness. I proposed to Korolev that he visit the laboratories. "No thanks," Korolev declined. He added in parting, "I'm returning to Nordhausen today. But I have a feeling that I'll be working with you quite a bit." He then shook my hand much more firmly than he had when we met.

Korolev went out into the reception room and lingered a bit, thoughtfully scrutinizing the secretary tapping away on her typewriter. Turning to me he asked, "She's German, of course?"

"Yes, of course."

He quickly descended the soft carpet runner down the stairway. I returned to my office and approached the window. Korolev was sitting behind the wheel of an Opel-Olympia. The car had not been washed for a long time and was therefore of indeterminate age. He turned sharply and zoomed down the road to the exit from town.

Now that I know about Korolev, based on everything that I have been able to learn from him personally, from his friends, acquaintances, and biographers, it seems to me that he had little interest in our institute and the details of our work that day. Yes, he had already thought over a plan of future actions and his meeting with me was meant to confirm certain ideas of his. But the main thing that excited, stimulated, and inspired him was the freedom of movement.

From the author's archives.

S. P. Korolev in Germany, October 1945.

Behind him were just over five years as a Soviet *zek*—at Butyrka, a transit prison, Kolyma, Butyrka again, the *sharaski* in Moscow, and then Omsk and Kazan.

And then suddenly Korolev was in Thuringia, the "green heart" of conquered Germany, in the uniform of a lieutenant colonel, with documents that opened every checkpoint, alone behind the wheel of a "trophy" car racing like the wind. He could drive wherever he wanted over such good roads. He could stop and spend the night in any town in the Soviet occupation zone. The military authorities would always give him refuge, and even if they weren't around, a good local *gasthaus* would take him in. Freedom! How beautiful! That is how, many years later, I imagined the feelings that must have overcome Korolev. He wasn't yet forty years old! He had so much yet to do! But he had the right now to take something from life for himself.

Soon after that memorable first encounter with Korolev in Bleicherode, Pobedonostsev arrived. He had spent a great deal of time in Berlin, and was now well informed about the organizational problems that troubled the Guards' Mortar Units command, the Party Central Committee military department, the industry People's Commissariats, and our institute. He reported that for the time being in Moscow, there was total confusion "at the top" as to who would become the actual boss of missile technology in the country. Meanwhile all of the power was in the hands of the military and the Central Committee apparatus. We would therefore be working according to the principle: "He who pays the piper calls the tune."

Pobedonostsev told Pilyugin and me in detail "who was who" at RNII and who Korolev was. Pobedonostsev was the first to tell us that Korolev had been Glushko's deputy in the "special prison." Glushko had been sentenced to eight years in a prison camp "for participating in the Kleymonov-Langemak sabotage organization." He was retained to work at the NKVD Fourth Special Department Technical Bureau. In 1940, they transferred him Aircraft Factory No. 16 in Kazan to develop aircraft jet boosters. In 1941, they appointed Glushko chief designer of an OKB staffed with imprisoned specialists. Over a three-year period they developed several types of boosters with thrusts from 300 to 900 kilograms. These boosters underwent tests on Petlyakov, Yakovlev, and Lavochkin aircraft. In 1942, at Glushko's request, they transferred prisoner Korolev from the Tupolev *sharashka* in Omsk to Glushko's OKB, where he was assigned as Glushko's deputy for testing.

According to all sorts of gossip, Glushko was supposedly delivered under guard to Stalin at the Kremlin in August 1944. There he not only reported on the work he had done, but managed to tell Stalin something about the future outlook. Glushko left the Kremlin a free man. Without a guard, he initially experienced some difficulty getting settled.

In 1944, the authorities decided to free Glushko and thirty-five other specialists who had worked in his OKB before their sentences had been served out, and their previous convictions were expunged. Korolev was also freed somewhat later in 1944. After both Glushko and Korolev were freed, they stayed on to work in Kazan, and Korolev tried to set up an independent rocket OKB there at Factory No. 22. Glushko, Korolev, and all the former NKVD prisoners were transferred to the jurisdiction of the People's Commissariat of the Aircraft Industry, which did not have a very good idea what to do with them.

POBEDONOSTSEV SAID THAT GAYDUKOV had tasked Korolev with organizing a service independent of the Institute RABE to study missile launch preparation equipment. Since we were the real bosses in town, Korolev would be asking us to render him all kinds of assistance, and Voskresenskiy and Rudnitskiy would become Korolev's subordinates. I agreed right away, but Pilyugin strongly objected. He had not yet met Korolev, and he believed that everything should be under RABE management. He felt that the existing Guard's regiment in Sondershausen should be reinforced to take on launch preparation. Regiment commander Chernenko and his officers were already working closely with us. But ultimately Pilyugin gave in.

From the author's archives.
S.P. Korolev in Germany, November 1945.

Several days later, Korolev arrived in Bleicherode with the plenary powers to create the *Vystrel* (shot) group. The tasks of this new service included the study of missile pre-launch preparation equipment, ground-filling and launch equipment, and aiming equipment; flight mission calculation; developing instructions for firing team personnel; and preparing all of the necessary documentation. Members of the *Vystrel* team included Voskresenskiy, Rudnitskiy, and several cadre officers. We allocated separate office space for them at the Institute RABE.

But soon Voskresenskiy was complaining that Korolev had left for Berlin on urgent business without notifying anyone, It turned out that Korolev had been included in a delegation which, at the invitation of the British military authorities, was traveling to Cuxhaven for V-2 demonstration launches near Hamburg. This news enraged Pilyugin and Voskresenskiy. They both blamed me, "What kind of chief are you if everybody there in Berlin decides for us and without us. We work so hard here! But when it comes to going to Hamburg, they forget all about us and send Korolev who only just got out of prison." But Pilyugin was a homebody by nature, and he quickly calmed down.

Soon thereafter Korolev returned along with Pobedonostsev. They were quite merry and excitedly told us about their visit to Cuxhaven, where the Brits, using

German prisoner of war missile specialists, had decided to conduct demonstration launches of the very V-2 missiles that had terrorized the Londoners.

All the members of our delegation, except for Korolev, had been sent to Cuxhaven with the ranks that had been conferred on them. But Korolev, on instructions from Moscow, had been ordered to change into a captain's uniform with artillery shoulder boards and "cannons."[1] Apropos of this, Pobedonostsev said that "this artillery captain" provoked considerably more interest among the British intelligence officers who were watching our delegation than General Sokolov, Colonel Pobedonostsev, and the other high-ranking officers. One of the Brits, who spoke excellent Russian, asked Korolev straight out what he did. In accordance with his instructions and his cover story Sergey Pavlovich responded, "You can see that I am an artillery captain." The Brit remarked, "Your forehead is too high for an artillery captain. What's more, you clearly weren't at the front judging by your lack of medals." Yes, for our intelligence services this disguise was a resounding failure.

The launches at Cuxhaven took place without a hitch. Filling us in on the details, Korolev commented ironically about the complete helplessness of the Brits, who themselves had not participated in the launch preparations at all, being entirely dependent on the German team. It had been impossible to determine the missile trajectories because it was overcast. But the launch made an impression. These missiles were, of course, a far cry from the GIRD rockets that Korolev had launched with Tikhonravov twelve years previously. At our officers' club at the Villa Franka, these first British launches served as an occasion "to celebrate" the start of training firing teams in the *Vystrel* group.

1. Shoulder boards with stitched renditions of cannons indicated service in the artillery forces.

Chapter 25

Engine Specialists

In the autumn of 1945, the Brits held V-2 demonstration launches for the Allies in Cuxhaven. How could we respond to show that we too understood this secret weapon, and furthermore, that we were already mastering this technology without the assistance of the Germans?

Our "Russian-sized" plan proved to be far more grandiose by propagandistic design. We were not ready for missile launches from German territory even with German assistance. We certainly weren't capable of doing that on our own territory in 1945. The Brits and Americans had managed to get their hands on fully tested missiles, an oxygen plant, and filling and launch equipment, along with a whole set of launchers and a troop detachment that had a great deal of experience firing against Britain.

I cannot recall who made the suggestion, and perhaps it was even a collective idea. It is more likely that it originated from the military leaders of GAU or the Guard's Mortar Units, since they were keeping abreast of our work and our resources to a much greater degree than the leaders of the industrial People's Commissariats or Party figures.

During the war, a large exhibition of all sorts of captured technology was set up on the grounds of Moscow's Central Park of Culture and Recreation. The exhibition was very successful and had tremendous propaganda value. It elevated the people's spirits during the most difficult years. After the victory, they added extensively to this exhibit. Someone came up with the idea of bringing a V-2 to put on display.

In September, our Institute RABE received an assignment to quickly prepare two missiles. I gave instructions to the factory in Kleinbodungen where Kurilo, one of the experienced production leaders at our NII-1, was now the director. The work there was in full swing. We naturally assumed that for the purposes of the exhibition it would be sufficient to assemble missiles without their internal instrumentation or electric devices, especially since the size of the propulsion system nozzle alone would make an impression.

But soon thereafter we received an absolutely staggering command from Moscow. The missiles had to be ready for firing tests on a rig that would be set up on the Lenin Hills. The fire plume would come crashing down with a horrific roar from a height of 80 meters along the bank of the Moscow River, to the delight of all the Muscovite spectators and numerous foreign guests who would be coming to the capital to celebrate the twenty-eighth anniversary of the October Revolution. This festive fireworks display would be quite an addition to the already customary victory salutes!

Presumably Stalin himself would want to feast his eyes on such an unusual firing performance. And after that decisions regarding the development of missile technology, despite all the postwar difficulties, would pass quickly through the Politburo. They would of course be interested in who organized all of this, and the organizers of the firing spectacle would be commissioned to head the development of this new type of weaponry. This assignment immediately shifted the primary responsibility from the assemblers to the engine specialists.

By that time, September 1945, the primary group of engine specialists had also settled down as an independent organization in Thuringia, near the town of Saalfeld at the Lehesten firing test base for series-produced A-4 engines.

In order for the reader to better understand the problems that engine specialists faced at that time, I will return to the events of May 1945 in Berlin.

On 25 May 1945, an American B-25 bomber landed at the airfield that serviced our "residence" in Adlershof. An assault landing force of "trade union" officers, headed by "Lieutenant Colonel" Aleksey Mikhaylovich Isayev, disembarked from the airplane. The commander of the B-25 was not an American but our test pilot Boris Kudrin. In 1942 and in 1944, he had tested the now legendary BI-1 rocket airplane in glider mode. With the end of the war, the Armed Forces command transferred several combat airplanes received through the Lend-Lease Program to our institute for our transport needs.

Aleksey Isayev's team consisted of liquid-propellant rocket engine specialists. It also included one of the first testers of liquid-propellant rocket engines, Arvid Pallo, whose face still showed traces of severe burns from the nitric acid that had splashed on him when an engine exploded during firing rig tests in the winter of 1942 in Bilimbay. The rest of the team had become involved in liquid-propellant rocket engine technology during the development of the BI airplane, and they had already mastered the engine terminology that Isayev had very successfully introduced. This group was the nucleus that would later diversify into specialized engine OKBs in the postwar era.

During the first years of work on missile technology in the Soviet Union, and during the same period in Germany, the creators of liquid-propellant rocket engines did not have any theoretical works at their disposal that were suitable for practical application to enable them to design a liquid-propellant rocket engine on a reasonably scientific basis.

A rocket designer used the sciences developed for the needs of artillery and aviation, such as ballistics, the flight theory of a variable mass body, aerodynamics, the theory of elasticity, and the resistance of materials—all of which depended on the classic works of famous scientists. The developers of missile control systems relied on the firm theoretical foundations of electrical engineers and radio engineers, who already had experience with aircraft automatic pilot technology and gyroscopic technology which had been very successfully applied in ship navigation.

Engineers who had taken on the risky work of developing liquid-propellant rocket engines did not have any special training, and during the first years they functioned almost blindly, using "hit-or-miss" methods. Fires and explosions on the test rigs and during rocket launches were not uncommon. We were not averse to good-naturedly teasing the engine specialists—we would ask them how many times a liquid-propellant rocket engine needs to explode before it's ready for its first flight.

As I recall my dealings with engine specialists—and most of all my work with Isayev, Glushko, and their engine specialist compatriots—I remember that they were always encountering problems when setting up the work process to obtain the highest possible specific thrust and combustion stability. All kinds of pulses occurred, both low-frequency and high-frequency. The pulses gave rise to vibrations, which would first destroy the equipment around the engine, and then cause the engine itself to explode. So as not to frighten the top brass, they replaced the word "explosion" with the term "breach of the combustion chamber."

Injectors, their arrangement, spray action, carburetion, temperature fields, cooling—all of these problems required study without the aid of any textbooks or manuals. And with the use of turbopump assemblies and gas generators in the makeup of engines, a multitude of highly complex problems of mechanical engineering for power plants emerged that were alien to traditional turbine construction.

Unfortunately, even now our literature does not have suitable works on engine science analogous to the numerous works that have been published on control systems and instrumentation. Contemporary rocket engines, in terms of their concentration of power per unit of volume, are a unique product of engineering. Perhaps this is why the majority of rocket accidents during the launch phase are caused by irreproducible failures in the engine systems.

MORE SO THAN ENGINEERS in other fields, engine specialists had to have intuition, engineering common sense, and courage to compensate for the lack of mathematical theories; they had to be able to recognize their mistakes and tirelessly conduct hundreds of tests rather than rely on calculations. Aleksey Isayev had been endowed with these qualities like a gift from God. He had tried many professions and specialties before he fell in love with liquid-propellant rocket engines, which remained his passion until the end of his days.

I made his acquaintance at Viktor Fedorovich Bolkhovitinov's design bureau long before the war. We worked together on the DB-A aircraft and participated in the preparations for Levanevskiy's flight to the United States. We vacationed together in Koktebel. There, during strolls on the stony beaches and rocky shore of Kara-Dag, Isayev told my wife Katya Golubkina and me about his adventurous attempt to escape from Koktebel to the Hawaiian islands with his boyhood friend, future writer Yuriy Krymov (Beklemishev).[1]

In 1941, when the Germans were advancing on Moscow, we were working together at Factory No. 293. At that time, without leaving the factory to seek outside assistance, Isayev and Bereznyak developed their idea for a rocket-propelled interceptor, which would later be called the BI-1. On one of those October days, Aleksey made me give my word that I would not divulge his secret. He brought me to one of the basement rooms of our factory and showed me an entire arsenal of aircraft weaponry. There were *ShKAS* rapid-fire aircraft machine guns, 20-millimeter *ShVAK* aircraft guns, dozens of solid-propellant rocket projectiles, and many cases of ammunition.

"Look, Boris, I have no intention of high-tailing it into evacuation somewhere in the east. We can send our families off there and set up a guerrilla detachment. These are the weapons that we must protect. You must help me pull together a detachment of loyal people."

We spent some time developing an operation to protect the bridge over the Moscow River, across which the Germans could force their way into Moscow, advancing along the Leningrad Highway. But our adventure was quickly found out as soon as we began recruiting volunteers to join our guerrilla movement. All the weapons were taken away, and under the supervision of real weapons specialists (including young engineer Vasiliy Mishin, future first deputy to Korolev and future chief designer and Academician), they were mounted on open railroad flatcars. The flatcars were coupled into the special train that evacuated us during the last days of October 1941 from Khimki to the Urals. During the movement of the special train, we took turns manning the makeshift anti-aircraft gun mountings.

Isayev was constantly generating all kinds of ideas—and not only technical ones. Sometimes he expressed thoughts that were risky in those days about the need for political restructuring. Especially interesting for that time were the projects that we discussed during our joint stay at Villa Franka in Bleicherode. Debates often flared up between us—for example, why in the world did they have it so good in the "lair of the fascist beast"? After all, all of this luxury and comfort was evidently generated before the war. How did they get such resources in Germany after losing World War I? Now Germany had also lost World War II—a myriad of damned issues needed to be discussed. But we were afraid to plunge too deeply into the heart of politics and history because there was the danger of "shaking the foundations." We agreed that ultimately, engineers and scientists should rule the world. Then there would be no national socialism, chauvinism, and anti-Semitism; no racism, and no national enmity at all. In our conversations we jumped from the Dora camp to the Russian camps, about which we knew a great deal less, and firmly concluded that it couldn't continue like this any longer. Power should be in

1. Soviet writer Yuriy Solomonovich Krymov (1908–1941) wrote under the pen-name "Yu. S. Beklemishev."

the hands of wise, talented, and honest scientists. Scientists and engineers were upright intellectuals and enthusiasts devoted to their work. They couldn't care less about your nationality. They had no use for concentration camps, special prisons, and crematoria. Scientists and engineers of the world, unite! But let's return once again to Berlin in May 1945.

ISAYEV, PALLO, AND THEIR FELLOW TRAVELERS, including Raykov, Tolstov, and other engine specialists who would become Isayev's future compatriots, had not been in Berlin very long before they set out for the *Walter* factory in Basdorf. This was the factory laboratory base where the Germans had developed and fabricated the liquid-propellant rocket engine for the Me 163 rocket-propelled fighter plane, which was functionally very similar to our BI. I will annotate my previous discussion of this factory with some details from Arvid Pallo that I believe are pertinent.

Having settled in Basdorf, Isayev and Pallo's team began searching for specialists who could restore documentation and hardware. Our troops had not fought with anyone here and hadn't damaged anything. But, in contrast to the Berlin enterprises, everything here had been carefully cleared away. The spindles and chucks had been removed from the machine tools. They couldn't even find everyday tools. They were not able to find engines or valves or any technical documentation. After a thorough inspection of the grounds, they finally discovered well-camouflaged aboveground storehouses containing concentrated hydrogen peroxide, *tonka* rocket fuel, and concentrated nitric acid which had been used as an oxidizer.

After lengthy questioning of the local inhabitants, they managed to find out the home addresses of several engineers from this firm, and using the successive approximation method, they managed to find the chief engineer in Berlin and one of the lead designers. They did not "crack" immediately. But after a certain amount of time they wrote a report about the firm's work and drew diagrams of the engines' construction. They claimed that the SS *Sonderkommandos* had taken the missing equipment and materials.

Nevertheless, one of the Germans, a test rig mechanic, directed Pallo's attention to a pipe sticking out of the ground and hinted that someone should dig there. They dug and found carefully packed and conserved liquid-propellant rocket engine combustion chambers, turbopump assemblies, and propellant component feed mechanisms with all of their valves. After they had laid this entire windfall out on tables in the shop and called the former directors to provide explanations, the latter announced, "Everything that we told you and wrote for you had nothing to do with reality. Everything was false. Now we see that we are not dealing with army officers but with real specialists, and we are ready to collaborate with you."

With the assistance of the German mechanics, our "real specialists" installed the engines they had found on test rigs and conducted a series of firing tests, taking down all the necessary specifications. When I arrived in Basdorf from Berlin, Isayev was distressed that he had previously (in Khimki) made me invent the most exotic methods of electrical ignition to set alight mixtures of kerosene and nitric acid.

The Germans had used a fuel called *tonka* instead.[2] When it was injected into the combustion chamber and mixed with nitric acid there was instant spontaneous ignition. "And it doesn't require any of your 'horns and hooves.'"

In 1943 in Khimki, I had indeed proposed an electric-arc type of discharge system for reliable ignition. This required a special mechanism that set up two arc-shaped "horns" in the engine nozzle before launch. At the required moment, a blinding electrical arc appeared between the electrodes, or "hooves", secured to these horns. Theoretically, the kerosene and nitric acid mixture had to ignite and then form the steady plume characteristic of a liquid-propellant rocket engine. We achieved ignition only half the time. Under my command, engineer and physicist Larisa Pervova worked on getting this capricious system to work right. The mechanics on the test rig joked in this regard, with some vexation, "Larisa still hasn't learned how to give anybody horns."

A threatening telegram addressed to Isayev arrived from Moscow from the deputy People's Commissar of Aviation Industry, accusing him of dawdling, and demanding that he return to make a report. In response, Isayev invited the directors of the aircraft industry to Basdorf in Germany. Makar Lukin, who was in charge of engine production in the aircraft industry, flew in and attended several firings. The engines generated 1.5 metric tons of thrust. Lukin praised them, "You guys are champs, what you're doing is fine, and for the time being you don't have to go to Moscow. Keep up the good work."

Each firing rig was provided with a technical description, then disassembled, carefully packed, and shipped to our institute, NII-1. And what about the dismantled machine tools and drawings? Eventually the Germans showed us that steel barrels with machine tool parts and instruments had been buried in a small grove under young birch trees. All of the technical documentation was discovered in an aluminum tube that had been buried in a bomb crater.

Having made a complete study of liquid-propellant rocket engines designed for aircraft, Isayev and Raykov headed by way of Magdeburg to Nordhausen, where we had come together to start joint work. The team of the remaining engine specialists headed by Pallo traveled to the A-4 engine firing test base in Lehesten near Saalfeld. On the way Pallo visited Kümmersdorf. The history of German liquid-propellant rocket engines began in Kümmersdorf. Here, von Braun and other future leading specialists of Peenemünde began their work and developed a series of engines with various combustion chamber geometries. The Germans expended a great deal of effort on developing combustion chamber head designs with injectors that provided the best mixture of alcohol and liquid oxygen.

In Lehesten, our engine specialists easily tracked down the directors of this unique engine testing base, including engineer Schwartz, director of the Haase

---

2. *Tonka* was a common term used by the Germans to refer to a mixed hydrocarbon fuel. Its most common form, *Tonka-250*, consisted of 57 percent crude oxide and monoxylidene and 43 percent triethylamine.

From the archives of Arvid V. Pallo.

A-4 rocket firing tests in Lehesten, September 1945.

oxygen plant, and many other specialists whom the Americans had not deemed necessary to take into their zone.

Lehesten was an enormous sand quarry with a firing rig set up on one of its slopes. In this same quarry was a subterranean plant that produced liquid oxygen for tests and a subterranean ethyl alcohol storage facility. Each engine slated for a V-2 missile underwent a preliminary firing test run using the propellant components fed under pressure from enormous, thick-walled tanks. In order not to disrupt series production at the subterranean *Mittelwerk* factory, they "lit up" more than thirty engines per day in Lehesten. All of the firing test equipment was in very good repair. The Americans, for unknown reasons, did not take anything away from the site.

The engines that had undergone firing tests at Lehesten, after being dried out and subjected to preventive maintenance procedures, were sent to *Mittelwerk* in Nordhausen. Turbopump assemblies that had undergone tests on a special test rig were also sent to Nordhausen, where they were connected with the engine during missile assembly. This process was forced on them and was not fruitful. In our production process, the entire propulsion system—the engine coupled with the feeding system (the turbopump assembly)—was provided for assembly from the very beginning of missile production. In essence, the propulsion system output parameters as a whole are determined during their joint operation.

Arvid Pallo became the Lehesten "firing boss" in July and began to debug the firing tests. Over fifty brand new combustion chambers ready for testing had been discovered in the subterranean warehouse. They had also found fully preserved freight cars on spur tracks containing property transported from Peenemünde. There were fifteen railroad cars containing A-4 engines; flatcars with ground equipment, including *meillerwagen* erectors; carriages for transporting missiles; tanks for transporting and filling liquid oxygen; alcohol filling units; and many other items from the ground facilities. These were very valuable finds. Pallo arranged with the military authorities to have them guarded. In Lehesten, they started having regular visits from high-ranking guests who did not mind tasting rocket fuel since it was ethyl alcohol of the highest purity.

Our engine specialists spent the entire period from July through September studying and mastering the technology for testing and adjusting the engines. They conducted over forty firing launches in various modes. To the Germans' surprise, our testers proved to be more daring and went far beyond the limits of the thrust regulation modes that had been permitted. In doing so, they discovered that the A-4 engine could be substantially boosted—to a thrust as high as 35 metric tons. They worked out new processes for the firing tests. These included processes to measure thrust parameters, calculate and select blends, flow-test oxygen and alcohol injectors, and perform quick analyses of the chemical and physical properties of both the fuel for the combustion chamber and the propellant components for the steam-gas generator.

The command to prepare for the firing launch on the Lenin Hills in Moscow interrupted the testing work. The documentation of the testing work had filled twenty-two report files. All of the prep work for the Moscow firing was supposed to be completed in one month. Arvid Pallo made the right decision—which was then approved by Valentin Petrovich Glushko, who appeared in Lehesten in October—to equip the missile that was being assembled in Kleinbodungen with the combustion chamber that had undergone firing tests in Lehesten. A special rig was being designed and fabricated in Lehesten for the missile's installation in Moscow. It would be outfitted with the necessary equipment for pre-launch preparation and launch: high-pressure tanks, tanks for alcohol and oxygen, pipelines and valves, and remote control panels to control the engine startup.

With the assistance of the Soviet Military Administration in Thuringia and the local authorities (using rocket alcohol as an incentive), they managed to design, fabricate, and test everything within one month in Saalfeld. We sent two assembled missiles without engines to Lehesten. There they completed work on the missiles, fully equipped them, and fit them to the rig. The special train carrying all of the equipment to set up firing tests in Moscow consisted of sixteen railroad cars. Pallo himself was in charge of this crucial expedition and heroically made his way through Poland and Brest, which was jammed with railroad cars. Finally they made it to Moscow's Byelorusskiy train station. Here the military authorities took over the entire train and relieved Pallo and those accompanying him of their sensitive cargo.

While the expedition was moving out of Germany, navigating through dozens of obstacles, some Politburo member reported the undertaking with the engine firings on the Lenin Hills to Stalin. It was not approved, and on that note everything ended.

WITH THE DEPARTURE OF ISAYEV, and Pallo's subsequent journey to Moscow, Glushko took on the directorship of all the engine-related work, and Shabranskiy, who would later become his deputy for testing, became the immediate chief of the Lehesten base.

Valentin Petrovich Glushko flew into Berlin with Korolev. They had worked together at the NKVD special prison in Kazan. There Glushko had been chief designer of aircraft power units, and Korolev had been his deputy for testing. Neither of them liked to recall this period of their joint work. After the prohibitions placed on historical aerospace technology publications were partially removed, Glushko exerted a great deal of effort to publish historical and popular literature. In one of his most respectable works, the *Kosmonavtika* (Cosmonautics) encyclopedia, he recalled: "In 1942–1946, Korolev worked at the KB (see (GDL-OKB) as deputy chief engine designer, involved with equipping series-produced combat aircraft with liquid-propellant rocket boosters."[3]

There are two inaccuracies here. First, the GDL did not exist at that time. They were imprisoned specialists, who under the supervision of chief designer Glushko, worked at a KB that was officially called the "NKVD special prison," but in common parlance was known as the *sharashka* attached to Factory No. 16 in Kazan. Second, in late 1945 Korolev was freed and was in Germany. During the spring of 1946, he occupied the post of chief engineer at the Nordhausen Institute. As such, Korolev was not Glushko's subordinate after 1945, Glushko was his. In 1946, Korolev was transferred to NII-88 and appointed chief designer of long-range ballistic missiles. Glushko's attempt in the later publications to lengthen Korolev's subordination to him by another full year to a certain degree characterizes the attitudes that formed between these two talented and very complex directors of our domestic aerospace industry.

My first encounter with Glushko in Germany took place soon after I had first made Korolev's acquaintance. Pobedonostsev had also informed me from Berlin of Glushko's arrival. I was convinced that Yuriy Aleksandrovich was bent on avoiding possible complications in connection with his former RNII colleagues' release. Before Glushko's arrival I invited Pilyugin to participate in the meeting with this most renowned engine specialist. In contrast to the incident with Korolev, I did not have to guess with whom I would be meeting.

First, Langemak and Glushko's book, *Rockets: Their Construction and Use*, which I had studied thoroughly, had declared in the foreword that Glushko was the most outstanding specialist in rocket technology. Second, the opinions of people whom I considered high-ranking authorities in that field, such as Bolkhovitinov, Isayev, and Pobedonostsev, were in and of themselves sufficient to instill my respect for the most important specialist in the field of rocket engines.

But there was one more reason why Glushko was for me a great authority. During 1943 and 1944, as I mentioned earlier, I developed an electrical-arc ignition system for Isayev's engine. In the process of this work, we decided to use an electrical arc not for ignition, but for generating thrust. Using the appropriate shape of electrodes, and by creating a magnetic field around them, it could have been possible to "shoot off" blobs of plasma at high speed and with a considerably

---

3. V. P. Glushko, ed., *Kosmonavtika entsiklopediya* (Cosmonautics Encyclopedia) (Moscow: Sovetskaya Entsiklopediya, 1985).

higher specific thrust than in engines using chemical propellant. In literature searches at the NII-1 library, I found secret Leningrad Gas-Dynamic Laboratory reports from which it followed that as early as 1929, V. P. Glushko had invented electric rocket engines. My passion for electric rocket engines ended there at that time. But since Glushko was involved with this, I was convinced that he couldn't be a dilettante in electrical engineering. It was yet another argument in his favor.

Two officers entered my office. I immediately recognized the colonel—it was Valentin Petrovich Glushko. The other officer, a lieutenant colonel, introduced himself simply as List. Both wore high quality uniform jackets, well-pressed trousers rather than tunics, jodhpurs, and boots. Glushko smiled slightly and said, "Well, it seems you and I have already met." Evidently he remembered our meeting in Khimki. Nikolay Pilyugin dropped in and I introduced him as the institute's chief engineer. I proposed that we all sit down and drink some tea or "something a bit stronger." But Glushko, without sitting down, apologized and said that he needed emergency automobile assistance. "We were driving from Nordhausen and our car was pulling badly and smoking severely. We were suffocating from the smoke inside. They say that you have some good repair specialists."

Nikolay Pilyugin went up to the window and announced, "Yes, it's still smoking. Did you turn the motor off?"

Suddenly in a calm, quiet voice List began to speak. He removed his service cap, revealing a shock of completely gray hair, and sank demonstratively into the armchair.

"Don't worry. The brake pads of the handbrake are burning out. We drove from Nordhausen with the handbrake engaged."

Pilyugin and I were flabbergasted.

"Why didn't you release it?"

"You see, Valentin Petrovich stipulated that if he was at the wheel, I was not to dare suggest anything to him."

Later we found out that before his arrest in 1938, Grigoriy Nikolayevich List had been the deputy chief designer at the I.V. Stalin Automobile Factory (ZIS).[4] In terms of outward appearance, his manner of speaking and holding himself, List was a typical old generation intellectual. Nevertheless, he understood automobiles in all their subtleties and drove them beautifully. He drove the car from Berlin to Nordhausen. But in Nordhausen, Glushko demanded that the wheel be turned over to him. And this was the result.

Pilyugin and I didn't know whether to laugh out loud or to sympathize. But there was not a trace of indignation or surprise on Glushko's face. He too sank calmly into his armchair, pulled out a pristine handkerchief, and wiped off his forehead. I called our repairs department, and after explaining the situation, requested that they quickly replace the Olympia's handbrake. And that was our first meeting with Glushko at the Institute RABE in Germany.

---

4. ZIS—*Zavod imeni Stalina.*

This incident was characteristic both for Glushko, who sometimes exhibited incomprehensible obstinacy and did not tolerate suggestions if he had set a certain goal for himself, and for List, who had worked at the Kazan *sharashka* under Glushko's authority. List now dreamed of breaking free from him, but Glushko was still his boss and would not let him go. Later, under the pretext of checking and revising the technical documentation on a propulsion unit, List nevertheless did break free from Glushko's daily surveillance. I set up a workstation for him at the Institute RABE, and here he serenely worked until the end of our activity in Germany. But upon his return to Moscow, List nevertheless returned to Glushko and worked in his Energomash design bureau, OKB-456, until he retired.[5]

Glushko and List did not stay long in Bleicherode. Glushko complained that Isayev had left for Moscow without waiting for him.

"The thing is, I was given great authority, and I wanted to enlist all of the engine specialists for prospective developments in this field." With the hint of a smile he added, "I was appointed chief of the engine section of a special governmental commission."

When we sent Glushko and List off to their hotel to rest until their trip to Lehesten, Pilyugin muttered, "So over in Moscow they are setting up special governmental commissions that will be boasting about our work, and we are going to start working for some other chief of the control section. And if it weren't for us, this special commission wouldn't even have anything to do."

We lit up our favorite Kazbek cigarettes and moved on to pressing business.

When Glushko and his people arrived in Lehesten, the work there got a new boost. The large amount of experimental material for the optimization of the engines and their series testing, plus the statistical data on thrust, temperature, and flow rate parameters, was extremely valuable for our engine specialists. In addition to the combustion chambers that had been stored in Lehesten itself, they found fifty-eight railroad cars in the surrounding area containing A-4 combustion chambers, five railroad cars carrying A-4 launchers and transporters, and nine liquid oxygen transport tanks. This wealth of equipment gave the engine specialists tremendous advantages over specialists in other fields, including control specialists such as ourselves.

---

5. Glushko's design bureau, OKB-456, was renamed KB Energomash in 1966.

Chapter 26

The Institute Nordhausen

In early 1946, with Ustinov's support, General Gaydukov managed to reach an agreement in the Party Central Committee in Moscow and in the Soviet Military Administration in Berlin for a significant expansion of operations in Germany. This had not been easy to do. A considerable portion of the Party and state apparatus involved with policy in Germany had demanded that the work in occupied Germany to restore German technology be curtailed and all Soviet specialists be called back to the Soviet Union no later than January or February 1946. Gaydukov and Ustinov, as well as Artillery Marshall Yakovlev, who supported them, did not agree—they insisted on expanding operations. At the same time, the Institute RABE was becoming the foundation for a significantly more powerful organization.

I should mention that the aircraft industry, using the Institute RABE as a model, had gathered German aircraft specialists in the Soviet occupation zone for work in Dessau, using the facilities of the Junkers factories.

Only the atomic experts immediately brought Professor Manfred von Ardenne and a small group of specialists to the Soviet Union. (The British had captured the primary developers of the German atomic bomb, headed by Nobel laureate Werner Heisenberg.)

The Institute RABE had a clearly pronounced emphasis in the field of electrical control systems because the institute management (Pilyugin and I from the Russian side and Rosenplänter and later Dr. Hermann and Gröttrup on the German side) consisted of specialists in electrical equipment and control. Korolev, who had taken charge of the *Vystrel* group; Glushko, who was directing the study and testing of engines in Lehesten; Kurilo, who was assembling missiles in Kleinbodungen; and other smaller groups were functioning more or less independently, often duplicating rather than complementing each other. Korolev accused Pilyugin and me of not paying proper attention to the general matters of missile construction—to its warhead and operation. Pilyugin replied harshly to Korolev.

"Sergey Pavlovich, you don't know the first thing about the control system, so go ahead and organize the work for all the other systems. Any aircraft factory in the Soviet Union will rivet the hardware for the body just as well as the Germans, but we still don't know how to make the instruments for the rocket. And even if we reproduce the instruments, we still have to learn how to control the flight so that the body and the tanks will fly where they are supposed to, instead of into the nearest garden. And as for which explosive to put in the warhead, it's better for you and the artillerymen to look into that without us." This skirmish had far-ranging consequences.

In February 1946, Korolev was summoned to Moscow. He returned in early March, cheerful, hale and hearty, radiating exuberant energy, and now a colonel. Thus, Korolev, if only outwardly in terms of military insignias, was now equal to Glushko, Pobedonostsev, Ryazanskiy, Pilyugin, and Kuznetsov, who had flown to Germany wearing colonels' shoulder boards. A day or two later, General Gaydukov arrived, also in an excellent, jubilant mood, and he asked me to assemble all the civilian specialists. He would be responsible for military specialists. At a large meeting of Soviet specialists, Gaydukov announced the decision to create a single organization based on the Institute RABE and all of the various and sundry functioning groups. It would be called the Institute Nordhausen.

Gaydukov was commissioned as director of the institute and Korolev was appointed his first deputy and chief designer. Next they reviewed and approved the general structure of the new institute.

Our Institute RABE became a part of the new conglomeration as an institute for control systems. Pilyugin, Ryazanskiy, Boguslavskiy, and I remained in charge,

From the author's archives.

Entrance to the Institute Nordhausen directorate.

but we were advised to prepare for a large number of new specialists who would be arriving from the Soviet Union shortly. They tasked me with helping to organize the Institute Nordhausen until the new staff was formed.

We agreed that RABE would make room—the new institute's headquarters and its management would also be located in Bleicherode. To do this we needed to commandeer some baron's private residence, which was located next door. This was no problem for the local authorities.

In addition to the Institute RABE, the following organizations became part of the Institute Nordhausen and were directly subordinate to Gaydukov and Korolev:

- The Montania factory in the vicinity of Nordhausen, which was used as a production base for engines and turbopump assemblies and a base for engine firing tests in Lehesten near Saalfeld. Glushko was put in charge of the general management of Montania and Lehesten, and Shabranskiy was appointed chief of Lehesten in place of Pallo, who had gone back to Moscow;
- The production facility in Kleinbodungen, which was officially named Factory No. 3 (*Werk Drei*). Kurilo was appointed director. The factory's objective was to restore the production process and assemble as many missiles as possible from everything that remained at *Mittelwerk*;
- The Olympia Design Bureau for the restoration of A-4 documentation and processing equipment that had been fabricated in Sömmerda at the Rheinmetall-Borsig factory. First Budnik was the head, and then Mishin. Mishin traveled with Bereznyak to Prague in search of technical documentation and got lucky. They found and brought back a large amount of design documentation which facilitated the beginning of work in Sömmerda;
- *Sparkasse* (savings bank), our term for the computational-theoretical group in Bleicherode. The team was established at a municipal savings bank which under the new authorities had been left without any monetary deposits. Colonel Tyulin, who had transferred to us from Berlin, headed the group. It included Lavrov, Mozzhorin, Appazov, and Gerasyuta. German theoreticians from RABE were transferred to the group, and more new specialists arrived, in particular, the chief ballistics expert from the firm Krupp, Dr. Waldemar Wolff, and aerodynamics specialist, Dr. Werner Albring, the former deputy director of the aerodynamics institute in Hannover.

GRÖTTRUP'S BUREAU BECAME an independent subdivision of the Institute Nordhausen, but at Gaydukov's insistence I was tasked with monitoring its activity. Taking advantage of that responsibility, I instructed Gröttrup to go beyond describing the history of A-4 development at Peenemünde and begin concrete work on proposals for longer-range missiles and high-precision control systems. Ryazanskiy and Boguslavskiy participated directly in these operations in the radio-engineering sphere.

The *Vystrel* group had expanded considerably. Now Voskresenskiy headed it in place of Korolev. This same group included Rudnitskiy, who had been tasked with searching for and restoring the ground-based filling, transporter-erector, and launch equipment. Here I should mention that at the same time the Institute Nordhausen was created, the Institute Berlin was established in Berlin, with the task of restoring anti-aircraft guided missile technology. Barmin was appointed chief engineer of the Institute Berlin. He had been the chief designer of the Kompressor factory where over eighty rocket projectile salvo-firing systems were developed during the war. Later Barmin was named chief designer of the entire complex for missile ground equipment while Rudnitskiy, who had worked with me at the Institutes RABE and Nordhausen, became his first deputy.

From the author's archives.

Vladimir Pavlovich Barmin, chief engineer of the Institute Berlin from 1945–46 and future academician and chief designer of rocket launching systems.

At the Institute Nordhausen, the Main Artillery Directorate set up its own representative office headed by Colonel Mrykin. He was a very demanding boss, who on first impression seemed extremely stern. He knew how to make sluggards and slovenly types tremble, but all the berating was in the interest of the issue at hand. The large body of cadre military specialists subordinate to him even suggested a unit of measurement to quantify the magnitude of his rebukes—the *mryk*. Leaving the office of Colonel Mrykin, officers would explain, "I received a one-*mryk* dressing down." A two- to three-*mryk* dressing down might be cause for an officer to be detached to another place of service. Later, Mrykin and I developed a very good relationship. I was convinced that his outward sternness and his not-always-pleasant, exacting nature did not prevent him from being a wise, objective, and keen manager who respected every honest specialist, both military and civilian. More than once I had the opportunity to witness Mrykin's indubitable decency.

The number of personnel at the Institute Nordhausen—composed of German specialists and Soviet military and civilian specialists—increased rapidly, requiring the creation of a new office to service the entire contingent. There was a shift from "quantity to quality" with regard to problems involving transport, accommodations, food, and receiving documentation and equipment and shipping them to the Soviet Union. At the Institute RABE, First Lieutenant Chizhikov, who had worked amicably with the services of the division, the commandant's office, and the German commercial director, coped with all of this excellently and single-handedly. I did not give up Chizhikov to Gaydukov's staff. To compensate, Gaydukov acquired a deputy for general work as well as a commercial unit and "rear services" that included a personnel department headed by Lieutenant Colonel Aleksandr Kaplun. Gaydukov asked that problems concerning light automobile transport be left to RABE and handled personally by Chizhikov.

As a rule, the stream of civilian specialists flying in from Moscow now without any sort of military uniform passed through Kaplun, whose responsibility it was to provide each of them with housing and food and direct them to the proper division of the institute. He asked for my assistance to accomplish the latter task. Thus, I became acquainted with many young specialists, still quite green, for whom fate had prepared the historic mission of humankind's breakthrough into space.

Among this stream of young people, my attention fell on the Kozlov family. Dmitriy Kozlov had gone through the war, lost an arm, graduated from the Leningrad Military-Mechanical Engineering Institute and married a coed. They showed up everywhere in tandem. I advised Kaplun to send Dmitriy and Zoya together to Sömmerda to work for Mishin. Fifty-six years later, at a meeting of the Russian Academy of Sciences, I met a corresponding member of the aforementioned institution, Dmitriy Ilyich Kozlov, two-time Hero of Socialist Labor, general designer and general director of the Central Specialized Design Bureau (TsSKB) and the Progress Factory in the city of Samara. He told me that he had been in Berlin at the aerospace exhibition, and he had taken a day trip to Bleicherode to recall the beginning of his career in Germany.[1]

"I was pleasantly surprised," said Kozlov. "More than a half century had passed, and there in Bleicherode they still remembered Chertok, they pointed out the building of the Institute RABE and were proud that Korolev had lived and worked among them."

Due to his new appointment, Korolev, as before, devoted a great deal of time to the subject matter of the *Vystrel* group and to organizing work on documentation in Sömmerda. There with Mishin and Budnik he began the first studies of a missile

---

1. TsSKB—*Tsentralnoye Spetsialnoye Konstruktorskoye Byuro.*

From the author's archives.

Evening gathering of Institute RABE officers at the Villa Franka. Sitting (from left to right): L. A. Voskresenskiy, unidentified person, V. A. Bakulin, V. I. Kharchev, V. P. Mishin, Yu. A. Pobedonostsev. Standing (from left to right): N. A. Pilyugin, A. G. Mrykin, S. G. Chizhikov, V. S. Budnik, S. P. Korolev—Bleicherode, Germany, 1946.

variant with a range of up to 600 kilometers—the future R-2 missile. In place of the Opel-Olympia, which lacked prestige, he acquired a powerful, sporty, dark red Horch. This car clearly suited Korolev's taste, and he never missed a chance to praise it. He offered to take his friends out for a drive, but did not trust any of them behind the wheel. I didn't pass up the chance to go out for a drive in the Horch. I already had a driver's license and had coped well with the Mercedes and the Opel-Olympia. Korolev did not trust me behind the wheel of the Horch because he knew about the "loop the loop" that I had executed in the Olympia with three passengers in the car.

Korolev drove through the narrow streets of Bleicherode and the village of Kleinbodungen in a way that made me plead with him, "Sergey Pavlovich! Your Horch is beautiful, but it's not a fighter plane, and we are in a populated area, not the sky."

Korolev's fitting response to that statement was, "But I have both a driver's license and a pilot's license."

"But I don't want to end up in heaven before my time, even in a Horch."

Soon thereafter, on the square in front of the Institute RABE, the Horch collided with an old Opel driven by a German driver who worked at our institute. The magnificent red Horch sustained a sizeable dent, but the small Opel was severely smashed. Korolev flew into my office extremely upset and demanded that I immediately fire the German driver and send Chizhikov, who was in charge of the institute's transportation, into exile in Moscow for not keeping order in his motor pool. Poor Chizhikov and the German skilled workmen spent the night in the workshop, and by morning the Horch looked better than it had before the ill-fated collision.

This was probably Korolev's first meeting with Chizhikov. Korolev was convinced that the former foundry pattern maker really did have the magic touch. Three years later, Korolev managed to bring Chizhikov into his creative team, which had received the Stalin Prize for the development of instrumentation for temperature measurement during R-1 missile flight tests. Chizhikov, who had been the first to receive a multi-*mryk* rebuke from Korolev for vehicular disorderliness, loved to boast "I'm not afraid of anyone in the whole wide world." And then he would pause and add, "except Korolev."

In order to relieve the Soviet specialists of concerns about the hardships their families were suffering, the military command decided in March 1946 to send the families of the "trade union" and cadre officers to their service posts on a semi-compulsory basis. For some of the cadre officers this was not very convenient, since they had acquired girlfriends during the war years, or as was customary to say at

From the author's archives.

S. P. Korolev, Institute Nordhausen's chief engineer, conferring with Ye. M. Kurilo—Bleicherode, 1946.

the time, "field wives" or PPZh.[2] But we civilian specialists immediately organized a service for the "re-evacuation" of our families from Moscow, Leningrad, and other cities, this time not to the east on the far side of the Urals and beyond as was the case in 1941, but to the west—to Berlin and then to the places where the heads of households had been posted in occupied Germany.

In May in Berlin, Chizhikov and I met our wives and children who had flown in on an Li-2 military transport plane. The meeting with our wives in Germany was not without incident. Chizhikov and I arrived in Berlin in two automobiles and waited for the plane from Moscow to land at Schönefeld airfield. Evening was already approaching when they told us that they were expecting no more airplanes that day.

From the author's archives.

In the vicinity of Bleicherode, Boris Chertok and Yekaterina Golubkina with son Valentin—Germany, 1946.

"How can that be? We have a message saying they departed from Moscow."

"Maybe they had to make a stopover along the way," suggested the dispatcher. "In any case, we'll radio Adlershof although the service aerodrome is already closed."

The dispatcher finally managed to get through to Adlershof and gladly reported that some women and children were there on the airfield. In the twilight we found our wives, who were exhausted by the difficult flight and incensed by the many hours of waiting, and our hungry little boys on the grass of the airfield next to the catapult used to launch V-1 cruise missiles. Pobedonostsev gave up his apartment in Berlin for us. There, Katya managed for the first time in twenty-four hours to feed some farina to our sick, one-and-a-half-year-old son.

After their arrival in Bleicherode, our wives, who had grown accustomed to the hardships of Moscow, were dumbfounded by the comfort of the Villa Franka. Nevertheless they demanded that we kick out the German maid: "We will cook and clean by ourselves." The officers' mess hall, our evening club at the Villa Franka, and also the German lessons all had to be closed down. Three meals a day and recreation were arranged for all the officers without families at *Japan*, the newly opened restaurant in Bleicherode.

---

2. PPZh—*Polevaya pokhodnaya zhena* (field camp wives).

Providing living quarters and food for the large number of German specialists and workers enlisted to work at the Institute Nordhausen was also a serious problem. When the work to restore the documentation and fabricate missiles was in full swing in the summer of 1946, the Germans themselves estimated the total number of German personnel at six thousand. Taking into account the personnel that worked at subcontracting firms, the number exceeded seven thousand. The assessment of this period of our joint activity in Germany by contemporary German historians is interesting. I shall quote excerpts from Manfred Bornemann's book *Geheimprojekt Mittelbau: Die Geschichte der Deutschen V-Waffen-Werke* (The Secret Mittelbau Project: A History of the German V-Weapons Factory), published in Munich in 1971.

> Material support for (German) specialists was on a level that had not existed in Germany for many years. Thus, for example, a degreed engineer received what were called Category-1 rations to last for fourteen days: sixty eggs, five pounds of butter, twelve pounds of meat, unlimited bread, plenty of vegetable oil, flour, cigarettes, and tobacco. For other categories of workers, these norms were lower, but for the situation at that time, still comparatively very high.

The German specialists were also paid relatively high salaries. Bornemann writes:

> One should also mention the rapport between the Russians and Germans during the missile project. The atmosphere was exceptionally amicable. The Russians showed their best side. Nevertheless, a certain distrust developed on both sides. If the Soviet specialists sometimes displayed reserve during work, fearing secret sabotage on the part of the Germans, this depressed the German missile specialists, who were worried about their future. As the rocket undergoing restoration acquired a more distinct shape, the Russians strove to obtain more documentary data on rocket technology during the production process.

The Germans' fear for their future is evident in these citations. Indeed, if the Russians were to understand and master everything, then what would become of the German specialists? Some of them hoped that the Russians would at least entrust the Germans with a field of activity that had never been attempted in Nordhausen, such as the very process of launching the missiles. But it turned out that the Russians also had already envisioned that. And not only in their small *Vystrel* group.

The 13 May 1946 decree of the Party Central Committee and Soviet People's Commissariat called not only for the creation of a missile industry, but also for the creation of a special missile State Central Firing Range (GTsP) and specialized troop units. In parallel with our military-industrial organization, which encompassed the Institute Nordhausen, they created a purely military system that was tasked with mastering the field operation of rocket technology.

From the author's archives.

L. A. Voskresenskiy (left), S. P. Korolev and V. K. Shitov (right) in Peenemüde, 1946.

From the author's archives.

General A. F. Tveretskiy and S. P. Korolev—Germany, 1946.

Using Colonel Chernenko's Guards Mortar Unit, which was stationed in the village of Berga near the town of Sondershausen, they began to form the reserve Special Purpose Brigade (BON) of the Supreme Commander-in-Chief.[3] Combat General Aleksandr Fedorovich Tveretskiy was appointed brigade commander.

Along with Korolev, Voskresenskiy, and Pilyugin, we set out for Sondershausen, where the entire BON officer staff was located, to familiarize ourselves with the new military organization and its commander. Korolev was afraid that the new complex technology would fall into the hands of martinet commanders—our work might be discredited at the very last stage. But our fears were unfounded. General Tveretskiy proved to be an uncommonly intelligent, benevolent, and prepossessing individual. We were soon convinced of this, interacting not only in the line of duty, but also socializing with our families.

But in one aspect Tveretskiy displayed firmness right from the beginning. Military specialists visited BON every day. They were officers with a great deal of frontline experience from various branches of the armed services. Tveretskiy announced that he did not intend to take up their time with drill and physical and political training. He categorically insisted that we grant them access to work in the institute's laboratories and subdivisions and admit them to missile tests at the

---

3. BON—*Brigada Osobogo Naznacheniya.*

From the author's archives.

Standing (from right to left): Major B. Chertok, Major Musatov, Colonel N. Pilyugin, and officers of the Seventy-fifth Guards' Division—Bleicherode, Germany, 1945.

facility in Kleinbodungen and to the *Vystrel* group's work. Korolev and Pilyugin were not enthusiastic because we were already thoroughly saturated with Soviet specialists, i.e. those engineers and military personnel who later were supposed to be transferred to the GAU central office to support the powerful military acceptance work forces.

We somehow fulfilled all of Tveretskiy's demands, and the officers, who in contrast to us were decorated with many combat medals, began to master their new field of work. Among the many BON officers who on orders from the army personnel department ended up serving in Sondershausen during those spring days of 1946, I must say some kind words about those whose subsequent work had a substantial impact on the development of our rocket, and later, space technology.

Nikolay Nikolayevich Smirnitskiy went from being assistant chief of the new formation's electrical firing group to very difficult service at the GTsP at Kapustin Yar. He eventually became a lieutenant general and headed the Main Directorate of the Rocket Armaments (GURVO) for nine years. Later, he was appointed deputy commander-in-chief of the Strategic Rocket Forces.[4]

Yakov Isayevich Tregub, a major of the special sapper troops, was in charge of the first launch team. Later, as a major general, he directed the testing of anti-

4. GURVO—*Glavnoye Upravleniye Raketnogo Vooruzheniya.*

aircraft guided missiles. He was the deputy to Chief Designer Mishin for the testing of spacecraft, and until recently he was working productively on new automatic space instruments for meteorology and to study Earth's natural resources.

After Sondershausen, Aleksandr Ivanovich Nosov served at Kapustin Yar and then was in charge of the directorate for testing the famous R-7 intercontinental missile at the firing range at Tyuratam (the Fifth Scientific-Research and Testing Range which would later become Baykonur). He did a great deal to optimize Korolev's rocket technology, but died tragically on the launch pad in the explosion of an R-16 intercontinental missile developed by Mikhail Yangel, another chief designer. This was not the only instance of tragic death among the first missile specialist officers.[5]

Boris Alekseyevich Komissarov arrived at BON with the rank of major. Later he specialized in testing automatic stabilization control instruments. He was in charge of military acceptance at missile factories and rose to a high governmental post: deputy chairman of the Commission on Military-Industrial Affairs under the Presidium of the USSR Council of Ministers. For departmental reasons, way back in 1946, Pilyugin and I had had to turn down Komissarov's requests to hand over gyroscopes, *mischgeräte*, and control surface actuators that we had gathered crumb by crumb and restored with difficulty. No, we could not foresee that many years later we would be coming to this unassuming major, petitioning for resolutions costing many millions of rubles.

I shall not list the many other military specialists with whom we worked jointly in Germany. Except for some random, minor exceptions, they all subsequently proved to be worthy fighters in the vanguard of the scientific-technical revolution in weapons technology and later in the peaceful field of cosmonautics.

I CANNOT SAY at whose personal initiative it took place—whether it was Korolev, Voskresenskiy, or one of the military specialists from the *Vystrel* group—but in early 1946 somebody came up with the idea of developing and constructing a special missile train using the workforces of German railroad car building firms and bringing in any other special rolling stock. The realization of this idea was beyond the capacity of the Institute RABE. With the creation of the Institute Nordhausen, however, the idea acquired many powerful advocates. The government allocated the necessary sizeable resources, the Soviet military administration drew up a top-priority order to railroad car and instrumentation firms, and the feverish activity began. The project called for the creation of a special train that could support the entire process of missile testing and launch preparation from any uninhabited location so that the only construction required would be railroad track.

---

5. During pre-launch preparations for the first R-16 launch in 1960, the rocket exploded and killed more than a hundred soldiers, administrators, and engineers.

The train was to consist of at least twenty special freight cars and flatcars. Among them were laboratory cars for off-line tests of all the onboard instruments, cars for the *Messina* radio telemetric measurement service, photo laboratories with film development facilities, a car for tests on engine instrumentation and armature, electric power plant cars, compressor cars, workshop cars with machine tools, cars containing restaurants, bathing and shower facilities, conference rooms, and armored cars with electric launching equipment. The train would have the capability to launch a missile by controlling it from the armored car. The missile would be mounted on the launch platform, which along with the transporter-erector equipment would be part of a set of special flatcars. Five comfortable sleeping cars with two-bed compartments, two parlor cars for high-ranking authorities, and a hospital car would make it possible to live in any desert without tents or dugouts. In the heat of the construction of this marvel of railroad technology, Tveretskiy convinced his superiors to approve and fund the construction of a second special train, but not for industry, just for the military. The program's doubling resulted in numerous conflicts due to the shortage of special testing and general-purpose measurement equipment to outfit the railroad cars.

But these were probably the only conflicts between industry and the military at that time. It was absolutely miraculous that both special trains were constructed and completely fitted out by December 1946. During the first years of the rocket era we simply could not imagine living and working at the firing range at Kapustin Yar without the special train. Not until the early 1950s, when, through the diligence of GTsP chief Vasiliy Ivanovich Voznyuk, hotels, an assembly and testing facility with workshops, domestic services, and much more were built, did life in the trans-Volga steppe become possible without the special trains.

WORK AT THE INSTITUTE NORDHAUSEN and correspondingly at the Institute RABE reached its most arduous level in August 1946. At *Werk Drei* in Kleinbodungen, they had managed to gather a sufficient number of parts to assemble more than twenty missiles. They were all provided with engines that had undergone firing tests at Lehesten and with turbopump assemblies that had been completed and tested at the Montania factory.

But an absolutely disastrous situation developed when it came to obtaining the necessary number of onboard instruments and test equipment for off-line and integrated tests. Nikolay Pilyugin, Viktor Kuznetsov's deputy Zinoviy Tsetsior, and I, again with the assistance of the SVA in Thuringia, visited the Karl Zeiss factories in Jena. There our Soviet optics experts were already lording over it as customers. Nevertheless, we arranged to place an order for the basic gyroscopic systems Gorizont, Vertikant, and Integrator. Karl Zeiss was a world-renowned manufacturer of optical instruments. These included glasses, binoculars, microscopes, telescopes, periscopes, all sorts of optical sights, and many other devices that were manufactured in large quantities to fill orders from Moscow. The engineers in Jena were not intimidated by the new orders and they claimed that, "Everything that Siemens did, we are capable of reproducing."

And they did! They received our instructions in March or April, and in September they returned the last of twenty sets of gyroscopic instruments. Our German specialists, Dr. Magnus and Dr. Hoch, grumbled that we wouldn't get the same precision with the Zeiss instruments that Siemens had guaranteed, but this didn't bother us very much at the time. The Germans' misgivings proved justified. But we did not understand this until we were performing flight tests after returning to the Soviet Union.

We tracked down the onboard electronic instruments—the *mischgerät* (mixing instrument), main distributor, time-domain current distributor, and the relays needed for their assembly and testing in the ruins of *Mittelwerk* and in trips made by the Germans over the border into the Western zones. But this proved to be clearly insufficient, and we had to open a special electronic instrument factory in Sangerhausen.

We also ran into a critical situation with the Askania control surface actuators and the graphite control surfaces. We did not have a single Soviet control surface actuator specialist at the Institute RABE until Georgiy Aleksandrovich Stepan appeared at my disposal, having first passed through Korolev. He was one of the young specialists whom Ustinov had ordered sent from various instrumentation factories to NII-88, which was being set up in Podlipki back in the Soviet Union. In the spring and summer of 1946, many of those who had not yet begun working in Podlipki departed immediately for Germany. They were no longer given military ranks. For that reason they got significantly less respect from the Germans than engineers with officers' ranks.

Under Pilyugin's direction and my supervision, Stepan, who had no prior knowledge of electrohydraulic control surface actuators, learned their structure and the minimum theory required to begin working independently in Podlipki. With his assistance, we managed to equip the missiles with control surface actuators and electromechanical trim motors to control the aerodynamic control surfaces. All of these drives were installed on a special load-bearing frame in the tail section and were tested for performance and correct polarity before the rocket was completely assembled.

The missile's electrical equipment consisted of a special 27-volt lead storage battery and two DC-to-AC converters to power the gyroscopic instruments, *mischgerät*, sequencing current distributor, and lateral correction radio system.

In late 1945, not far from Berlin, we met our electric machine specialists headed by NII-627 Director Andronik Gevondovich Iosifyan. They were busy with completely different business. Iosifyan was already a well-known specialist in the field of slaving systems and devices for electric synchronous communications. Nevertheless, he was included in our co-op, and soon thereafter we gained a powerful and loyal ally—not only in Germany but also for many years in the Soviet Union—in the solution of diverse electrical engineering problems of rocket and space technology.

Captain Kerimov was in charge in the well-lit, dry, and comfortable basements of the Institute RABE. He would later become the chief of the Main Directorate of the Ministry of General Machine Building and then the permanent chairman of the State Commission on Manned Flights. In 1946, he was responsible for restoring six onboard sets of extremely scarce Messina telemetry equipment and a ground receiving-recording station. G. I. Degtyarenko, a specialist from the Moscow radio institute NII-20, assisted him. Subsequently, fate separated these first telemetry experts. Kerimov made his career in a purely military and later, ministerial field. At NII-20, Degtyarenko attempted to restore and perfect the German Messina, but strong competition from the new work forces of Boguslavskiy at NII-885 and Bogomolov at MEI forced him to capitulate.

THE MAIN NUCLEUS OF SOVIET DIRECTORS of the Institutes Nordhausen and RABE consisted of a comparatively small group of specialists who first became intimately acquainted in Germany. With the arrival of their families, the business relationships were supplemented with familial relations and friendships between their wives. We tried to set aside time to take trips together to nearby towns. Tonya Pilyugina, Lesha Ryazanskaya (her name was Yelena, but we all called her Lesha as her husband did), and Katya, who had quickly become friends, were extremely active. It wasn't possible for us husbands to take part in all their sightseeing trips, but we tried to quench their thirst for knowledge of the Western world by making our service vehicles and German drivers available to them. After seeing points of interest in Weimar, Erfurt, and even Leipzig, our wives would report the expenses that the drivers had presented for the kilometers driven. We obtained gasoline, which was in short supply, from the military administration or illegally from the black market. Conventional filling stations were not in operation.

Family "receptions" were other forms of social mingling. We contrived anniversary dates for these. I organized the first such banquet-reception at the Villa Franka under the pretext of the first anniversary of the Institute RABE's founding. Its success encouraged a chain reaction of "comradely dinners" on the occasion of whomever's birthday it happened to be. The birthday boy or girl would arrange with one of the very modest local restaurants to provide service for these dinners. It was the responsibility of the hospitable host to provide all the requisite beverages and *hors d'oeuvres*, and the German staff provided the service, including the dishes, the table settings, and so on. We usually paid not in marks, but in kind—foodstuffs and schnapps. For our wives, these activities at first seemed sacrilegious in the face of the horrors of the war and of our relatives and friends who had perished. After five dreadful years they had grown unaccustomed to such revelry, but they quickly adapted.

Korolev's wife, Kseniya Vintsentini, and their eleven-year-old daughter Natasha arrived in Bleicherode after the other officers' families. Korolev tried very hard to show them all around Germany as much as possible. Perhaps because Korolev's wife and daughter left Bleicherode almost immediately and for a long period of time, they did not fit in with the "club" of officers' wives, who exchanged information almost daily about where and what to buy, what to see during the day if they didn't have a car, and what was playing in the evening at the town's movie theater. Korolev never told any of my close comrades at work anything about his family. We only learned from Pobedonostsev that his wife Kseniya Maksimilianovna Vintsentini was a leading surgeon at the renowned Botkinsk hospital. His daughter Natasha had not seen her father for the last six of her eleven years. Kseniya Vintsentini was the only wife of a former "enemy of the people" in Bleicherode. Katya met her for the first time at the children's swimming pool. It turned out that Natasha Koroleva and Nadya Pilyugina, who had already learned to swim, helped my seven-year-old Valentin master the technique. That evening when she was telling me about her encounter at the pool she said, "She's an amazing woman. We talked as if we had already known each other for years. It's too bad that she and her daughter are getting ready to leave soon. She said that she is needed at work and Natasha must be back in time for the beginning of the school year."

Late one evening in the first week in August, the field telephone rang in the bedroom. "Something's happened again at your institute," said Katya anxiously. Korolev had telephoned, and to my surprise, he excused himself for calling so late and asked me to hand the phone to Yekaterina Semenovna. At the end of the conversation Katya explained that the Korolevs had invited us to their fifteenth wedding anniversary party on 6 August. The next day I found out that the entire officer elite of the Soviet colony of the Institute Nordhausen and the military administration had been invited.

At the anniversary party there were many traditional toasts "to the health of the newlyweds" and long speeches about what an uncommon gathering the "renowned *dzhigit* Sergey and his beautiful Ksana" were holding "in the land of the conquered enemy."[6]

Korolev began his toast with the words, "Do you all know what it means to live fifteen years with Korolev? No, you cannot imagine such a thing. But my Ksana has endured this heroically. And for that I drink this toast to the last drop!"

We all stood up at once and drained our glasses. Having refilled my glass, I began with difficulty to elbow my way over to the happy couple. When I had finally managed to squeeze into an empty space among those who wanted to clink glasses with the Korolevs and had begun to utter something, Kseniya Maksimilianovna interrupted me, saying that she knew about me as the discoverer of this town, that she was acquainted with my wonderful wife, and that if we were going to drink, let's drink to our friendship. I expected to see a radiant, almost sainted woman, who after many long agonizing years was reveling in domestic bliss. But her face did not have a happy smile. She looked at me with dark, sorrowful eyes. Korolev was not next to his wife, and I went out into the garden to light up a cigarette.

---

6. A *dzhigit* is a Caucasian horseman.

Our hussars, as we jokingly referred to the three bachelor lieutenant colonels Boguslavskiy, Voskresenskiy, and Rudnitskiy, were already out there smoking. The three of them lived amicably in a single residence, the Villa Margaret, and shared a tiny Fiat. The hussars were surprised and concerned by Korolev's unexpected conduct. Voskresenskiy explained, "Somebody really set Sergey off. He needed to unwind. While you were paying court to Ksana he invited your Katya into his Horch and then tore off into the night." About ten minutes later Korolev's Horch illuminated us with its powerful headlights and screeched to a halt. Korolev jumped out of the car and like a true gentleman, throwing open the door on the passenger's side, he gave Katya his hand, led her over to me, and—wobbling slightly—returned to the festive, noisy banquet hall. At home during our post-party "flight analysis" Katya said that Korolev suddenly approached her and offered to demonstrate the performance of his sports car. "When we took off, he apologized, saying that he needed to unwind and that speed was the only thing that let him unwind. By some miracle we avoided hitting a bicyclist. Somewhere we struck what was probably a milk can but we didn't stop. He thanked me and said that if he had been alone, he might have crashed the car." And that was that.

IN AUGUST 1946, after traveling around many towns and enterprises in the Soviet zone, a high-level government commission headed by Artillery Marshall Nikolay Yakovlev arrived in Bleicherode.

The following individuals were members of the commission: Minister of Armaments Colonel General Dmitriy Fedorovich Ustinov; Major General of the Artillery Lev Robertovich Gonor, who had already been appointed director of head missile institute NII-88; Chief of the Main Directorate of the Ministry of Armaments Colonel Sergey Ivanovich Vetoshkin; Director of the *Gosplan* Department of the Defense Industry Georgiy Nikolayevich Pashkov; and Deputy Minister of the Communications Systems Industry Vorontsov. The only old acquaintance on the commission was Pobedonostsev. Officially he—like Pilyugin, Mishin, several others, and me—was still considered to be affiliated with NII-1 of the Ministry of the Aviation Industry.

We immediately understood that all of the commission's primary decisions actually came from Ustinov and Pashkov. This duo had to decide how to distribute the cadre of specialists gathered at the Institute Nordhausen among the ministries and agencies, along with the material and intellectual wealth that had been accumulated. Despite the fact that the distribution of duties in rocket technology had been stipulated in principle by the governmental decree dated 13 May, many design and production issues, and especially personnel issues, had not yet been resolved. Ustinov informed us that his ministry was officially in charge and that he had already arranged, with Korolev for Korolev to transfer to NII-88, the new head institute, as a chief designer. Pobedonostsev would be the chief engineer at the new institute. Ustinov then acknowledged Gonor and announced that he would be the future director of NII-88.

Ustinov felt that he needed to dwell on a subject that appeared to greatly disturb him, and on which a firm position had not been reached in the commission or at some higher level.

"Very great and vital work has been done here. Our industry will have to begin not from square one, not from a void, but by first learning what was done in Germany. We must accurately reproduce the German technology before we begin to make our own. I know that some of you don't like this. You also found many deficiencies in the German missile and you are burning with the desire to go your own way. Initially we will forbid that. To begin with, prove that you can do as well. And to anyone who alludes to our experience and history, my response is that we have every right to this, we paid for it with a great deal of blood!

"But we will not force anyone. Whoever doesn't want to do this can look for other work."

Ustinov continued, "We cannot take it upon ourselves to develop and manufacture engines, and therefore Glushko will transfer with this problem to the aircraft industry, which will allocate a special factory for him in Khimki. As far as the control systems are concerned, except for control surface actuators, basically this has been entrusted to the Ministry of the Communications Systems Industry (MPSS), and Ryazanskiy has already agreed to be in charge there, but under the condition that Pilyugin and Chertok transfer with him as his deputies."[7]

Apropos of that, Ustinov decided that one of us would be enough for Ryazanskiy because a leading specialist on the whole complex of control problems would also be needed at the NII-88 head institute. This specialist would also be responsible for starting up the production of control surface actuators, which MPSS had rejected.

Thus, Pilyugin and I were going to have to part company and decide who went where. One way or the other, we both would be leaving our old home in the aircraft industry. Soon thereafter I understood that our respective assignments had already been decided in Moscow in the Party and state offices, and they had been approved at all security and cadre levels. That explained why Nikolay Dmitrievich Yakovlev, a Marshal with the outward appearance of a typical simple-minded peasant, had been smiling slyly while looking at Pilyugin and me when Ustinov was giving his impassioned speeches regarding who needed to go where to work. It was clear that it was all a show—the decisions had already been made.

For purposes of decorum, we were given twenty-four hours to think it over. After many hours of debate in the commission, Pobedonostsev divulged to me that he had already made an agreement with Ustinov. Under this agreement, Pilyugin would go with Ryazanskiy as his deputy, and I must transfer to NII-88 as deputy to Pobedonostsev, the chief engineer. Pobedonostsev felt it necessary for persua-

---

7. MPSS—*Ministerstvo Promyshlennosti Sredstv Svyazi.*

From the author's archives.

S. P. Korolev, last photos at Bleicherode, 1947.

siveness to add that he knew Ustinov a lot better than I did. "Believe me. He is a very powerful man. One can work with him. I don't know Gonor, but I have been told that, in any case, he is a decent man and we will be able to get along with him, especially since he is Ustinov's man. It will be difficult for Sergey of course, but we will help. Give your consent!"

And I did. I must confess that I liked the young, energetic Ustinov. And I wasn't the only one. Ryazanskiy said, "You know, I regret this assignment for only one reason—instead of having a smart, energetic minister such as Ustinov, I will have some spineless windbag or some indifferent bureaucrat over me." It remained for me to meet with Pilyugin. He waved it off and came to the following conclusion: "We're not getting a divorce. We're still going to be in the same field. The main thing and the saddest thing is that we're both leaving the aircraft industry. They didn't want to take on the problems of missile technology. Ustinov is taking that on. That means we have to help him."

After the commission had dealt with personnel allocation and distribution, the division of laboratories and production property went rather smoothly. Ustinov demanded that we increase documentation so that no one would be denied the necessary number of sets. But the originals, copies, and "copies from copies" had to be stored in the NII-88 central archives.

Gaydukov and Korolev were ordered to prepare detailed reports concerning the work conducted at the Institute Nordhausen, bearing in mind (this was the first time this was said officially) that operations in Germany would be curtailed no later than the end of that year. They did not name a specific date. Gonor felt the need to clarify that he would insist that Pobedonostsev, as well as Korolev's deputy for technical documentation and several other leading specialists, arrive in Podlipki no later than September. To fulfill this completely legitimate demand as early as August and September, we began to gradually send our specialists to Moscow and "its environs," as the military used to joke, alluding to Podlipki, Bolshevo, and to the as yet unknown site of the missile test range.

After the relatively peaceful division of the laboratories, problems once again arose. Who would get what in terms of the technical wealth that had been accumulated over the course of more than a year? The two special trains were equipped in large part with the laboratories' apparatus and testing equipment. The institutes in Moscow must have their own laboratories, but it would be impossible to outfit them there.

After all of the emotional experiences, to the surprise of the Germans, we began a hectic, redoubled effort to manufacture two more sets of special laboratory equipment. Each set included the massive test benches for the control surface actuators; the so-called "Häuserman pendulums," the first primitive electromechanical simulator for the adjustment of *mischgeräte*; all sorts of panels for testing gyroscopic instruments; a central distributor; a time-domain current distributor; and finally, panels for the integrated tests of the entire missile.

The reviving German industry fulfilled our orders readily and rapidly. The directors of the enterprises which contracted with the Institute Nordhausen were already accustomed to impossible deadlines and they joked, "This again. 'Let's go, let's go.'"

We paid them generously, almost without haggling, and by October we had turned out and procured a sufficient amount of equipment for the initial period.

By October we had also manufactured twelve missiles and had completed their horizontal integrated testing. The horizontal tests proved to be the most complicated technological process. From the first try, something always went counter to the procedure and instructions. Display lights lit up, stayed illuminated, and went out regardless of the instructions. We had to develop a good understanding of the operating logic behind the the general "ground-to-air" circuit in order to quickly figure out the causes for the malfunctions. As a rule, the malfunctions were caused either by the inexperience of the operator/tester or by equipment failures.

The horizontal test process provided us with a graphic demonstration of the low degree of reliability of the A-4's electrical system as a whole. Out of the twelve missiles, not one underwent testing without receiving ten negative remarks for such reasons as "no contact when there should have been one" or "contact present where it shouldn't have been." The latter cases were due to a short circuit—smoke came out of the missile and all of the power supply sources shut down. A technical council began to ask the eternal questions: "Who is to blame?" and "What is to be done?"

In addition to these twelve missiles, we put together other assemblies and put the final touches on them for the purposes of rocket assembly training at the factory in Podlipki. The factory accumulated enough assemblies and performed off-line tests for ten missiles.

IN EARLY OCTOBER, all the main directors of the Institute Nordhausen were assembled for a closed meeting in Gaydukov's office. Here we saw Colonel General Ivan Aleksandrovich Serov for the first time. All we knew about him was that he was Lavrentiy Beriya's deputy for counterintelligence and an authorized representative in that field in Germany who purportedly had no direct relation to the NKVD office of internal repression.

Addressing all of us, Serov asked us to give some thought to making a list with brief descriptions of those German specialists who in our opinion might be of use working in the Soviet Union. To whatever extent possible, we would not take superfluous specialists. We were to hand over our lists to Gaydukov. The German specialists that we selected would be taken to the Soviet Union regardless of their own wishes. The Americans had designated the German specialists whom they needed "prisoners of war." We would not act like that. We would allow the specialists to bring along their families and all of their household effects. We would know the precise date very soon. There was already a resolution to that effect. All that was required from us were fully-verified lists without errors. Specially trained operations officers would carry out the operation. They would each have a military interpreter and soldiers to help load the personal effects. The German specialists would be told that they were being taken by decision of the military command to continue the same work in the Soviet Union because it was no longer safe to work in Germany.

"We will allow the Germans to take all their things with them," said Serov, "even furniture. We don't have much of that. As far as family members, they can go if they wish. If a wife and children wish to stay, then by all means. If the head of the family demands that they go, we will take them. No action is required of you except for a farewell banquet. Get them good and drunk—it will be easier to endure the trauma. Don't tell anyone about this decision so that a brain drain won't begin! A similar action will be underway simultaneously in Berlin and Dessau."

We left this meeting with mixed feelings. It was difficult to meet and work with the Germans and to seriously discuss future projects knowing that one night soon they and their families would be "seized."

Three days later the date was announced—from the late night to early morning hours of 22–23 October. On the evening of 22 October we held a banquet at the *Japan* restaurant with a completely open bar for the Germans and a strict prohibition against drinking for all Soviet specialists who were acting as hosts. The banquet had been supposedly organized to commemorate the successful completion of the assembly and testing of the first dozen missiles. In all, around two hundred people were "enjoying themselves." This, of course, meant only the

Germans—the Russians were in a gloomy mood in light of the ban on drinking in the presence such a beautiful spread of *hors d'oeuvres*. The party broke up at about one o'clock in the morning. After returning home, I informed Katya for the first time about the operation that would be staged that night and asked that she wake me up at three o'clock.

At four o'clock in the morning, hundreds of military Studebakers began to rumble through the streets of the quiet, soundly sleeping town. Each operations officer had previously located the home that he was supposed to pull up to, so that there would be no confusion and needless fuss. At each home, the interpreter rang, woke up the head of the household, and explained that she had an urgent order from the Supreme Commander-in-Chief of the Soviet Army. The dazed, half asleep Germans didn't immediately grasp why they needed to go to work in the Soviet Union at four o'clock in the morning, much less with their families and all their possessions. But the discipline, order, and the unquestioning subordination to authority that had been drilled into them and under which the entire German people had lived for many decades did the trick. An order is an order. They proved to be much more quick on the uptake, obedient, and submissive than we had assumed. There were no serious incidents or hysterics—with one minor exception.

At five o'clock, Pilyugin called me and, stuttering from emotion, said that operations officers had come for him and asked him to drive over to Dr. Rule, who had poisoned himself to make a point, and on his deathbed was requesting to see Pilyugin. I told him, "Go—just request a doctor to render first aid." When Pilyugin entered Rule's apartment, Rule was lying there with a military doctor bustling about him trying to determine how many pills he had taken. The doctor told Pilyugin that the pills were harmless, that they were not life threatening, and that there was nothing here for her to do. Pilyugin asked Rule what he wanted from him. Slurring his words, he demanded a guarantee that in the Soviet Union he would be offered work in his specialty together with Pilyugin—whom up until now he had trusted—and not be sent to Siberia. Pilyugin gave his word, and with that the incident was settled. Pilyugin really did appreciate Dr. Rule, and unbeknownst to the other German specialists, the two of them were designing a longitudinal acceleration integrator using new principles.

The second hitch took place at the Villa Franka. Frau Gröttrup announced that she could not starve her children. Here she had two beautiful cows, and if she were not allowed to take them with her, she would refuse to go. Helmut Gröttrup declared that he would not go without his family. A communication with the operational leadership ensued. They sent an immediate response: we guarantee that we will hitch a goods wagon for the two cows to the special train and fill it with hay. Only who would milk them? Frau Gröttrup thanked them and announced that she was prepared to milk them herself.

Now this incident was also settled. The owners supervised as the soldiers loaded their possessions—everything that they possibly could have wanted—into the Studebakers. There wasn't much furniture because almost all of the German specialists lived in strangers' apartments and the furniture did not belong to them. The cars loaded with people and possessions headed for the Kleinbodungen train station. There on a sidetrack stood a special train made up of sixty cars. The people found places in the sleeping cars, and their possessions were loaded into the freight cars under their supervision.

In the morning as I walked along the now quiet streets to the institute, solitary Studebakers and military Jeeps were still scurrying about the town. Someone forgot something; someone wanted to say one last farewell to his beloved. Without complaint, Serov's staff complied with these requests.

When I appeared in my office, the institute's most beautiful woman, Frau Schäfer, who managed our archives and our Photostat, came flying in. She was indignant because she had not been arrested and was not being taken to work in the Soviet Union. Her husband was a prisoner of war there and if she were in the Soviet Union she would certainly find him. "Why haven't they taken me?" I explained that they had taken only engineers and scientists and that Russia had enough archivists, Photostat operators, and typists.

But she did not give in and demanded that I report to Gaydukov. Instead of Gaydukov, I telephoned the commandant's office where the temporary operational headquarters was located. There, after some brief confusion, they made the following decision: "Give this Frau a car and let her go home immediately, gather her things, and set off for the special train."

Thus, Frau Schäfer turned out to be, perhaps, the only one who left on that special train for Russia as a complete volunteer.

The fuss continued around the special train for another entire day until everyone was arranged, everything that had been forgotten in the nighttime confusion was fetched, all the abundant rations were passed out, and the two Gröttrup cows were loaded. The Institutes RABE and Nordhausen switched over into liquidation mode.

All of the technical and maintenance personnel remained, the commercial office did not move, and having received assurance that none of them would be taken away, they set about the work of settling debts. We still had the big job ahead of us of copying and collating sets of documentation, disassembling and packing up the laboratory and production gear, and collecting the orders that remained unfulfilled from subcontractor enterprises. The BON officers who also remained in Germany for the time being were very helpful during this time.

Wrapping up operations on this scale took almost three months, and it was not until January 1947 that the entire primary staff of the Institute Nordhausen arrived with their families in Moscow at the Byelorusskiy train station. The assembled missiles, missile parts, machine tools, instruments, equipment, and automobiles that we had acquired as personal property, including Korolev's Horch, arrived in Podlipki ahead of us, and by the time we arrived, they had already been partially "distributed." The BON personnel and the hardware that had been transferred to them—several A-4 missiles and ground equipment—did not depart Thuringia for the State Central Firing Range that was under

From the author's archives.

By the entrance of the Villa Franka: German journalist Karl Heinz, Ursula Gröttrup (Helmut Gröttrup's daughter), and B. Ye. Chertok—Bleicherode, Germany, 1992.

construction until the summer of 1947. And that was the end of our almost two years of activity in Germany.

To this day the debate continues as to what significance the German achievements during World War II had in the development of our domestic missile building industry.

I will attempt to briefly formulate my responses to that question. First, we and the Americans, British, and French, who had worked on the new types of weaponry, were convinced that long-range automatically guided missiles were not a thing of the distant future, not fantasy, but a reality. There was a shared belief that in the future this type of weapon would certainly be used on much broader scales than the Germans had been capable of achieving.

Second, we had the capability, based not on literature but on our own experience, to study the shortcomings and weak points of the German technology and think about substantially improving it while we were still in Germany.

Third, missile technology attracted the attention of the all-powerful hierarchy of the Party, state, and military leadership. The resolution dated 13 May 1946 was the direct reaction to our activity in Germany, and of course also to a certain extent a response to the work being conducted in the United States on the basis of the same German technology.

Fourth, we acted correctly, having organized the study and restoration of the technology on German soil and taking advantage of the Germans' still powerful technical potential, with the participation of German specialists. It would have been impossible to provide similar working conditions on that scale during the first two postwar years in our own country.

Fifth, and this is perhaps one of the most crucial results, our work in Germany produced more than a reconstruction of German technology. Korolev once gave a very sound assessment of the work of the Soviet specialists in Germany: "The most valuable thing that we achieved there was forming the basis of a solid creative team of like-minded individuals."

Index

B

C

D

E

F

G

H

I

J

K

L

M

N

O

P

R

S

T

V

W

Y

Z

OIL CITY LIBRARY
3 3535 00207 0452
DISCARD